ESSENTIAL ENTOMOLOGY

ESSENTIAL ENTOMOLOGY

SECOND EDITION

George C. McGavin

Honorary Research Associate of the Oxford University
Museum of Natural History and Honorary Principal Research Fellow,
Imperial College

Leonidas-Romanos Davranoglou

Leverhulme Trust Early Career Fellow,
Oxford University Museum of Natural History

OXFORD
UNIVERSITY PRESS

Great Clarendon Street, Oxford, OX2 6DP,
United Kingdom

Oxford University Press is a department of the University of Oxford.
It furthers the University's objective of excellence in research, scholarship,
and education by publishing worldwide. Oxford is a registered trade mark of
Oxford University Press in the UK and in certain other countries

Text © George C. McGavin and Leonidas-Romanos Davranoglou 2022

Illustrations © Richard Lewington 2001 except those listed on page xi

The moral rights of the authors have been asserted

First Edition published in 2001
Second Edition published in 2022

Published in the United States of America by Oxford University Press
198 Madison Avenue, New York, NY 10016, United States of America

British Library Cataloguing in Publication Data
Data available

Library of Congress Control Number: 2022921538

ISBN 978–0–19–284311–1 (hbk.)
ISBN 978–0–19–284312–8 (pbk.)

DOI: 10.1093/oso/9780192843111.001.0001

Printed and bound by
CPI Group (UK) Ltd, Croydon, CR0 4YY

Preface

Life began in the seas, and salt water is still home to most of the earth's **phyla**. On land, however, two groups dominate: the plants and the insects. The word insect comes from the Latin, *insectus*, the past participle of *insecare*, to cut or incise, and refers to the major divisions of an insect's body.

Insects have been around for well over 400 million years. They were the first animals on land and the first to get into the air. They are the most species-rich and abundant group of multicellular animals the Earth has ever seen. More than half of all described species and about 75% of all animals are insects. But exactly how many species share the planet Earth with us? The truth is: we don't have the foggiest idea. Some early guesses of 30 million or even 100 million have been replaced in the last decade with more realistic estimates of somewhere between 5 and 10 million species. Despite this massive uncertainty, one thing is indisputable: the majority of Earth's inhabitants are insects.

In terrestrial ecosystems, green plants are the primary producers, trapping the sun's energy and using it to convert carbon dioxide to organic matter. Without photosynthesis the planet would be a very different place. Autotrophs like plants form the base of terrestrial and freshwater food chains, but every **trophic level** above is dominated by insects.

For example, it may come as a surprise to many that all the heaving, snorting herds of grazing ungulates on the African savannahs are entirely 'out-munched' (perhaps by a factor of ten to one) by myriads of tiny mandibles. What about the meat eaters? Again, insects consume many times more animal flesh than all the vertebrate carnivores put together and ants alone are the major carnivorous species in any habitat you could mention. If this sounds implausible, consider that, although insects are individually small, there are an awful lot of them.

Insects pollinate the vast majority of the world's quarter of a million or so species of flowering plants. Pollination is one of the most essential symbioses ever to have evolved. This plant–insect version of 'I'll scratch your back if you scratch mine' has lasted for 100 million years and it has generated a rich diversity of species. Twenty thousand species of bee are, to a large extent, responsible for the continued survival of the angiosperms, which includes a very long list of the things we eat: fruit and vegetables from pumpkins, plums, and peas to cherries, cucumbers, and cocoa.

Without the protein provided by insects many groups of higher animals would simply not exist. Insects feature exclusively or very largely in the diets of a whole beastly bestiary from the aardvark to the zorilla (an African striped polecat). Birds, the descendants of the dinosaurs, are mostly insectivorous. A brood of nine great tit chicks will consume around 120,000 caterpillars while they are in the nest. A single swallow chick may consume upwards of 20,000 bugs, flies, and beetles before it fledges, and even species such as

Insects   56.3%

Arachnids   4.5%
Crustaceans   2.4%
Other Arthropods 1.2%
Molluscs   4.2%
Nematodes 0.9%
Other Invertebrates   4.0%
Vertebrates   2.7%

Plants   14.3%

Fungi   4.2%
Algae   2.4%
Protozoans   2.4%
Bacteria and Viruses
0.5%

> Vertical bar chart showing the approximate proportions of all species on Earth.

> The image shown here is a composite of photographs of individual insects attracted to a sea holly flower over the course of 15 minutes. It is obvious that without insects, pollination would be impossible for most plants, and the world's terrestrial ecosystems would collapse.

© Ed Phillips

buntings, which are seed feeders as adults, rear their young on a nutritious diet of insects. The earliest primates were insectivores, and many primates still are today. In countries where insects are large or very abundant, they form an important source of protein, vitamins, and minerals for humans. Locusts and grasshoppers, beetle grubs, caterpillars, and termites feature regularly in tropical diets and their nutritional value exceeds that of western-style fast foods.

At least a quarter of all insect species are parasites or predators of other species of insect. Insects recycle nutrients, enrich soils, and dispose of carcasses and dung. Apart from these essential ecosystem services, insects provide humans with useful products such as silk, honey, waxes, medicines, and dyes. We use them to control pest organisms and as model systems to help us study many aspects of biology from genetics and behaviour to physiology and ecology.

Of course, insects have a dark side. Herbivorous insects may eat up to 15–20% of all crops grown for human consumption, and locally, the percentage may be much higher. Although, contrast this with the 30% of all food grown that is wasted by humans. In addition, insects carry the viruses, bacteria, spirochaetes, rickettsiae, protozoans, roundworms, and fungi that are responsible for innumerable plant, animal, and human diseases. About one in six human beings alive today is affected by an insect-borne illness of some kind. Besides stings and allergies, which can be fatal for some, the list of diseases carried by insects includes plague, typhus, sleeping sickness, river blindness, Chagas' disease, yellow fever, Zika virus, epidemic typhus, trench fever, loiasis, filariasis, and leishmaniasis. In the early twentieth century, Sir Ronald Ross showed conclusively that mosquitoes were responsible for transmitting malaria. Although much more is known about malaria today, it remains a deadly disease responsible for hundreds of thousands of deaths every year. To complete the destructive side of their activities, insects cause great damage to wooden structures and a wide range of natural materials and fabrics. There are even, much to the aggravation of museum curators, beetles whose larvae have a voracious appetite for collections of dried insects.

Being much smaller, insects may not have the immediate appeal of the 'furries or featheries', but if you look closer, you will find that they are a lot more varied and interesting. What animals other than humans build air-conditioned condominiums, use an underwater breathing apparatus, construct acoustic equipment, cause explosions, make paper, cultivate gardens, and farm other animals? The world of insects is endlessly fascinating, and they have extraordinary relationships with other organisms. Some of their lifecycles beggar belief. There are moths that suck mammalian blood, flies that breed in pools of crude petroleum, flies whose larvae develop only in the tracheal passages of red kangaroos, and lice that solely inhabit the throat pouches of certain cormorants and pelicans.

Despite all the talk of the need to quantify biodiversity in recent years, taxonomy and systematics are still less fashionable today than they once were and there is a lack of funding for basic taxonomic research. How are we going to save biodiversity, if we do not know what we are trying to save in the first

place? Perhaps we have lost sight of the fact that, without a system of classification, our investigations can only be a confusion of unconnected facts. We
survive in a complex world by defining what things are, naming them, and
thus identifying them. Imagine the colossal task of having to remember each
and every novel object individually by its unique characteristics. Classification is a central process of the human brain, and it, not mathematics, as has
sometimes been suggested, is the foundation and cornerstone of all science.
As British biologist and beetle taxonomist Roy Crowson pointed out, you
need to know what you are going to count or measure before your numbers
are going to make any sense.

Sex, violence, and a cast of trillions—the study of insects is exciting and
intellectually satisfying, but where do you start? With classification of course.
In any study, whether ecological, physiological, or behavioural, the first thing
you need to know is what sort of insect you're dealing with. Textbooks sometimes emphasize systems biology at the expense of systematics with the result
that students are familiar with the broader picture of insect physiology or
behaviour but are confused when it comes to recognizing a specific insect.
Whether there are 2 or 10 million insect species, you do not need to know
them all individually. At first glance, a beetle may resemble an ant, and some
flies are excellent mimics of bees, but ants do not have wing cases and bees
have two pairs of wings. Within each order there are conspicuous features
that we use to aid correct identification.

The amount of biological information available, especially concerning insects, is increasing rapidly and newcomers to the study of insects can be
put off simply by the enormity of the subject. Some species have received
a huge amount of attention for various reasons. The American Cockroach,
Periplaneta americana, and locust species, such as *Schistocerca gregaria*, are
enduringly popular as models for research, and fruit flies (*Drosophila* spp.)
are extraordinarily useful in genetics. Particular researchers have had their
favourites. Wigglesworth's pioneering work on insect cuticle used a South
American, cone-nosed, reduviid bug, *Rhodnius prolixus*, and Dethier's choice
of a particular blow fly, *Phormia regina*, for his work on insect feeding and
sensory biology resulted from a lunchtime chance observation of a female of
this species laying her eggs on a liverwurst sandwich. These, and a handful of
other species, are rare exceptions. The rest have yet to enjoy their 15 minutes
of fame. The majority of the world's living insects are completely unknown
and the biology of many of those we know about are obscure. There is still
plenty of work to be done.

But there is a problem. Human activities have brought about a biodiversity crisis. The conversion of more and more of the planet's natural capital
into producing human beings and serving their needs and desires has come
at an enormous cost. We seem to have developed a serious disconnect with
the natural world and this disconnect may be our downfall. Populations of
major groups of organisms, for example, birds, reptiles, and mammals, have
declined by an average of 68% in the last 50 years. Almost every recent
study undertaken points to a marked slump in insect species' richness and
abundance. The decreases seen in well-studied insect groups like bees and

butterflies are occurring in many other groups. The causes of these declines are very clear: the accelerating loss of natural habitat, mostly due to agricultural intensification, coupled with the prodigious amounts of pesticides used, are taking a heavy toll. On top of this we are facing global climate change caused by our use of fossil fuels, which affects every species and every habitat on Earth.

It has long been thought that the world's tropical forests are home to more than half of all extant species. If these complex habitats continue to be felled for timber, cash crops, and ranching at even the slowest rate suggested, it will still only be a matter of a few hundred years before they are completely lost. It is therefore an inescapable conclusion that our planet could lose more than half of all its living species in the time it takes for an acorn to become a veteran oak tree. We know that no species lasts forever. Of all species that have ever lived, 99% are now extinct and it was the numerous mass extinction events that paved the way for the appearance of those creatures that led to human beings. The difference is that now we know enough to do something that might prolong our own survival.

There can be little doubt that humans are the most intelligent and capable species yet to evolve on Earth. In the very short time since our appearance, we covered the entire globe and established colonies wherever it was possible to survive. A few of us have walked on the surface of the moon and visited the deepest abysses of the oceans. We spend vast sums of money to probe the very make-up of matter and examine the universe. We want to understand the science of everything from the infinitesimally small to the astronomically large but seem to be paying little heed to the health of the very ecosystems that support us.

The aim of this book is to provide a readable introduction to the most abundant multicellular life forms on Earth. It is not a field guide but, should you need to know the essential facts about a particular order of insect (e.g. where they occur, how many species are known, what makes them different and interesting from all other orders), this book will provide them and direct you to specialist texts and sources of information to guide you further along any particular path. If you really want to understand how the natural world works, you need to know about insects.

The book is laid out in an uncomplicated fashion. The first section gives a brief introduction to the insects and covers topics such as the evolution of the group and the factors that made them such successful organisms. It discusses the importance of their role in terrestrial ecosystems and outlines the features of structure, function, and physiology that they share. The second section provides a semi-pictorial key to the insect orders. Unambiguous text, coupled with clear drawings designed to highlight key features, allow the reader to assign most adult insects to the correct order. The bulk of the book is devoted to the essentials of the 28 living insect orders. The final section deals with collecting techniques.

Acknowledgements

We thank the following people for providing the photographs that decorate the new edition of our book: Nicky Bay, Paul Brock, John Gausas, Martin Gore, Matt Doogue, Piotr Naskrecki, Ed Phillips, Mike Picker, Hans Pohl, Anne Riley, Gilles San Martin, John Smit, Zestin Soh, Rupert Soskin, and Manfred Ulitzka.

Figure Acknowledgments

All illustrations not listed below are © Richard Lewington, 2001 and 2023

Preface page vi Vertical bar chart © Oxford University Press 2001.

Pages 4-7 Diagrams of the arthropod-like groups © Oxford University Press 2001.

Page 13 Diagram from Wilmer, P., Stone, G. and Johnston, I. (2000) *Environmental Physiology of Animals*. Blackwell Science, Oxford.

Page 15 Diagram from Gullan, P. J. and Cranston, P. S. (2000) *The insects: an outline of entomology*, 2nd edition, Blackwell Science, Oxford.

Page 17 Diagrams from Chapman, R.F. (1998) *The insects: structure and function,* 4th edition. Cambridge University Press. © After Hepburn, H.R. (1985) Structure of the integument, in *Comprehensive insect physiology, biochemistry and pharmacology,* Vol. 3 (eds G.A. Kerkut and L.I. Gilbert), Pergamon Press, Oxford.

Page 19 Diagram © Oxford University Press 2001.

Page 20 Diagram redrawn from Kukulova-Peck, J. (1991) *The insects of Australia*, 2nd edition. Melbourne University Press, Australia. © CSIRO Australia 1991.

Page 21 Diagram after Chapman, R.F. (1998) *The insects: structure and function,* 4th edition. Cambridge University Press.

Page 22 Diagram © Oxford University Press 2001.

Page 36 Diagram of the phylogeny of hexapods. Adapted from © Hans Pohl (2014).

Page 286 Diagram © George C. McGavin 2001.

Contents

Section 3 **Fieldwork** 277

Biographies

George McGavin studied Zoology at Edinburgh University, followed by a PhD in entomology at Imperial College and the Natural History Museum in London. After 30 years as an academic, mostly at Oxford University, he became an award-winning television presenter. George is an Honorary Research Associate of the Oxford University Museum of Natural History and an Honorary Principal Research Fellow at Imperial College. He is a Fellow of the Linnean Society an Honorary Fellow of the Royal Society of Biology, and an Honorary Life Fellow of the Royal Entomological Society. In 2019 George became the President of the Dorset Wildlife Trust.

Leonidas-Romanos Davranoglou studied Zoology at Imperial College London. He then undertook his DPhil thesis in entomology at the Department of Zoology, University of Oxford, funded by the prestigious Oxford Natural Motion and the Onassis Foundation scholarships. In 2021, Leonidas was awarded with both a Leverhulme Trust Early Career Fellowship and the John Fell OUP Fund at the Oxford University Museum of Natural History to study the evolution of insect communication. Leonidas will be researching how insects communicate using sounds by examining the morphology and biomechanics of living species, as well as their extinct relatives preserved in 100-million-year-old Cretaceous amber.

Glossary

Aedeagus The organ of copulation in male insects.

Ametabolous Development without metamorphosis, such as in bristletails and silverfish.

Apodous Without legs.

Apodemes Internal tendon-like extensions of the cuticle, which serve as muscle attachment sites.

Apolysis The separation of the old and new cuticles.

Aposematic Having bright colouration to serve as a warning to predators. Usually red, yellow, and black.

Apterous Lacking wings.

Apterygota (adj. apterygote) Primitively wingless insects. The smaller subclass of the Insecta.

Arrhenotoky Parthenogenesis where haploid males are produced from unfertilized eggs.

Asexual Of reproduction without separate sexes. Reproduction occurring by parthenogenesis.

Autotomy The casting or shedding off of a part of the body, for example, when under threat.

Batesian mimicry Where an edible or palatable species mimics the colours and patterns of an inedible species. The mimics are rare relative to the models.

Biological control Use of natural predators, parasites, or disease producing organisms to effect a reduction in the populations of pest species.

Biramous Having two branches.

Bisexual Having two separate sexes.

Caste Any group of individuals in a colony of social insects that is structurally or behaviourally different from individuals in other groups, that is, soldiers, workers, and reproductives as seen in termites, bees, and ants.

Cerci (sing. cercus) A pair of terminal abdominal appendages, which typically have a sensory function.

Chitin A high molecular weight, unbranched, amino-sugar polysaccharide made up of $\beta(1-4)$ linked units of N-acetyl-D-glucosamine. Chitin is the main structural element in arthropod exoskeletons and is also found in fungal cell walls.

Chrysalis The alternative name for a lepidopteran pupa.

Class A major taxonomic subgroup, above Order and below Phylum.

Coelom The main body cavity surrounding the gut in animals such as annelid worms, starfish, and vertebrates. In arthropods and molluscs, the coelom is much reduced, and the main body cavity is a haemocoel.

Colony An aggregation of social insects sharing a nest.

Corbiculum (pl. corbicula) The pollen basket of honey bees, which is a concave, shiny area on the hind tibiae, fringed with stiff hairs.

Cosmopolitan Occurring throughout most of the world.

Coxa (pl. coxae) The first segment of an insect's leg, joining the rest of the leg to the thorax.

Crop An area of the foregut in insects where food is stored.

Cuckoo A term used for an insect that uses the food stored by another to rear their own young.

DDT Dichlorodiphenyltrichloroethane.

Dimorphic Occurring in two distinct forms. The sexes of some insects are differently coloured or shaped.

Ecdysis The last stage of moulting, which is the casting or sloughing of the old cuticle.

Elytra (sing. elytron) The rigid front wings of beetles, modified as covers for the hind wings and not used in flight.

Endite A lobe of a leg segment that is directed inwards towards the midline of the body.

Endopterygota (adj. endopterygote) Insects where the wings develop internally. Metamorphosis is complete and there is a pupal stage (holometabolous).

Entognathous Having eversible mouthparts contained inside the head within a small pocket.

Epidermis In insects, the single layer of cells that secretes the cuticle.

Exite A lobe of a leg segment that is directed outwards away from the midline of the body.

Exopterygota (adj. exopterygote) Insects where the wings develop outside the body. Metamorphosis is incomplete and there is no pupal stage (hemimetabolous).

Family A taxonomic group below that of Order and comprised of subfamilies, genera, and species.

Fat body An aggregation of metabolic and storage cells in the haemocoel.

Femur (pl. femora) The portion of an insect's leg that corresponds to the mammalian thigh. The order of segments is coax–trochanter–femur–tibia–tarsus.

Fungivorous Eating fungi.

GABA Gamma-aminobutyric acid. Typically an inhibitory transmitter of neuromuscular junctions. Gamma is often shown symbolically: γ-aminobutyric acid.

Gall An abnormal plant growth caused by a virus, bacterium, fungus, mite, or insect.

Genitalia The hard parts of insect reproductive systems that engage between males and females during mating.

Genus (pl. genera) A low taxonomic group composed of species.

Haemocoel The haemolymph-filled internal body cavity of arthropods, such as insects, which is essentially an enlarged part of the circulatory system.

Haemolymph A fluid in invertebrate animals that is the equivalent of blood.

Haltere(s) The greatly modified hind wings of Diptera, which serve as balancing organs.

Hemimetabolous Developing by incomplete or gradual metamorphosis, for example, in the Orthoptera and Hemiptera. Immature stages are called nymphs (see exopterygota).

Hemimetabola Those orders with incomplete metamorphosis.

Holometabolous Developing by complete metamorphosis, for example, in the Diptera and Lepidoptera. Immature stages are called larvae (see endopterygota).

Holometabola Those orders with complete metamorphosis.

Homeotic genes Genes involved in the regulation of body plan.

Homology Character states acquired by direct inheritance from an ancestral form.

Homoplasy Character states acquired by convergent evolution.

Honeydew The carbohydrate-rich excrement of plant sap feeders such as aphids.

Hyperparasitoid A parasitoid of another parasitoid.

Idiobiont A parasitoid that kills or paralyses its host when laying an egg, and thus preventing the host from developing any further.

Inquiline A species that lives in the nest, gall, or home of another.

Insect The word insect comes from the Latin, *insectus*, past participle of *insecare*, to cut into, and refers to the major divisions of an insect's body.

Integument The cuticular and cellular covering of insects and other arthropods (i.e., the epidermis and cuticle combined).

Koinobiont A parasitoid that does not kill the host when egg laying, thus allowing the host to develop further.

Lobopods Protoarthropods with antennae, annular body segmentation (not a jointed exoskeleton), and soft, fleshy appendages bearing claws.

Meconium The accumulated solid wastes of the larval stages of some insects, which are expelled when the adult stage is reached.

Metamorphosis The transformation between the immature form and the adult form, with the forms being different.

Monophagous Restricted to eating a single species plant or type of foodstuff.

Monophyletic Of organisms that have descended from a single common evolutionary ancestor or ancestral group.

Mullerian mimicry Where several inedible or unpalatable species converge in colour pattern. As all the species are non-palatable, mimicry rings, as they are known, are enhanced by the frequency all ring members and even distantly related species can be involved.

Mycetocytes Special cells that permanently house intracellular bacterial symbionts.

Myrmecophilous Greek for ant-loving. Any species that lives with and depends on ants for food, protection, and care.

Neotenic Retaining features of the immature stages in the adult.

Ocelli Simple light-receptive organs on the head of many insects.

Oligophagous Restricted to eating a small range of related plants or foodstuffs (see polyphagous, monophagous).

Ommatidia Individual light receptive units in the compound eyes of insects. Each ommatidium comprises a clear lens system and light-sensitive retinula cells.

Omnivorous Having variable eating preferences.

Ootheca Protective egg case. Can contain a single egg in some orders, and up to several hundred eggs in other orders.

Order A taxonomic rank below a class and above a family.

Oviparous Producing young by means of eggs expelled from the body before they are hatched.

Ovoviviparous The young are produced by means of hatching eggs inside the body.

Palaeozoic (Paleozoic) The first of the three eras of the Phanerozoic (having visible life). The Palaeozoic includes the Cambrian, Ordovician, Silurian, Devonian, Carboniferous, and Permian periods. The Palaeozoic is followed by the Mesozoic and Cenozoic eras.

Parasite (adj. parasitic) A species living off the body or tissues of another, but not causing the death of the host.

Paraphyletic Of a group of species that includes some, but not all, of the species that share a single common ancestor.

Parasitoid A species that develops by consuming the body tissues of a single host, eventually causing its death (see idiobiont and koinobiont).

Parthenogenesis (adj. parthenogenetic) Reproduction without the need for fertilization.

Perineurium A layer of glial cells inside the neural lamella, which acts as the blood–brain barrier.

Pheromone A volatile substance produced by insects to communicate with others for the purposes of reproduction, aggregation, defence, etc.

Phloem The main transport vessels of plants taking nutrients from the leaves to other parts.

Phyla (sing. phylum) The major groupings of the animal kingdom, comprised of superclasses and classes, for example, Mollusca, Arthropoda, Chordata, etc.

Physogastric When the abdomen (usually in females) becomes greatly distended and membranous, typically after a meal, when pregnant, or when storing fat or food.

Polyembryony A single egg giving rise to many embryos by subdivision.

Polymorphic Having more than two forms.

Polyphagous Eating a wide range of plants or foodstuffs.

Polyphyletic Of a group of species that have different ancestors. Such taxa are 'unnatural' or incorrect.

Predacious (n. predator) Eating other animals.

Proleg A fleshy abdominal limb of insect larvae.

Pronotum The dorsal cover of the first segment of the thorax.

Pruinescent (colouration) Of a powder, or powdery.

Pterygota (adj. pterygote) Winged insects. The larger subclass of the Insecta.

Rostrum The prolonged part of the head of weevils and scorpionflies. Also the tubular, slender, sucking mouthparts of insects such as the Hemiptera (also known as the labium in the latter).

Saprophagous Eating decayed organic material.

Sclerite A hardened chitinous body plate surrounded by flexible membranes or sutures (Gk. *skleros*—hard).

Sister groups Groups of the same taxonomic rank resulting from the splitting of an ancestral lineage.

Solitary Occurring singly or in pairs, not in social groups.

Somite A single body division of a segmented animal.

Species A group of living organisms consisting of related similar individuals capable of exchanging genes or interbreeding. A taxonomic rank below a genus.

Spiracle(s) The breathing holes of insects leading to the tracheae occurring at the sides of the abdomen and thorax.

Stridulation The act of producing sound, usually by rubbing two parts of the body together.

Subfamily A taxonomic subdivision of Family.

Subimago Pre-adult winged stage of Ephemeroptera.

Suborder A taxonomic subdivision of Order.

Suture A groove on the outside of an insect body segment that indicates the fusion of two sclerites.

Symbiosis Different species living in an association that brings mutual benefit.

Synanthropic Associated with humans and their habitations.

Synonymy Where a single species has been described and named more than once

Tagmata (sing. tagma) A group of body segments that together form a major functional unit such as the head, thorax, or abdomen.

Tarsus (pl. tarsi) The foot of an insect. It is attached to the end of the tibia and made up of a small number of subsegments or tarsomeres.

Thelytoky Parthenogenesis where only female offspring are produced.

Tibia (pl. tibiae) The lower leg segment of insects, corresponding to the mammalian shin.

Tracheae (sing. trachea) The internal airways of insects.

Trochanter The small segment of the insect leg between the coxa and the femur.

Trophallaxis The exchange of food between colony members in social insects.

Trophic level The levels of food or energy transfer in food chain. Food producers or autotrophs form the first level, herbivores form the second level, and primary carnivores the third level. Other carnivores form the fourth and fifth levels.

Vector An intermediate host that carries a disease organism in contact with its target organism.

Introduction to insect evolution and biology

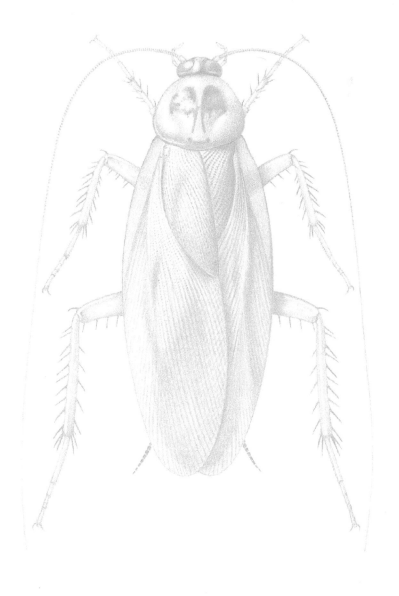

The evolution of the arthropods

> **radiodonts**
> Early arthropods. The front section of the elongate body carried a pair of appendages and a ventral mouth with distinctive radiating teeth. The trunk of the segmented body had lobe- or flap-like appendages. Some were predators while others were filter feeders

> **lobopods**
> early arthropods with antennae, annular body segmentation (not a jointed exoskeleton) and soft, fleshy appendages bearing claws. Earlier studies suggested that velvet worms (Onycophora) and water-bears (Tardigrada) are direct descendants of lobopods, although recent works suggest that their similarities are likely only superficial

> **homeotic genes**
> genes involved in the regulation of body plan

> **monophyletic**
> (of) a group of species that share a common ancestor

> **polyphyletic**
> (of) a group of species having different ancestors. Such taxa are 'unnatural'

The Earth formed from gas and debris around 4,600 million years ago (mya). The first traces of life appear as fossils of blue-green algae around 3,600–3,400 mya, and yet for more than 2 billion years, nothing more complicated than bacteria, algae, and minute, unicellular planktonic organisms existed. Then evolution went into overdrive. Known popularly as biology's 'big bang', the time around the beginning of the Cambrian period (543 mya) saw the 'sudden' appearance in the fossil record of an incredible assemblage of multicellular animals, some of which were the ancestors of all species on Earth today. Chief among these animals were early arthropods, such as **radiodonts** and **lobopods**.

Although the Cambrian explosion is characterized by some exceptionally well-preserved material, such as those found in the Burgess Shale deposits of British Columbia, it is often difficult to say exactly which might have been the direct ancestors of modern forms. The process of fossilization is unpredictable, and fine detail is often recorded only under special conditions. Evolutionary intermediates may never be found. The segmented, worm-like forebears of these creatures had certainly arisen by the late Precambrian (the Vendian), for the early arthropods had already evolved into a range of clearly recognizable groups with distinct body plans by the early to mid-Cambrian.

Arthropods differ from each other chiefly in the number and design of their appendages. The discovery of **homeotic genes** and the function of some of these genes has provided some insight into the early diversification of the group.

We now know how certain homeotic genes may have determined the evolution of segment diversity, as well as how they led to the evolution of the dazzling range of body plans exhibited by arthropods as a whole.

Just as the invention of simple and then compound microscopes in the seventeenth century advanced the biological techniques of the day, and therefore made it inevitable that old classifications had to be abandoned and new ones adopted, molecular techniques are producing data that sometimes contradict views based on traditional morphology. While the phylogeny (evolutionary history) of any group may be inferred from morphological or molecular studies, both approaches can suffer from being unable to distinguish whether or not particular character states shared by two organisms are due to common evolutionary descent (acquired by direct inheritance from an ancestral form—homology, or by convergent evolution—homoplasy). Very different animals might evolve features that make them look superficially similar (e.g. moles and mole crickets; see figure on next page), and it is not unusual for natural selection acting over a long time to make closely related species look very different from each other (e.g. elephants and hyraxes).

Arthropods are a **monophyletic** group, that is, they have a single common ancestor. The staggering morphological diversity found in today's arthropods led some authors in the past to suggest a **polyphyletic** origin, where arthropods evolved from four different types of worms that independently gained an exoskeleton. This view is now obsolete. We know that arthropods

Essential Entomology. Second Edition. George C. McGavin and Leonidas-Romanos Davranoglou, Oxford University Press.
© George C. McGavin and Leonidas-Romanos Davranoglou (2022). DOI: 10.1093/oso/9780192843111.001.0001

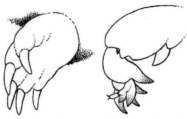

> Front legs of the mole and
the mole cricket,
demonstrating their
similarity.

Foreleg of mole   Foreleg of mole cricket

originate from a single worm-like ancestor, which gained a chitinous, segmented exoskeleton that was regularly shed to allow the organism to grow and to heal from injuries. The cousins to all other arthropods, the tardigrades or water bears (Tardigrada) and the velvet worms (Onychophora), do not possess a rigid exoskeleton, but a soft cuticle also made of chitin, which they regularly shed, therefore giving us an idea of how early arthropods may have looked. Arthropods, tardigrades, and velvet worms all belong to a likely monophyletic lineage known as the Panarthropoda.

Features that characterize arthropods

- Possession of a chitinous cuticle, which acts as a rigid exoskeleton for the internal attachment of striated muscles and is periodically moulted to allow for growth.
- Segmental paired legs.
- Body segments aggregated and/or fused into tagmata (discrete functional units) of which the most universal is the head. Besides the head there may be a trunk, as in the myriapods, or a separate thorax and abdomen, as in the crustaceans and hexapods.

The arthropod-like groups (Panarthropoda)

Onycophora

Velvet worm

Onycophora (velvet-worms) > 200 species
Onycophorans are soft-bodied, worm-like animals that are confined to tropical soil and litter.

Tardigrada

Water bear

Tardigrades (water-bears) > 1,300 species
Tiny (less than 1 mm from head-to-tail) and typically live in the water films associated with mosses.

NB: Illustrations not to scale.

The true arthropods (Euarthropoda)

Trilobita

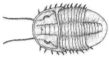

Trilobite

Trilobita (trilobites) > 20,000 species
Extinct marine, bottom-living, and pelagic arthropods that ranged from
1 mm to 1 m in length with the body divided into three regions: a head or
cephalon, a mid-body or thorax, and a hind-body or pygidium. The head
had antennae and eyes, the latter similar to the compound eyes of modern
arthropods.

Chelicerata

Mygalomorph spider

Merostomata (horseshoe crabs) 4 species
Horseshoe crabs are large, marine, bottom-living scavengers. Species of the
genus *Limulus*, which belong to the Xiphosura, a group little changed since
the Silurian, can reach up to 60 cm in length. The Eurypterida, predatory
giant sea scorpions, were very abundant in the Silurian and up to 2 m long.

**Arachnida (spiders, scorpions, harvestmen, ticks, mites, psuedoscorpions,
palpigrades, sunspiders, whipscorpions, whipspiders, and ricinuleids)** >
100,000 species
A largely terrestrial group, although some mites and a few spider species
are found in freshwater. Most arachnids are predators, but some mites are
parasites or herbivores.

Pycnogonida

Sea spider

Pycnogonida (sea spiders) > 1,300 species
Highly specialized, marine, inconspicuous, leggy arthropods which do not
have an obvious trunk or body. Sea spiders are predators of hydroids and
polyzoans.

Pancrustacea (crustaceans and hexapods)

Krill

Cephalocarida (cephalocaridans) 12 species
These are small (< 4 mm), marine, bottom-living, filter-feeding crustaceans
first described in the 1950s.

Branchiopoda (water-fleas, fairy shrimps, and brine shrimps) ~ 1,100 species
Branchiopods are small, filter-feeding, freshwater and marine crustaceans.
Unlike most crustaceans, they have flat, leaf-like body appendages called
phyllopodia.

Oligostraca (mussel shrimps and seed shrimps) > 13,000 species
Ostracods are marine and freshwater crustaceans mostly around 1 mm long
(although extinct species were ten times longer). The entire body is enclosed
in a bivalvate shell or carapace.

Malacostraca (mantis shrimps, crabs, shrimps, krill, lobsters, and hermit crabs)
> 40,000 species
Most living Crustacea, including all the larger shrimp- and crab-like
species, belong to this group. They are primarily marine, although some are
freshwater or terrestrial.

Maxillopoda (copepods, fish lice, and barnacles) > 14,000 species
These small, filter-feeding crustaceans are major elements of marine plank-
ton. Some species are parasites of fish and crabs, and some live in freshwater.
This is a polyphyletic group, and some lineages are related to the Ostracoda.

Remipedia (remipedians) 28 species
The first representative of this marine group (*Speleonectes* sp.) was found in a
sea cave in 1980. Remipedians are the sister group to the Hexapoda.

Hexapoda

Dragonfly

Collembola (springtails) > 9,000 species
Springtails are mainly terrestrial, but a few occur on the surfaces of fresh and
saltwater pools. They are mostly detritivores but also eat pollen and algae,
and a few are pests of crops, such as sugarcane. They are < 4 mm in length
and have a special abdominal 'spring-and-catch' jumping mechanism.

Protura (proturans) > 800 species
Proturans are soft-bodied, minute (< 2 mm), soil-dwelling, blind, fungus
feeders.

Diplura (diplurans) 800 species
Diplurans are small to quite large (5–50 mm), slender, blind, terrestrial
hexapods. Most species live under stones and in soil and litter. The abdomen
has a pair of conspicuous terminal cerci.

Insecta (insects) > 1,000,000 species
Insects are mainly terrestrial but there are a significant number of aquatic
species. Ranging in size from minute parasitic wasps at around 0.2 mm to
stick insects measuring 30 cm in length, insects have evolved very diverse
lifestyles. The head bears external mouthparts, antennae, and compound
eyes. The thorax bears three pairs of legs and (typically) two pairs of wings.
Insects have a well-developed tracheal system.

Myriapoda

Pauropoda (pauropods) > 800 species
Pauropods are soft-bodied, soil-dwelling, blind, scavenging myriapods that are < 2 mm long.

Diplopoda (millipedes) > 12,000 species
Millipedes are terrestrial, slow-moving, mostly herbivorous, burrowing myriapods with elongate cylindrical bodies. They vary considerably in size (< 2 mm to > 25 cm) and are typically round in cross section. The first three segments of the trunk have no legs, but the remaining segments are fused in pairs, each bearing two pairs of legs.

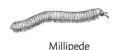

Millipede

Chilopoda (centipedes) 8,000 species
Centipedes are terrestrial, carnivorous myriapods with venom-bearing claws derived from the first pair of legs. Some are elongate, thread-like, slow-moving, soil dwellers, while others are long legged and fast running. Centipedes can range from 5 mm to 20 cm in length. The trunk segments each bear a single pair of legs.

Symphyla (symphylans) 200 species
Symphylans are typically small (1–8 mm), soft-bodied, blind, soil- or litter-dwelling, herbivorous myriapods.

NB: Illustrations not to scale.

The origins of insects and other hexapods

Insects evolved on land from arthropod ancestors that had long before made their way from the sea. However, the lack of fossil material has meant that it has been challenging to find details about this sea-to-land transition.

Most of the fossils of the earliest purported insects from the Devonian have now been shown to either be an entirely different organism altogether, or they are too poorly preserved to enable a reliable identification. *Rhyniella praecursor*, for example, was initially thought to be a larval insect, and is now known to be a springtail (Collembola). One of the most important of these fossils undoubtedly is *Rhyniognatha hirsti*, which, for more than a decade, was considered as the earliest example of a flying insect, and consequently was of essential importance for disentangling the complex origins of flight. However, a 2017 study suggested that *Rhyniognatha* more likely represents an early centipede, confirming previous studies indicating a myriapod origin for this fossil. It is clear that the deep origin of insects is clouded by much debate and will likely remain so in the near future unless spectacular new fossils come to light and change the game.

Based on this consideration, one could be forgiven for taking the rather pessimistic view of John L. Cloudsley–Thompson, who, in 1988, wrote the following:

> Time spent on insect phylogeny is often wasted. The origins of most insectan orders is a mystery, and is likely to remain so into the foreseeable future— certainly until existing fossils have been studied more extensively and further palaeontological evidence has been collected.

> **entognathous**
> having eversible mouthparts contained inside the head within a small pocket

> **cerci** (sing. *cercus*)
> a pair of terminal abdominal appendages that typically have a sensory function

Nevertheless, Cloudsley–Thompson concluded that they evolved from terrestrial ancestors, which had antennae, **entognathous** labiate mouthparts, probably 14 post-cephalic segments, tracheal respiration, 12 pairs of legs, coxal styles, eversible vesicles on most segments, and a pair of terminal **cerci**. The validity of his assertions is highly debated, since, as one can imagine, they are highly speculative.

> **sister groups**
> groups of the same taxonomic rank resulting from the splitting of an ancestral lineage

A revolution in our understanding of hexapod evolution came not from fossils, but from studying the DNA and structure of extant insects. Phylogenomic and morphological studies have convincingly suggested that the Remipedia, a group of elusive, marine cave-inhabiting crustaceans are the **sister group** of insects and hexapods as a whole. Although we may have not yet found the closest relatives of hexapods in the fossil record, these findings clearly demonstrate that insects are nested deeply within crustaceans, and they could be described as flying prawns!

Five factors in a winning formula

The staggeringly high insect diversity on Earth today is the result of a combination of high rates of speciation and the fact that many insects are very persistent, that is, they show relatively low rates of extinction. Most insect species may survive, little changed, for 2 million years or so, but a few species, found in 35-million-year-old Baltic amber, are morphologically identical to living species. A number of central factors that have led to the success of the Insecta can be identified as crucial.

Five factors contributing to the success of insects

- Small size
- A protective cuticle
- An efficient nervous system (the blood–brain barrier and sensory neuro-motor refinement)
- The evolution of flight
- High reproductive rate

Size

Your size determines how the environment will affect you. The range of sizes of all organisms covers eight orders of magnitude, from the smallest bacterial cell at around 0.3 µm (micrometres) long to a blue whale at 30 m long. In the middle of this range, organisms of around 3 mm—the insects—dominate. In rain forests, where, intuitively, large species might be expected to abound, 90% of the beetles collected were < 5 mm in length. To really understand the insect world, you need to imagine how the physical environment affects small

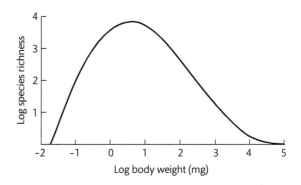

> A typical relationship between log body weight and insect species richness.

creatures. Take falling for example. On Earth, the acceleration due to gravity is a constant, but its effects vary according to the size of the falling object. Falling pollen grains and similarly small objects, for example, may be blown around for weeks before they reach the ground, and no matter the height from which an insect is dropped, it will not be damaged.

J. B. S. Haldane takes up the story:

> *You can drop a mouse down a thousand-foot mine shaft and, on arriving at the bottom, it gets a slight shock and walks way. A rat is killed, a man is broken and a horse splashes.*

It is due to the ratio of surface area to volume. Area is proportional to the square of the length (l^2), whereas volume is proportional to the cube of the length (l^3). Drag forces cause the terminal velocity achieved by falling objects to be proportional to their frontal surface area-to-mass ratio. Thus, a 1-kg lead ball will attain a much higher terminal velocity than a 1-kg lead sheet, and a 1-cm diameter stone will attain a higher terminal velocity than a 1-cm diameter sponge. For most organisms, however, the density of the tissues is much the same, and volume is thus a good measure of mass. A petit pois has a much smaller volume (and therefore mass) in relation to its greater surface area than a ripe water melon, which has a large volume (and therefore relatively greater mass) in relation to its small surface area. If you are still not convinced, ask yourself which you would rather drop on your kitchen floor. More massive objects have greater momentum; or put simply, the bigger they come, the harder they fall.

Small animals are also relatively more powerful than large ones. The cross-sectional area of insect muscles is relatively large compared to the mass they are supporting. Humans are able to carry a load roughly equal to their body weight, whereas the load-carrying ability of an ant is many times its own body weight, and earwigs are able to move more than fifty times their own weight.

A chrysomelid beetle hanging on to a leaf in a gale may resist a force ten or more times greater than its mass because the adhesive structures on its feet have a relatively huge surface area compared to the rest of the body.

What other consequences are there to being small? The world is a much finer-grained environment to small animals and offers a lot more places to live. Insects have simply evolved to occupy many more ecological niches than larger organisms ever could. An oak tree during the course of its life may support a few mammal and bird species, but many hundreds of different insect species. Dozens of species of gall wasp attack the oak's buds, leaves, flowers, acorns, and roots. The leaf-mining larvae of flies and moths neatly graze inside its leaves, secured between the upper and lower epidermal layers. Scuttling, flat bugs move about under flakes of its bark, feeding on fungal threads, while many more closely related species insert slender, needle-like mouthparts into the oak's xylem or phloem vessels, or systematically empty the contents of individual cells. The larvae of longhorn, bark, and ambrosia beetles devour the very fabric of the tree. In the height of summer in northern latitudes, the sound of falling caterpillar droppings can be heard quite clearly. Each of these herbivorous insects may fall prey to a number of carnivores and they all may act as hosts to one or more species of parasitic wasps or flies. Nests, fungi, lichens, and the rain-filled hollows left when branches have fallen all support rich and varied communities. When the tree ages and finally dies, its decaying structure will be alive with insects for months or years to come. Our oak tree is just one of 600 species of oak found growing in northern temperate regions of the world, each with a distinctive insect fauna.

That there are no really large insects today does not mean that it would be impossible for them to exist. It is often stated that the upper limits to the size of an insect are constrained by the mechanics of moving with an exoskeleton and the difficulties of getting enough oxygen to all the tissues via **spiracles** and diffusion alone, but this is not really why insects are not bigger than they are.

Gaseous exchange by means of air-filled tracheae is quite an efficient system and, in any case, suitable modifications, such as sac-like air pumps, would most likely evolve. It is now known that caterpillars possess quite a sophisticated gaseous exchange system equivalent to the vertebrate lung (see Lepidoptera). If larger, the weight of an insect's body at moulting might be greater than the soft cuticle could bear, and being vulnerable to predators, they would probably have to hide. However, they could get round the mechanical problem by moulting in water or by maintaining their body shape with hydrostatic pressure (although falling any distance at this stage might be a disaster). Fossil evidence shows that, in the late Carboniferous and early Permian, most insects were bigger (some very much bigger) than they are today. *Meganeura*, a carboniferous dragonfly relative, had a wingspan of 75 cm; the wingspan of the biggest odonate today, *Megaloprepus coerulatus* (a beautiful damselfly of South and Central America, which oviposits in rain-filled bromeliad tanks and tree holes), is a quarter of this. *Ramsdelepedion schusteri*, a carboniferous silverfish, was a chunky 6 cm long; today, the very biggest silverfish are approximately 1 cm in length. It has been suggested that large size evolved as a defensive strategy against predators such as proto-dragonflies.

> **spiracles**
openings to the tracheal system of insects, located on thoracic and abdominal segments

They, in turn, got bigger and the process continued until it was halted, either because they were no longer mechanically viable or, more likely, because of the appearance and rise of vertebrate predators, such as reptiles, who ate them. From that time on, insects got smaller and smaller to avoid predation and were thus able to radiate into all the available empty niches. The lower limit of the insect size range is determined by a number of physical constraints.

The heaviest Goliath beetles weigh about 60–100 g, making them heavier than many birds; at only 1.6 g, the bee hummingbird is lighter than an adult locust. A few butterflies and moths have very large wingspans (up to 280 mm), but their bodies are not very massive. The longest species, a stick insect, has a body length of around 30 cm. The smallest insects are species of parasitic wasp, being a little more than 0.2 mm from head to tail; this is no more than the thickness of the cuticle of large insect species and about half the size of some protozoan organisms. The smallest beetle in the world, a tiny feather-winged ptinid beetle, could sit comfortably on the very claw tip of a Goliath beetle.

So what are the constraints of size, and how can insects be so small and still function appropriately?

Oxygen supply   All animal cells need a supply of oxygen and organisms that are very minute might be able to get enough oxygen by means of simple diffusion. This is because a small body has a proportionally greater surface area than volume. Active insects need to get oxygen to all their cells and although oxygen diffuses 10,000 times faster in air than in water, it will not pass through waterproofed cuticle. The tracheal system of insects is a simple and effective means of gaseous exchange. Pairs of spiracles, the openings of the system, are located laterally on various body segments and are connected to air sacs and large tracheae, which then branch into even smaller branches, getting finer and finer. The finest branches, the tracheoles, are in intimate contact with the tissues and are especially abundant in tissues where oxygen demand is high. Interestingly, no matter how big or small the insect, the finest tracheoles always have a diameter of 0.2 μm. This is because if the tubes were any narrower, oxygen molecules would simply not diffuse any further along them. Evolution has produced a tube diameter that is 'just right'. A small insect can have a small number of tracheoles, but they cannot be any finer than 0.2 μm.

Cell size   Another reason why insects may have reached their minimum potential size might be due to the fairly constant size of animal cells. For various reasons concerning internal transport and operation dictated by relationships between surface area and volume, animal cells tend to be the same size. From this, it follows that larger animals are made up of more cells and smaller animals of fewer cells, and there must be a minimum number of cells needed to make a functioning insect. An adult locust, for instance, may have a million or so neurons to serve all its various sensory and motor needs. Although axons can vary in length greatly, nerve cell bodies tend to be similarly sized. A flying insect the size of a full stop might be able to fit in between one and ten thousand neurons, but any fewer would be inadequate for it to function. To overcome these difficulties, insects have very economical wiring and have shifted the job of integration from the central nervous system (CNS) to the peripheral nervous system.

Vertebrate muscles have a single motor neuron innervating each separate muscle fibre. To increase or decrease the force exerted by a muscle requires the recruitment of more or fewer motor neurons. Such nervous extravagance is impossible in tiny creatures. Instead, insects have polyneural, multiterminal innervation. In other words, only a few motor neurons innervate an entire muscle by branching extensively and by having multiple terminals. There are three types of motor neuron: fast, slow, and inhibitory. Fast motor neurons innervate most of the fibres within a muscle and their terminals release the excitatory transmitter L-glutamate, causing fast 'twitch' responses. Slow motor neurons run to a smaller subset of fibres and also release L-glutamate, causing slow, graded contractions. Inhibitory motor neurons innervate to all muscle fibres and their terminals release the neurotransmitter GABA, which inhibits contraction. In addition, neuromodulatory neurons are used to 'fine-tune' the system by setting the gain and threshold of motor innervation and muscle response. This arrangement allows simplification of the nervous system while still retaining flexibility. In vertebrate animals, the CNS is heavily involved in the integration of motor patterns. However, in insects, the motor neurons themselves play an important part in the integration of walking, flying, and reflex leg movements. There are also ways in which insects can shift the early stages of sensory integration to their peripheral nervous system. A good example comes from the African migratory locust; the sensitivity of chemosensillae on the maxillary palps is affected by the nutritional state of the animal and it is measured by the levels of nutrients in the **haemolymph**.

> **haemolymph**
the body fluid in invertebrate animals that is the equivalent of blood

In other words, if a locust has fed on carbohydrate-rich food, its sensitivity to carbohydrate is reduced and other nutrients, such as protein, will be phagostimulatory. The locust can thus regulate its diet selection in quite a sophisticated manner without the need for high-level CNS integration.

Temperature regulation   The large surface area-to-volume ratio of insects can make it difficult to keep cool and expensive to keep warm. Insects are ectothermic, but this is not to say that insects cannot regulate their temperature at all. Bees and moths, for example, need to warm up their flight muscles before take-off, and, when flying in hot conditions, need to lose heat by increasing haemolymph flow to the abdomen, where excess heat can be lost. Even in the hottest of environments, insects can survive. Desert beetles and ants often have very long legs and run fast to minimize contact with the scorching sand. The important point to remember, however, is that in any habitat there will always be a great range of temperatures available. Cooler, moister microhabitats will exist near plants or under stones, and burrowing underground, even a small distance, will ameliorate the hottest desert conditions. Small animals like insects are much more able to take advantage of a thermally heterogeneous and fine-grained environment. An elephant cannot move from an air temperature of 50°C to a much more bearable 30°C simply by hiding under a leaf; put another way, microclimates matter much more to insects (see figure below).

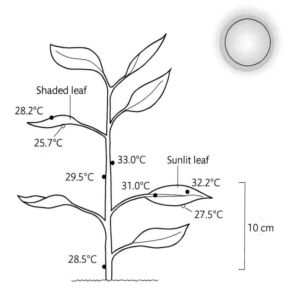

Shaded leaf
28.2°C
25.7°C
33.0°C   Sunlit leaf
29.5°C
31.0°C   32.2°C
27.5°C
10 cm
28.5°C

> **Microclimate matters. The relation between body temperature of a dead 50-mg insect, measured with an inserted thermocouple, and position on a plant.**

Cuticle

The common analogy of a medieval suit of armour for the protective and waterproof outer covering of insects is as much of a hindrance to understanding as a help. The only similarities are that both are the interface between the living animal on the inside and its environment, and, just like armour, the cuticle provides mechanical protection for the body and its appendages.

A conventional suit of armour was made of a single type of metal that was tough enough to deflect a mace blow or stop an arrow. However, armour has progressed somewhat in several hundred years. Nowadays special alloys can stop knives (but not bullets) and semi-flexible polymer materials can stop bullets (but not knives). Composite tank armour made from many layers of metal and ceramic sandwiched together has increased the effective range of these fighting machines by greatly reducing their weight. Although we have realized the great potential of studying natural designs, no biomimetic product has yet matched the versatility and effectiveness of the insect cuticle.

> **integument**
> the cuticle, together with the epidermis that produces it

Unlike man-made armour, the **integument** is, to a large degree, capable of self-repair. Mechanical damage to larval cuticle, such as a small tear or puncture, is quickly sealed.

If penetrated, the basement membrane of the integument closes up and new epidermal cells are produced on either side, and full repair by the production of a completely new cuticle takes place at the next moult. The ability of most insects to regenerate legs after amputation, known as autotomy, is well known.

Once thought of as a purely inert mechanical barrier, the integument also provides protection from the biological warfare of most viruses, bacteria, and fungal pathogens. In many cases, it is also able to mount an immune response by producing antibacterial proteins.

The cuticle is not simply protective—it also serves as an exoskeleton that anchors muscles and allows movement. Muscles are attached to the inside of the integument and in places the cuticle is extended to form **apodemes**.

The cuticle also bristles with chemo- and mechanosensory devices of all kinds and has built-in strain gauges to signal loads and tensions within its structure.

The outward appearances of insects are much more variable than the insides. Most insects have the same kinds of internal organs, even though they may differ in size by three orders of magnitude. The cuticle has been moulded throughout the course of evolution to provide all kinds of devices for running, burrowing, jumping, flying, swimming, catching food, and so on. Cuticle may be cryptically or brightly coloured, or it may smooth, spiny, or hairy, but it can be tough enough to cut through metals and hardwood or so delicate that gases will diffuse through its surface.

> **apodemes**
> the internal extensions of the cuticle, for the attachment of muscles

Cuticle structure   We need to look closely at the structure of cuticle to understand just how it serves all its various functions. Cuticle is produced by the *epidermis*, a single layer of cells supported by a basement membrane. It covers the whole of the outside of an insect, also extending inside where it lines the foregut, the hindgut, and the tracheal system. Cuticle is made up of two major layers: a chitin-containing *procuticle*, which can be nearly 0.20 mm thick (200 μm), and a much thinner outer layer, the *epicuticle* (< 5 μm thick) (see Figure).

The procuticle is made up of chitin, a high molecular weight, unbranched amino-sugar polysaccharide made of β-$(1{\rightarrow}4)$-linked units of N-acetyl-D-glucosamine. Packs of chitin molecules are formed into microfibrils, and these are embedded side by side in protein to make sheets. These sheets are laid down in the plane of the cuticle and successive layers are rotated slightly so that a helicoidal pattern is produced, giving the cuticle a lamellar appearance in cross-section. This arrangement confers great tensile strength. The outer part of the procuticle becomes tanned or sclerotized, an irreversible process of stiffening caused by protein chains becoming chemically cross-linked to each other by quinones. This dark, sclerotized layer of the procuticle is now known as exocuticle, while the thicker, inner membranous layer is known as endocuticle. Very highly sclerotized cuticle is found in areas demanding great toughness, such as claws and the mandibles of chewing mouthparts. To give flexibility at leg joints and between adjacent body **sclerites**, the procuticle is not differentiated and only flexible endocuticle is present.

> **sclerites**
> a hardened chitinous body plate surrounded by flexible membranes or sutures

A special rubber-like protein called resilin is often incorporated into particular regions, for example, wing hinges, and other structures that require great elasticity and energy storage, for example, those associated with jumping (see Orthoptera and Siphonaptera).

The outermost layer of insect integument is the thin, brittle epicuticle (1–4 μm thick). The epicuticle is composed of distinct inner and outer layers, collectively known as the cuticulin layer. Cuticulin is a composite made from quinone-tanned protein and lipids and is covered by two further layers of

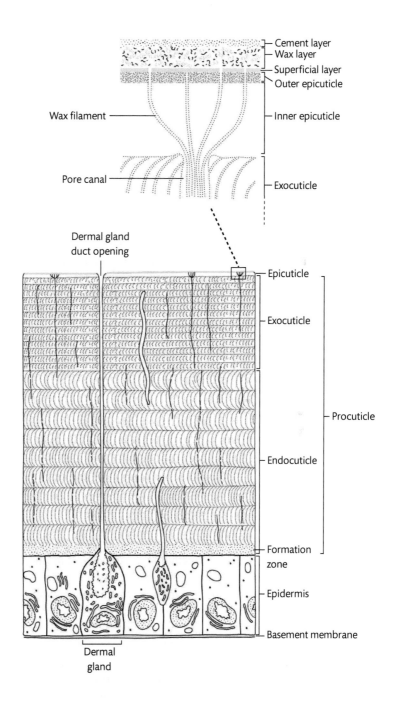

Cement layer
Wax layer
Superficial layer
Outer epicuticle
Inner epicuticle
Wax filament
Pore canal
Exocuticle
Dermal gland
duct opening
Epicuticle
Exocuticle
Procuticle
Endocuticle
Formation
zone
Epidermis
Basement membrane
Dermal
gland

❯ Cross section through a
typical insect cuticle.

hydrocarbon wax or lipid. The molecules of the inner wax layer are closely packed as a highly orientated monolayer, while those of the thicker, outer wax layer are not. It is the waxy epicuticle that is largely responsible for limiting water loss (of major importance in terrestrial environments). Minute pore canals run at right angles through the cuticle from the epidermal cells to the surface at densities that can be as high as a million per mm². These canals, whose function is to transport waxes (lipids) from the epidermis to the surface, branch into even finer wax canals as they travel through the epicuticle. The outermost layer in the complicated sandwich that comprises an insect's integument is a thin coating of lacquer-like cement.

Moulting   As insects grow, they need to periodically replace the rigid exoskeleton. Some insects moult as few as three or four times during their life, while others may moult 50 times or more. There are dangers in moulting because, for a short time, the protective function of the cuticle is compromised. Moulting is a complex process controlled by hormones secreted by the brain and neuroendocrine glands, which act on the cells of the epidermis and on the nervous system. Moulting involves two major processes. First, the old cuticle separates from the underlying epidermal cells. This is called **apolysis**.

> **apolysis**
> the separation of the old and new cuticles

> **ecdysis**
> the last stage of moulting being the casting or sloughing of the old cuticle

Between this and **ecdysis** (the action of shedding the old cuticle), unsclerotized parts of the old cuticle are digested and the new cuticle is formed (see figure below).

Stages in moulting

1. When moulting begins, the cells of the epidermis draw back from the inner surface of the old cuticle and increase in number, or, in some cases, in size. The new epidermal surface, having a greater area, is wrinkled. This is important as the surface area of the epidermis ultimately determines the surface area of the cuticle it will secrete.

2. Next, moulting fluid, a mixture of protein- and chitin-dissolving enzymes, is secreted into the space between the epidermis and the old cuticle. The enzymes are not activated until the epidermis has secreted the new cuticulin layer (inner and outer epicuticle). The outer epicuticle is made first, then the inner epicuticle, with its closely packed wax molecules. This layer will protect the growing procuticle beneath.

3. Once activated, the enzymes of the moulting fluid digest the old unsclerotized endocuticle, but do not affect the old sclerotized exocuticle. The digested material, comprising the bulk of the old cuticle, is reabsorbed.

4. The epidermal cells secrete the procuticle under the new cuticulin layer. The production of new cuticle may take place for many days and, in some species, endocuticle may continue to be produced many days after the old cuticle is shed.

5. When the old endocuticle has been fully digested and the new procuticle has formed, ecdysis occurs. As a result of muscular activity and/or increased haemolymph pressure (caused by the insect swallowing air or water), the old cuticle, now composed only of the thin epicuticle and the brittle exocuticle beneath, fractures along lines of weakness called sutures.

The main lines of weakness occur along the dorsal midline of the body and as a Y-shaped suture on the head.

6. Once free of the old cuticle (which includes the lining of the fore- and hindgut and the tracheae), the insect must expand the new, soft, wrinkled cuticle before the processes of sclerotization irreversibly change the outer layer of procuticle into exocuticle. In the case of insects with grub-like larvae, only the head capsule may be sclerotized and most of the cuticle remains soft and membranous. If the moult has been from the last immature stage to the adult of a winged species, the soft, crumpled wings must be fully expanded.

7. Soon after moulting, dermal glands associated with the epidermis secrete a cement-like substance, which travels to the surface through large dermal gland ducts and then covers the surface of the epicuticular wax.

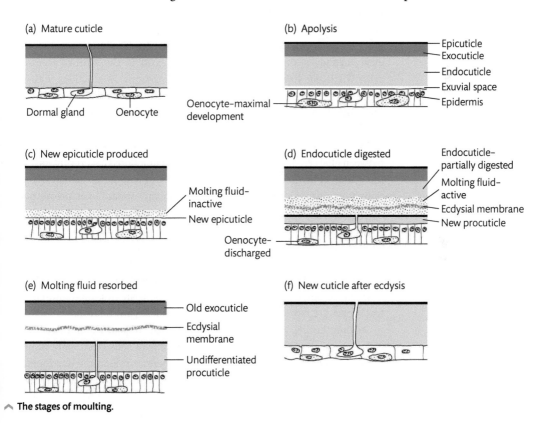

The stages of moulting.

The nervous system and the blood–brain barrier

Most animals have similar kinds of internal systems: a digestive system, a reproductive system, a circulatory system, and so on. Of primary importance, the nervous system integrates information about the environment outside the body and the physiological conditions inside. The internal organs of insects are constantly immersed in haemolymph, whose composition varies with food intake and environmental conditions. However, to work properly the nervous system needs an environment that is chemically stable within a narrow tolerance band. Other tissues can cope quite well with fluctuations in osmotic pressure and ionic concentrations, and normal homeostatic processes are able to maintain conditions within acceptable limits. A conspicuous feature of the CNS of insects and other terrestrial arthropods (and vertebrates) is the presence of a clearly defined blood–brain barrier. The nervous system needs a private pool, not communal bathing.

The insect CNS   The CNS of an insect is made up of a dorsally located brain connected on two sides to a large ventral ganglion lying beneath the oesophagus (the suboesophageal ganglion). From the suboesophageal ganglion, a pair of stout connectives (the paired ventral nerve cord) run backwards along the body cavity and join a series of thoracic and abdominal ganglia.

The brain, ganglia, ventral nerve cords, and the major peripheral nerves are sheathed by a supportive, corset-like layer of connective tissue called the neural lamella (or neurolemma) inside which the *perineurium*, made up of glial cells, acts as the blood–brain barrier (see Figure below). The degree to which this sheath reaches down the peripheral nerves varies, but, even beyond it, some sort of barrier layer exists in the form of accessory cells. Glial cells have different sorts of connections or junctions with the other cells around them. Some of these junctions are adhesive and structural, bonding the glial cells in the blood–brain barrier to each other and to the neural lamella above. Neighbouring glial cells communicate metabolically by means of gap junctions, but the most important sort of junction between glial cells in the perineurium are tight junctions that prevent ions and molecules in the haemolymph from diffusing in and out of the CNS. These tight junctions ionically isolate the fluid in the tiny extracellular 'private pool' immediately surrounding the CNS from the rest of the haemolymph. Active transport mechanisms pump ions in and out of the CNS to keep the ionic concentrations of this fluid constant. The nervous system can thus function efficiently at all times, unaffected by fluctuations in ionic concentrations in the haemolymph. Tracheae for gaseous exchange penetrate deep inside the CNS and branch so extensively that the finest branches, the blind-ending tracheoles, are never located more than 5 µm from any neuron.

Within the CNS, other sorts of glial cells, which greatly outnumber nerve cells, intimately wrap each nerve cell. These glial cells electrically insulate the nerve cells from each other, play an important role in the development of the nervous system, and provide metabolic support. Imagine the modern world without electrical insulation. Very careful routing of bare wires would be needed to avoid the continual shorting-out of circuits. If it were not for the insulating properties of the glial cells, electrical signals passing through the million or so neurons that make up an insect's nervous system would be grossly degraded.

A blood–brain barrier is found in other terrestrial arthropod groups, for instance, scorpions and spiders, whereas crustaceans only have a partial barrier. Myriapods do have a barrier of some sort, but there is no evidence yet that it is especially ion-tight. Though of variable complexity within insects, the possession of a highly efficient blood–brain barrier is a major contributory factor to their huge radiation on land and in the air, allowing them to successfully colonize even the driest and most inhospitable habitats.

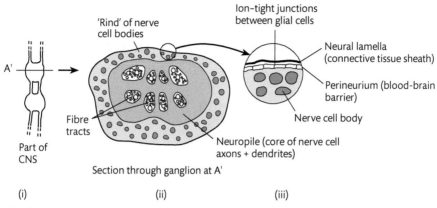

Diagram of the blood–brain barrier (perineurium).

Flight

The origin of wings The power of flight, another major chapter in the success story of insects, would have never been possible without lightweight, rigid cuticle. While there might be arguments about exactly what sort of arthropod first walked over dry land, there is no doubt that insects were the first animals to take to the air. Insects evolved wings in the Carboniferous around 300 mya, and today, only a minute percentage of the world's extant insect species belong to the orders (Archaeognatha and Zygentoma) that are ancestrally wingless. Insects with developed wings appear relatively suddenly in the fossil record, and as there are no fossils that show intermediate stages, there is a great deal of speculation as to how they might have evolved. One additional source of controversy is the debate around the purpose of the first protowings. One view has been that outgrowths from the top of the thorax would have enabled insects to have a longer, more stable glide path. This 'paranotal theory' envisages early insects climbing plants and launching themselves into the air as a more efficient, and perhaps safer, means of getting around or avoiding predators. These lobes might have had other functions, and experimental evidence suggests that, when small, they would have had a significant thermoregulatory function. Other ideas have involved small insects being carried around in air currents or larger insects jumping from the ground and taking to the air. There is a major problem with ideas that involve paranotal lobes and gliding, which is that, at some point in time, these fixed structures need to have developed hinge mechanisms and muscle systems to make them flap up and down. It is much easier to imagine structures that were already hinged, tracheated, and muscled fulfilling some prior function and then eventually taking on an aerodynamic role. The 'endite–exite theory' suggests that protowings developed from the fusion of inner and outer appendages (**endites** and **exites**, respectively) of basal leg segments, which were already articulated and might have been under some sort of muscular control (see Figure below).

The latest phylogenomic and developmental studies, as well as a critical reinterpretation of wing development in fossil insects, suggest that the origin of wings is more nuanced than previously thought. A dual model, which combines the contribution of both thoracic (paranotal) and leg structures for

> endite
a lobe of a leg segment that is directed inwards towards the midline of the body

> exite
a lobe of a leg segment that is directed outwards away from the midline of the body

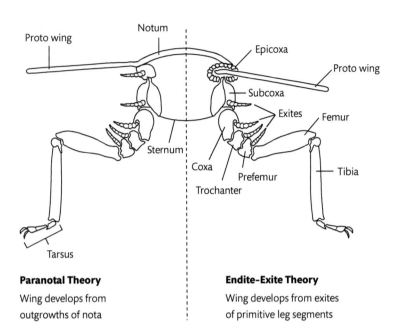

> **Comparison of paranotal theory and endite–exite theory.**

Paranotal Theory

Wing develops from
outgrowths of nota

Endite-Exite Theory

Wing develops from exites
of primitive leg segments

the development of wings, is now the most strongly supported theory. During their development, insect wings express a set of genes that otherwise form tergal (the dorsal surface of the thorax) and leg structures. Importantly, these sets of genes exist in Archaeognatha and Zygentoma, insects that predate the evolution of wings, as well as crustaceans, which demonstrates that they are part of two distinct ancient sets of gene networks involved in the formation of the dorsal body wall and appendages, respectively. The dual model suggests that these two distinct regulatory networks at some point collaborated to form an entirely novel structure: the insect wing. Indeed, in insects with incomplete metamorphosis, wings originate from primordial cells that are part of primitive leg segments, which, during early development, move upwards to the tergum. They then receive a signal from thoracic genes and become flattened and expand outwards to form early wing pads. Also, the direct muscles that are involved in the movement and articulation of the wings are originally leg muscles, further highlighting the participation of the latter in wing formation. Future developmental studies, and perhaps new fossil data, will help clarify the complex origins and timing of insect wings. However, we might never find the exact developmental intermediate steps and the evolutionary pressures that led to the evolution of protowings.

Flight certainly allowed insects to disperse, to colonize new habitats, and to escape from their enemies and the vagaries of the natural environment. Which of these aspects might have been of primary significance in the evolution of wings is a matter of debate, but predator avoidance is an excellent candidate—life in the Palaeozoic was dangerous! Pre-eminent among predator avoidance mechanisms must have been increasingly sophisticated early-warning systems composed of delicate vibration- or wind-sensing hairs. Abdominal cerci coupled to giant interneurons running to thoracic ganglia and the brain are common in some insect orders today and also existed in early insects (in most fossils, of course, the neural connections are educated conjecture). These systems elicit an escape response of rapid running or jumping. It has been suggested that this escape response, development of pro-towings, and a gradual rewiring of thoracic and abdominal ganglia might have been the route by which insects achieved powered flight so quickly. It is easy to see how being airborne for longer and/or being further away from danger would have conferred a major selective advantage.

Viscosity   As things get smaller, the effect of the density of the medium they are in becomes more important. Viscous forces become greater than the inertial forces. The Reynolds number is a ratio of these forces. At high Reynolds numbers (100 and above), flow of liquids or gases across solid structures is turbulent and chaotic. At Reynolds numbers of 10 and below, flow is laminar (follows the surface contours). At very low Reynolds numbers, the air to a very small insect is much more like a liquid than a gas, and not surprisingly, the wings of minute insects differ greatly from their much larger relatives. The wings are reduced to slender, hair-fringed paddle-like struts with which the tiny insect 'rows' through the relatively sticky air (see Figure).

⌃ The wings of a fairyfly.

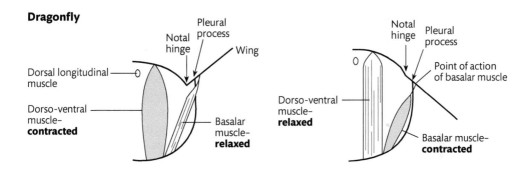

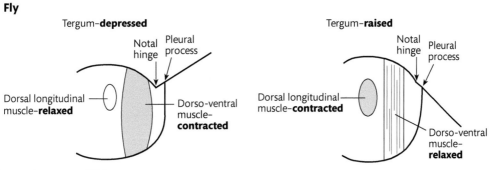

⌃ Diagram of two types of flight motor. Dragonfly: direct muscle for downstroke–neurogenic rhythm. Fly: indirect muscle for downstroke–myogenic rhythm.

Flight mechanics   Flying is a key adaptation of insects, but as insects get smaller, the faster they have to flap their wings to fly. There are limits to the rate at which muscles can contract to bring about wing movement. The limits are not set by the muscles' ability to contract but by the inability of motor neurons to fire above a certain rate. In large insects with long wings, the flight muscle can be attached directly to the wing bases and speed of firing the motor neurons is sufficient to flap the wings at a suitable speed. The rhythm produced by the 'flight motor' is termed neurogenic. Smaller insects have smaller wings and the frequency at which they need to beat is far above the speed at which motor neurons can fire. This problem has been solved in two ways. Firstly, indirect flight muscles are attached to a highly elastic, box-like thorax. Distortion of the thorax and the way in which the wings are joined and articulated to the latter generate the up and down stroke, as well as the angle of attack. Secondly, the flight muscles can be asynchronous and react rapidly to being stretched by contracting. The net result is that a high wing-beat frequency can be produced without the need for a similar firing rate in the flight muscle motor neurons. The rhythm produced by this type of flight motor is termed myogenic (see figure above).

We now have a clearer understanding of the way in which insect wings generate lift. For many years, insect flight was explained by comparing insect wings to simplistic models such as fixed-wing aerofoils and propellers. This led to confusion and to patently absurd statements such as 'according to the laws of aerodynamics, bumble bees should not be able to fly'. In reality, bees fly very well; the problem has been how additional, unpredicted lift is generated. Research has demonstrated that the leading edge of an insect's wing generates a strong helical airflow along its length during the downstroke. Vortices (rotating cylinders of air) move along the length of the wings and are shed from the wing tip (see Figure). We can now dispense with explanations involving steady-state aerodynamics and begin the work of discovering exactly how these vortices are generated and maintained along the wing. There is still much to do when we consider the flying abilities of the hover fly or the difficulty in swatting a house fly.

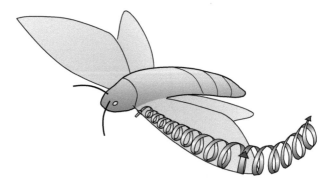

❯ Diagram of vortices being shed from wings.

Once wings evolved, there is no reason for them to then disappear unless having them incurred any kind of selective disadvantage. Secondary flightlessness is not uncommon. Loss of wings has occurred several times in insects, most dramatically among parasitic species. It is likely that the major radiation of groups, such as lice and fleas, took place only after the loss of wings. Wing loss seems to be more common in habitats such as mountains, deserts, and caves, which are very stable over long periods of time, or cold and windy habitats where flight would be very costly. There is also strong evidence of a trade-off between flight and fecundity. Wings and their muscles are developmentally and energetically costly appendages. By not developing wings, wingless individuals might be able to produce more offspring over a longer period of time than winged individuals of the same species, as they allocate their resources into developing larger reproductive organs instead of wings.

Reproduction

Many insects have remarkably high reproductive rates. This, coupled with typically short generation times, enable them to evolve faster, and to adjust to environmental changes much more rapidly than slower-breeding animals. Consider the folk tale about an old man who helped a vain King (the story likely originates in India, with variants told in European and Islamic cultures). The King had the man brought to court and told him that he could have anything he wanted as a reward. The old man asked for a chessboard. The King naturally agreed but, as he had wanted to show off his great wealth, demanded that the man asked for something else more valuable. The old man then asked for one grain of wheat to be put on the first square of the chessboard, two grains on the second square, four grains on the third, and so on, until the sixty-fourth square. Still not very pleased, the King agreed. It was not until they got to the last few squares that the King realized the trouble he was in. On the sixty-fourth square would rest more than 9×10^{18} grains—probably more than has been cultivated in the 6,000 years of human history. It is also, incidentally, nine times more than the estimated number of individual insects living on Earth.

A pair of small bruchid beetles can produce about eighty offspring with equal numbers of male and females every 21 days or so. Theoretically, the progeny from the second generation would number 3,200, and the third generation would number 12,800 beetles. By the eighteenth generation (a mere 432 days) the beetles would number 1.4×10^{29} individuals and occupy a volume equal to that of the Earth. This kind of population explosion is never seen due to a variety of regulatory factors, including weather conditions, disease, predation, and food availability, but it does illustrate the potential for population growth. Insect outbreaks are favoured by the coincidence of two or more of these factors, often in agricultural or other non-natural systems.

Interactions with other organisms

Apart from their obviously successful design characteristics, insects have a vast number of interactions with a multitude of other organisms from viruses, bacteria, and protozoa to fungi, plants, themselves, and other animals. A full description of these interactions, which can be facultative or obligatory, simple or complex, and beneficial to one partner or both, is well outside the scope of this volume. However, here we touch on relations involving some bacteria and fungi, and deal with the enormous subject of plant–insect interactions in slightly more depth.

Wolbachia are a widespread group of bacteria that are inherited cytoplasmically and cause a range of reproductive changes in arthropods, especially insects. Distortions of sex ratio such that mostly females are produced have been found in ladybirds, parasitic wasps, and many other taxa. Another effect of *Wolbachia* and related bacterial infections is reproductive incompatibility. If mating occurs between two individuals of strains or geographic isolates of the same or closely related species that do not carry the same infection, embryos will not develop. The effects are typically unidirectional but can be bidirectional. As yet, little is known about the actual mechanisms by which these microbes alter the reproduction of their hosts to their own advantage. A survey of 154 insect species in Panama found that nearly 17% were infected with *Wolbachia*. Given this prevalence, and the number of insects species in the world, *Wolbachia* must be very abundant organisms, and it is likely (although not yet conclusively proved) that they are a significant potential mechanism for speciation in insects by preventing the reproduction of otherwise compatible populations.

A widespread phenomenon in insects is the possession of symbiotic, non-pathogenic bacteria, protozoan, yeasts, or fungi. In some insects, these organisms are extracellular and may be found in the gut epithelium or free inside the gut. A much more intimate association and one pointing to co-evolution, concerns intracellular bacterial symbionts permanently housed inside special cells called **mycetocytes**, which are usually found within the **fat body**.

> **mycetocytes**
> special cells that permanently house intracellular bacterial symbionts

> **fat body**
> an aggregation of metabolic and storage cells in the haemocoel

These symbionts are thought to have evolved from extracellular bacteria. Mycetocytes may be grouped into bodies or organs called mycetomes. The presence of symbionts is strongly associated with insects whose diet is poor or lacking in particular nutrients, such as blood-feeding species (e.g. cimicid bugs and lice), wood-feeding beetles (e.g. cerambycid beetles and woodworm), and sap-sucking bugs (e.g. planthoppers and aphids). In the case of bacterial symbionts, studies using antibiotics have shown that the insects benefit greatly, as the former metabolize amino acids and vitamins of essential importance to the latter. Extracellular symbionts are transferred from generation to generation by the young eating either the adults' excreta or their own eggshells, which have been daubed with symbionts. Intracellular symbionts are transferred by transovarial transmission, that is, internally from the mycetocytes to the ovaries, and then to the eggs.

Insect–fungal interactions are numerous and varied. Fungi are eaten directly by many insects; leaf-cutter ants and termites culture fungal species as food. Wood-boring beetles, such as bark beetles (Scolytidae) and ambrosia beetles (Platypodidae), carry fungal species with which they infect dead or dying wood. The growth of the fungus 'pre-digests' the wood, making it more edible by the beetles and their larvae. Female *Sirex* wood wasps (Siricidae) carry spores of rot-producing fungus in their reproductive tract, which are passed into the wood at oviposition. Although the cuticle is an effective barrier to most species, at least 700 known species of entomopathogenic fungi are able to penetrate its defensive layers by being able to produce a number of cuticle-degrading proteinase, chitinase, and lipase enzymes. Besides killing and consuming insects, many fungal pathogens have interesting effects on their hosts. *Entomopthora muscae*, for example, infects house flies and yellow dung flies and kills very large numbers annually. Male house flies are sexually attracted to fungus-infected corpses, perhaps because the swollen abdomen of the infected flies might look like a fecund female with which they can mate. Oddly though, the males prefer fungus-infected corpses to similarly sized non-infected corpses, so there must be more to the story. In yellow dung flies the *Entomopthora* fungus changes the perching behaviour of the fly such that it dies in a suitable elevated position to ensure maximum spread of the released fungal spores. Within the Basidiomycota, species of a large genus, *Septobasidium*, are found only in association with armoured scale insects (Diaspididae), and, unlike many insect–fungus relationships, the interaction between the fungus and the scale insect is mutually beneficial. The fungus grows on the bark of trees and shrubs, such as citrus and tea, where it forms structured colonies, known as 'houses', inside which the scale insects live, feed, and reproduce. Only a certain fraction of the scale insects become infected by the fungus, and hyphae growing from their bodies join other hyphae and transfer nutrients from the fluids of the insect's body cavity to the main body of the fungus. The infected scale insects do not become adult or reproduce, although they may actually live longer than non-infected scales. The fungus receives food and the uninfected scale insects benefit from protection from the weather and natural enemies.

As more than half of all insect species are herbivorous, their relationship with plants must be considered the biggest single type of biological interaction on this planet. Initially saprophagous, some early insects might have avoided competition and predation by moving up to plants to feed on dead or decaying parts, or nutrient-rich structures. Insects faced several problems along the evolutionary path to become true herbivores. One of the difficulties of feeding on aerial parts of plants is that holding on requires special adaptations. Living above ground level also increases the risk of desiccation. Of these two obstacles, the first was overcome by the evolution of specialized tarsal and pretarsal structures. Sharp claws and adhesive hair pads allow modern leaf-feeding insects to cling to the shiniest leaves. Waxy cuticle, spiracular control, and highly efficient physiological regulation of water balance, together with the ability to use microclimatic variation in relative humidity, has allowed insects to tolerate very dry environments. Some herbivorous species have avoided desiccation problems by feeding internally as leaf miners, stem borers, or gall formers. The biggest problem of being herbivorous stems from the fact that plant tissue is not very nutritious, having low protein, sterol, and vitamin contents. Although some spring growth may, for short periods, have higher levels of nitrogen, the dry weight nitrogen content of most plant tissue lies at 3–4%. Leaf-chewing insects with dry weight nitrogen contents of 7–15% have a much harder time than their carnivorous cousins. Sap feeders fare even worse because phloem sap has a dry weight nitrogen content of no more than 0.6%, and xylem sap of no more than 0.1%, and mostly less than 0.01%. It should be remembered that the total nitrogen content of plant tissue is higher than the soluble nitrogen content, that which is actually available to herbivorous insects. There are a number of solutions to a nutrient-poor diet. Insects, such as sap-sucking bugs, can consume large amounts of liquid food and void the excess carbohydrate present. Insects can be very selective in their feeding, attacking only the most nutritious parts and timing their life cycles to coincide with rich resources. They can develop slowly or use symbiotic microorganisms to boost levels of essential nutrients. Cicadas, whose nymphs feed on xylem sap from roots, are famous for their long development: species can take 5–17 years to develop.

The relationship between herbivorous insects and the plants that they eat is often termed co-evolutionary, and what we see today is the result of an 'arms race' that began in the late Devonian or early Carboniferous. Adaptations in one species to overcome defence mechanisms evolved by another species are countered by novel defences in the first, and so on. Working on the principle of damage limitation, plants have evolved a number of defences to herbivores. Physical properties, such as shiny or waxy surfaces, hairs, and sharp or sticky trichomes make holding on difficult or dangerous, but are not much defence against internal feeders, such as miners or borers, once they are inside. Incorporated silica grains, or leaf toughness caused by high levels of lignin or cellulose, will deter most chewers but will not affect sap suckers much.

Of more interest are the chemical defences that plants have evolved to deter or poison insects. There are a very large number of these secondary plant compounds or allelochemicals, and more are being discovered. They fall into a number of classes, of which alkaloids, terpenoids, and flavonoids are the biggest, with more than 10,000 different active compounds between them. We should be grateful for the huge diversity of insect–plant interactions, because many of the compounds we value most highly, and may be addicted to, such as nicotine, caffeine, and tannins, as well as many useful drugs, are nothing more than plant-produced anti-insect compounds.

It is now clear that elegant tritrophic relationships have evolved between plants, insect herbivores, and the wasps that parasitize them. Locating suitable resources, whether plant tissue in the case of the herbivore or insect tissue in the case of the parasitoid, is of primary importance. Plant volatile chemicals may act as attractants or repellents allowing insects to orientate from some distance. Plants damaged by herbivores can produce volatile chemicals that attract natural enemies—predators and parasitoids. One study using corn seedlings and caterpillars of the beet armyworm showed that certain oral secretions of the caterpillars act as triggers for the release of plant volatiles—mechanical damage alone does not have the same effect. The production of these volatiles, induced through a pathway known as the octadecanoid pathway, can also be stimulated by applying jasmonic acid (jasmonates are a class of fatty acids). It remains unclear as to why herbivores should stimulate clear signals that alert enemies to their presence, but chemicals such as that found in the oral secretions of the armyworm caterpillars must be advantageous to the caterpillars in other ways. Natural selection acting on the plants and the response of natural enemies to volatiles has produced a winning strategy (as far as the plants are concerned). The stage may now be set for 'stealth herbivores' to evolve.

Insect–plant relationships are not one sided. Mutualisms abound, and at the other end of the spectrum, there are carnivorous plants, which trap and digest insects. Even here, things are not clear cut; some pitcher plants harbour many larvae of many flies that live unharmed on the plant's digestive juices. It turns out that these larvae feed on the thin soup of water and decaying insects and provide the plant with nutrients in a readily assimilable form.

Even the shortest discussion of insect and plants should deal with pollination. Using insects as 'flying genitalia' is a marked improvement over wind pollination. Winds are unpredictable in speed and direction, there is much wastage of pollen, and rare plants would stand little chance of ever being fertilized. Insects may have visited plants to feed on spores as early as the Devonian, and it is likely that the first step towards true pollination came about incidentally as a result of herbivorous feeding. Flying insects would have carried spores attached to their bodies, and any plant that, by chance, produced a structure, odour, or some other cue that attracted insects would be at a selective advantage. Plants evolved structures, such as carpels, to protect their seeds from being eaten by herbivores, while at the same time ensuring that pollinators could locate them and would continue to visit through the production of rewards such as nectar. Flying insects need a carbohydrate fuel for energy and nectar is typically rich in sucrose, glucose, and fructose. Pollen is a protein-rich food, but plants that are pollinated by insects unable to utilize pollen directly (e.g. butterflies) produce nectar with amino acids and proteins. Since the first flowering plants appeared in the Cretaceous, co-evolutionary processes have produced a wealth of flower size, shape, and colour, and the appearance of groups of insects that are tied to them, to a greater or lesser degree. Many groups of insects visit flowers, in particular, beetles, flies, moths and butterflies, and bees. Mutualistic adaptations are numerous. Plants that evolved long, tubed flowers were tracked by species that evolved similarly elongated mouthparts to cope. In this manner, non-specialist flower visitors are defeated to the mutual advantage of plant and pollinator. The end result of evolution in plants that ensure extreme pollinator constancy is that tightly co-evolved groups of plants and pollinators occur. Fig wasps (Agaonidae) and fig trees are totally dependent on each other for survival. The trees rely on the wasps for pollination and the wasps, which have complex life cycles, can only reproduce inside figs (see hymenopteran superfamilies). Bees (see Apoidea) are considered the most sophisticated plant pollinators and feed their young on a diet of nectar and pollen. Occasionally, the plants seem to benefit more than the pollinators do. Orchids of the genus *Ophrys* have evolved flowers that mimic the shape and sexual odour of certain female bees. Males attempting to copulate with the flowers get a pollinium or pollen packet stuck to their backs. When the bee is pseudo-copulating with another flower, the pollen in transferred. Very complex relationships exist between certain South American orchids and about 200 species of orchid bees (Apidae: Apinae: Euglossini). Only the males of these long-tongued, fast-flying bees visit the orchids. The bees collect the orchid's fragrance using pads on their front legs, and the scented oils are then passed back to the rear tibiae, which are filled with spongy storage tissue. While this is occurring, the bee gets a pollinium stuck to a part of its body. The exact location of the pollinium is important, and several closely related species of orchids might be reproductively isolated by this simple mechanism. It is not yet clear exactly how the bees use the scents, as they attract only males; perhaps females are attracted to aggregations of males and then select a mate on other criteria.

Insect structure

For all their diversity, insects are remarkably similar in overall design, both inside and out. The basic plan has been reworked many times over to produce a multitude of variants based on three main tagmata or body regions: the head, thorax, and abdomen.

The head is formed from six fused segments, and carries the compound eyes, secondary, light-receptive organs called **ocelli**, the antennae, and the mouthparts.

> **ocelli**
> simple, light-receptive, non-image forming organs on the head of many insects

The mouthparts may be modified according to diet, allowing the sucking or lapping of liquids, or the chewing and grinding of solid foods. The thorax is the powerhouse of the insect and is made up of three segments, each bearing a pair of legs. The posterior two thoracic segments usually each bear a pair of wings. The legs, which are each made up of a number of articulated components—coxa, trochanter, femur, tibia, and tarsus—can be greatly modified to serve a variety of functions. The abdomen, which is usually made up of eleven segments, contains the digestive and reproductive organs, and carries the internal and external genitalia. The latter are often species specific and, therefore, of great value to taxonomists.

On the inside, the major systems are the nervous system, the circulatory system, the digestive system, the respiratory system, and the reproductive system.

Sense organs

From the CNS, described earlier, arises the peripheral nervous system. This comprises sensory nerves bringing information from sensory receptors into the CNS, motor nerves that control the action of muscles, and the stomatogastric nervous system, which innervates the gut. The compound eyes are the main visual sense organs of adult insects. The eyes are made up of individual light receptive units called **ommatidia**, and there may be anything from only one or two in some ants to more than 10,000 in dragonflies.

> **ommatidia**
> individual light receptive units making up the compound eyes of insects

Each ommatidium comprises a clear lens system and light-sensitive retinula cells. In day-flying insects, the image received by the eye as a whole is made up of a mosaic of spots of differing light intensity from all the separate ommatidia. The more ommatidia present, the greater will be the insects' visual acuity. The eyes of night- or dusk-flying insects have a different internal construction and sacrifice visual acuity in favour of light sensitivity. Colour vision has been shown to occur in all orders of insects, and some can detect the plane of polarized light. In general, insects 'see better' at the blue end of the spectrum than at the red end, and, in some groups, the range of sensitivity carries on into the ultraviolet.

Hairs on the cuticle surface are responsive to vibrations, touch, and sound. Special hearing structures called tympanal organs may be present on various parts of the body (legs, wings, thorax, abdomen, and antennae). Depending on the species, these organs are responsive to sound frequencies ranging from less than 100 Hz to 240 kHz. Hair beds and arrays of small hair-like sensilla on the cuticle at joints of legs and segments provide information about the relative position of body parts. Strain gauges called campaniform sensilla respond to stresses in the cuticle, and internally, stretch receptors sense distension of muscles and parts of the digestive tract. Chemical sense organs or chemoreceptors are present on the mouthparts, antennae, tarsi, and other parts of the body. Insects smell by means of olfactory sensilla. These structures are found mainly on the antennae and can be present in very large numbers to detect remarkably low odour concentrations. In some cockroaches, there may be up to 100,000 of these sensilla. Many insects have temperature and humidity sensors, and some can detect magnetic fields and infrared radiation. Closely associated with the nervous system is a complex array of hormonal organs. Together, the nervous and hormonal systems regulate insect behaviour and physiology.

Circulatory system

The circulatory system is quite simple. The internal organs are bathed in haemolymph, which is moved around the body cavity by means of a combined heart and aorta lying just under the dorsal body surface. This dorsal organ has a number of openings called ostia along its length. In very simple terms, body fluids enter through the ostia and, as muscular contractions proceed from back to front, the fluids are passed towards the head end. The haemolymph does not move back through the ostia because they act as one-way valves. The main function of the haemolymph is concerned with nutrient and waste transport, although it also has a role in temperature control, protection against parasites and diseases, and as a hydrostatic skeletal element.

Digestive system

The digestive system of insects is essentially a long, open-ended tube with special regions for grinding and storing food, the production of enzymes, and the absorption of nutrients. The fore- and hindguts, being invaginations of the epidermis, are lined with cuticle. The unlined midgut is the main area for enzyme production and nutrient absorption. The nutrients (carbo-hydrate, fats, proteins, vitamins, etc.) needed by insects are essentially the same as for all other animals. The midgut epithelium of insects secretes a mucopolysaccharide layer called the peritrophic membrane, which envelops the food as it enters from the foregut. The membrane prevents bacteria, and perhaps parasites, from contacting the epithelial cells directly. However, it mainly works as a filter to prevent the passage of large molecules and food particles through its pores. Digestion takes place in a series of stages. The first stage occurs inside the peritrophic membrane, where large molecules are broken to smaller ones. The second stage occurs in the space between the peritrophic membrane and the gut wall, where small molecules are broken down. Lastly, final digestion and absorption of nutrients takes place on the surface of the midgut epithelial cells. The peritrophic membrane is produced continuously and eventually passes out of the anus with food residues. In some beetles, it continues to be produced after feeding has stopped just prior to pupation. It is issued from the anus as a thread-like ribbon and the larva uses it to make a cocoon. In most animals, the removal of the toxic waste products of metabolism requires their solution in water. For small terrestrial animals like insects, loss of water is critical, and a different system is employed. At the junction of the mid- and hindgut, there are a number of long, slender, blind-ended ducts called Malpighian tubules. These tubules extract the waste products from the haemolymph in which all the internal organs are bathed. In effect, they continually produce the insect's urine, which is moved to the hind gut. Valuable water and salts are reabsorbed by the hindgut and rectum, and the excrement is voided mainly as non-toxic and insoluble uric acid granules.

Respiratory system

The respiratory system of insects takes the form of an incredible network of branched tubes, which take gases directly to the cells of the body. The openings of this tracheal system, called spiracles, are located on the sides of each abdominal segment and on the thorax. No insect has more than ten pairs of spiracles, and the number is often reduced. In the aquatic nymphs of dragonflies, damselflies, mayflies, and stoneflies, which have tracheal gills, there are no functional spiracles. Spiracles can be closed by means of valves to reduce water loss. The main tracheae are spirally thickened like the flexible hose of a vacuum cleaner, to prevent them from collapsing. The tracheae are often swollen into air sacs, and insects can set up a flow of air through their tracheal system by means of body movements and the rhythmic flattening of these sacs. The tracheae branch continuously, getting finer and finer, until they become very fine intracellular tubes known as tracheoles, which are < 1 μm in diameter. Gases are exchanged by diffusion of oxygen into cells and carbon dioxide out of the cells. Several studies have shown that caterpillars have a special system analogous to the lungs of vertebrates (see Lepidoptera). Aquatic insect species show many interesting adaptations to life underwater, such as gills, air stores, and breathing siphons.

Reproductive system

The reproductive system of insects consists of a pair of ovaries in the female and a pair of testes in the male. Each ovary is joined by its own lateral oviduct to a main or common oviduct. The common oviduct leads to the single vagina or genital chamber. Each testis is joined by means of a duct, known as the vas deferens, to a common ejaculatory duct. This common duct takes the sperm to the external genitalia, of which the aedeagus (or penis) is the most important component. Eggs are fertilized by sperms stored inside the body of the female. The fertilization of eggs by sperm is internal and requires the male to introduce sperm by means of an aedeagus or penis via the female's vagina. Sperm is deposited in or near a sperm-storage organ, known as the spermatheca, and is subsequently used to fertilize the eggs. In some insects, a single mating will provide the female with all the sperm she will ever need, while in others, sperm cannot be stored indefinitely and the insects mate repeatedly to maintain a fresh supply.

Conclusion

Before we end this brief introduction, we would like to answer a question often asked by students: Why are there no truly marine insects? To our eyes, the sea can seem a dangerous and inhospitable place, but reasonably constant temperatures (certainly less of a range than found on land), coupled with fairly constant ionic concentrations, make the seas (physiologically speaking) about as benign an environment as there is on Earth. Life in freshwater is much more difficult osmotically, yet only a small fraction of 1% of all insect species live in marine habitats. Many insect orders contain a few species with marine associations, but the majority of species that spend at least part of their life cycle around the seashore, in brackish water as well as on the surface of the ocean, belong to the orders Hemiptera, Coleoptera, Diptera, and Trichoptera, among others. A few species of small predatory bugs with the evocative common name of Coral-treaders belong to the exclusively southern hemisphere family, the Hermatobatidae. The detailed biology of only one or two species is known, but they hide in coral and rock crevices and only emerge to feed at low tide. The most remarkable insects to have colonized saltwater are a few species of caddisfly species, whose larval and pupal stages are found in rock pools along the coast of New Zealand and south-east Australia (see Trichoptera).

Of the various types of marine environment, the least populated by insects are pelagic and coastal water habitats. Members of the hemipteran family Gerridae (pond skaters) are predacious bugs adapted to life on the surface of water. The vast majority live on freshwater, but five genera contain marine species. Of special interest are the 50 or so wingless species of ocean strider belonging to the genus *Halobates*. Most of these are confined to mangroves and coastal regions, but at least five species are only found far from land on the open oceans of the southern hemisphere. They are pelagic in as much as they live independently of land, but they live *on* saltwater, not *in* it. Lice in the anopluran family Echinophthiriidae are parasitic to seals, walruses, and sea lions. Although reproduction takes place when their hosts are on land, the lice are well adapted to long periods (many weeks or months) of feeding on their hosts while at sea. The lice are vertically transmitted (from mother to offspring) before the pups go to sea and, like all parasitic lice, there is no stage where they are independent of their hosts.

The greatest numbers of marine-associated insects are found in intertidal zones, salt marshes, mangroves, and mudflats, where they are phytophagous on salt-tolerant plants, predators, or scavengers. However, these inhospitable habitats are as much extensions of the terrestrial environment as of the marine. The question 'why don't insects live in the sea?' is easily answered. There is no physiological reason why insects could not live in saline water; in fact, it could make things a bit easier. Simply, most of the features that have made insects so successful on land would be of no great advantage relative to existing marine arthropods (the Crustacea), whose member species already fill every available niche. Put simply: insects just don't have a chance.

A final word must be on biodiversity. An increasing interest in cataloguing the Earth's biological diversity made it fashionable, for a while, to talk about larger and ever-larger numbers of undiscovered species in terrestrial and marine habitats. Some of the estimates for the total number of extant insect species, based on tropical forest samples, even reached the 100 million mark. Larger estimates might seem more impressive, but there came a point at which the brakes were applied. A number of studies suggest that there might not actually be quite as many species as originally thought. Undoubtedly, there are more species not described than described, but the figure probably lies somewhere in the range of 3–8 million, not 30–80 million. All species do not have an equal chance of being described. Small, rare species are much less likely to be named than larger common species. Not all things collected are ever described, and, if they are, the time from collection to description is getting longer and longer. Apart from the serious problem of the dwindling number of trained taxonomists capable of contributing to the immense task, or the lack of funds available to train new ones, one of the major difficulties in biodiversity studies is **synonymy**. This is where a species has been described and named more than once, sometimes by different authors and sometimes by the same author. Taxonomic revisions have shown that, in some taxa, species may have more than a dozen different names. Overall, it is estimated that about 20% of all species names are not valid. Studying the patterns of species description in different groups, especially the very large ones, would provide interesting insights. One such study on the Geometridae, a large and widely distributed family of moths, suggests that the actual number of species is nowhere near the estimation of an order of magnitude greater than the number of described species. Despite this, there are still plenty of places around the globe where every other species collected is likely not to have been described.

Further reading

Alborn, H. T., Turlings, T. C. J., Jones, T. H., Stenhagen, G., Loughrin, J. H., and Tumlinson, J. H. (1997). An elicitor of plant volatiles from beet armyworm oral secretion. *Science* **276**: 945–9.

Carlson, S. D., Juang, J., Hilgers, S., and Garment, M. B. (2000). Blood barriers of the insect. *Annual Review of Entomology* **45**: 151–74.

Chan, W. P., Prete, F., and Dickinson, M. H. (1998). Visual input to the efferent control system of a fly's gyroscope. *Science* **280**: 289–92.

Choe, J. C. and Crespi, B. (eds) (1997). *Mating systems in insects and arachnids*. Cambridge University Press, Cambridge.

Clark-Hachtel, C. M. and Tomoyasu, Y. (2016). Exploring the origin of insect wings from an evo-devo perspective. *Current Opinion in Insect Science* **13**: 77–85.

Clark-Hachtel, C. M. and Tomoyasu, Y. (2020). Two sets of candidate crustacean wing homologues and their implication for the origin of insect wings. *Nature Ecology and Evolution* **4**: 1694–1702.

Dudley, R. (2000) *The biomechanics of insect flight: form, function, evolution*. Princeton University Press, Princeton, NJ.

Ellington, C. P., van den Berg, C., Wilmott, A. P., and Thomas, A. L. R. (1996). Leading edge vortices in insect flight. *Nature* **384**: 626–30.

Ertas, B., von Reumont, B. M., Wägele, J.-W., Misof, B., and Burmester, T. (2009). Hemocyanin suggests a close relationship of Remipedia and Hexapoda. *Molecular Biology and Evolution* **26**: 2711–18.

Gaston, K. J. (1991). The magnitude of global insect species richness. *Conservation Biology* **5**: 283–96.

Giribet, G. and Edgecombe, G. D. (2012). Reevaluating the arthropod tree of life. *Annual Review of Entomology* **57**: 167–86.

Giribet, G. and Edgecombe, G. D. (2019). The phylogeny and evolutionary history of arthropods. *Current Biology* **29**: R592–602.

Grimaldi, D. A. and Engel, M. S. (2005). *Evolution of the insects*. Cambridge University Press, Cambridge.

Gullan, P. J. and Cranston P. S. (2014). *The insects: an outline of entomology*, 5th edn. Blackwell Science, Oxford.

Fanenbruck, M., Harzsch, S., and Wägele, J.-W. (2004). The brain of the Remipedia (Crustacea) and an alternative hypothesis on their phylogenetic relationships. *Proceedings of the National Academy of Sciences of the United States of America* **101**: 3868–73.

Haug, C. and Haug, J. T. (2017). The presumed oldest flying insect: more likely a myriapod? *PeerJ* **5**: e3402.

Herre, E. A., Jandér, K. C., and Machado, C. A. (2008). Evolutionary ecology of figs and their associates: recent progress and outstanding puzzles. *Annual Review of Ecology, Evolution and Systematics* **39**: 439–58.

Linz, D. M. and Tomoyasu, Y. (2018). Dual evolutionary origin of insect wings supported by an investigation of the abdominal wing serial homologs in *Tribolium*. *Proceedings of the National Academy of Sciences of the United States of America* **115**: E658–67.

Nilsson, L. A. (1992). Orchid pollination biology. *Trends in Ecology and Evolution* **7**: 255–9.

Osorio, D., Averof, M., and Bacon, J. P. (1995). Arthropod evolution: great brains, beautiful bodies. *Trends in Ecology and Evolution* **10**: 449–54.

Pellmyr, O. (1992). Evolution of insect pollination and angiosperm diversification. *Trends in Ecology and Evolution* **7**: 46–9.

Prokop, J., Pecharová, M., Nel, A., Hörnschemeyer, T., Krzemińska, E., Krzemiński, W., and Engel, M. S. (2017). Paleozoic nymphal wing pads support dual model of insect wing origins. *Current Biology* **27**: 263–9.

Regier, J., Shultz, J., Zwick, A., Hussey, A., Ball, B., Wetzer, R., Martin, J. W., and Cunningham, C. W. (2010). Arthropod relationships revealed by phylogenomic analysis of nuclear protein-coding sequences. *Nature* **463**: 1079–83.

Seimann, E., Tilman, D., and Haarstad, J. (1996). Insect species diversity, abundance and body size relationships. *Nature* **380**: 704–06.

Simpson, S. J. and Douglas, A. E. (eds) (2014). *R. F. Chapman—The insects: structure and function*, 5th ed. Cambridge University Press, Cambridge.

Walker, S. M. and Taylor, G. K. (2021). A semi-empirical model of the aerodynamics of manoeuvring insect flight. *Journal of the Royal Society Interface* **18**: 20210103.

SECTION 2

The insect orders

All six-legged animals belong to a single superclass, the Hexapoda of which the Insecta forms the largest class (circa 1 million spp). The three non-insect classes are the Protura (~ 800 spp.), Diplura (~ 800 spp.), and the Collembola (> 9,000 spp.). These three classes are known as 'entognathous hexapods', which have jaws inside an eversible pouch in the head. The figure below shows the subdivisions of the Hexapoda, and the latest insect tree of life.

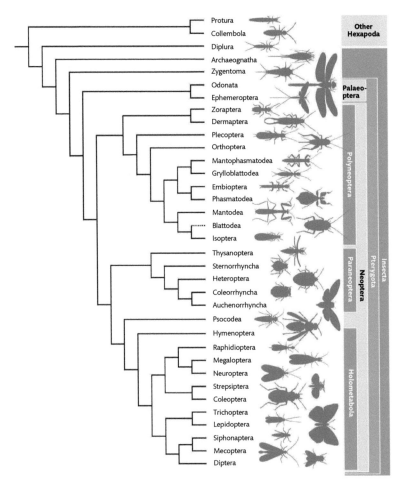

Phylogeny of hexapods as resolved from the most comprehensive phylogenomic study (Misof et al., 2014) at the time of writing this book. Note that the position of Psocodea as sister to the Holometabola is probably artefactual, and that termites (Isoptera) are included within Blattodea. Adapted from © Hans Pohl.

For practical purposes, the most commonly used level of classification is the order. These correspond roughly to familiar English names for different types of insects, flies, beetles, butterflies and moths, earwigs and so on, but also include less well-known groups like bristletails or ones without vernacular English names like the Strepsiptera. The 28 recognized insect orders are described here in a sequence that reflects their taxonomic grouping (i.e., Archaeognatha, Zygentoma, Palaeoptera, Neoptera, and so on). The classification in this book is based on the latest molecular and morphological systematic studies, which are provided as further reading at the end of each order.

Essential Entomology. Second Edition. George C. McGavin and Leonidas-Romanos Davranoglou, Oxford University Press.
© George C. McGavin and Leonidas-Romanos Davranoglou (2022). DOI: 10.1093/oso/9780192843111.001.0001

Insect groups that predate the evolution of wings

Ancestrally wingless insects, also sometimes known collectively as the apterygotes, do not undergo any sort of metamorphosis, that is, they are **ametabolous**. The young stages resemble the adults in all respects except size and sexual maturity and moulting continues until death. These are the most ancestral of all insects alive today, and the only descendants of a much larger array of wingless species that once existed. They have persisted in specialized cryptic microhabitats and have retained several ancestral features that likely date back to the Carboniferous times (300 million years ago, mya). The only two surviving orders are superficially similar in overall appearance but are in fact independent monophyletic lineages with subtle morphological differences.

Essential Entomology. Second Edition. George C. McGavin and Leonidas-Romanos Davranoglou, Oxford University Press.
© George C. McGavin and Leonidas-Romanos Davranoglou (2022). DOI: 10.1093/oso/9780192843111.001.0001

Archaeognatha
(bristletails–alternative name: Microcoryphia)

Common name	Bristletails	Metamorphosis	Ametabolous
Derivation	Gk. *archaeo*–ancient; *gnathos*–jaw	Distribution	Worldwide
		Number of families	2
Size	Body length 7–25 mm	Known world species	550 (0.05%)

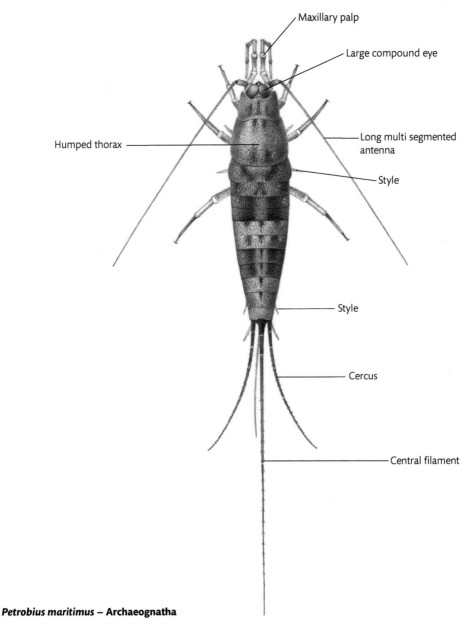

Maxillary palp

Large compound eye

Long multi segmented antenna

Humped thorax

Style

Style

Cercus

Central filament

Petrobius maritimus – **Archaeognatha**

Essential Entomology. Second Edition. George C. McGavin and Leonidas-Romanos Davranoglou, Oxford University Press.
© George C. McGavin and Leonidas-Romanos Davranoglou (2022). DOI: 10.1093/oso/9780192843111.001.0001

Key features

- ancestrally wingless
- elongate, cylindrical body, covered with small scales
- jumping ability
- scavengers
- in litter, soil, among coastal rocks and cliffs, some in dry regions

© Rupert Soskin

> Jumping bristletails are the most evolutionarily primitive of all living insects.

Archaeognathans are **elongate and cylindrical** in cross-section with a noticeably **humped thorax** and an exoskeleton covered with scales. The head has a pair of **long, multi-segmented antennae, large compound eyes that touch on top of the head,** and **three ocelli** (single-faceted, simple eyes). Beneath the **elongate, seven-segmented maxillary palps** there are **downwards-facing mouthparts.** The jaws, which have a single or two points of articulation, operate with a vertical, auger-like action. Single-point (monocondylic) jaws were considered as a primitive feature that defined Archaeognatha and distinguished them from all other insects, including the Zygentoma, all of which have two points of articulation (dicondylic). It is now known that many Archaeognatha have dicondylic jaws as well, which pushes the appearance of complex jaws even earlier in insect evolution. In addition to the three pairs of thoracic legs, there are **accessory appendages** called styles, which support the abdomen. These structures, which have their own muscles, are present on abdominal segments 2–9 and assist when running by supporting the abdomen over uneven or steep surfaces. There are also one or two pairs of eversible vesicles located on the underside of abdominal segments 1–7, which are involved in absorbing water and assisting with moulting. The abdomen has a pair of **multi-segmented cerci** and a **much longer central filament.**

Bristletails in the family Machilidae are often nocturnal, eating algae, lichens, and plant debris. Large populations can be found living in cliffs and crevices in old stone walls. The Meinertellidae are diurnal and scavenge sometimes in very dry environments, for example, deserts and dunes. For only a handful of species is a detailed life history well known. *Petrobius maritimus,* a common bristletail species living on rocky shores in the northern hemisphere, grazes at night on lichens and algae-growing rocks and is active even in near-freezing conditions.

Immature archaeognathans may moult up to ten times and may take up to two years to reach sexual maturity after which they may live for another two years. Some species have been known to moult more than 50 times during the course of their lives and many species use their adhesive coxal vesicles to stick themselves safely to vertical or inclined surfaces during moulting. Courtship and mating, which is quite simple, may involve some sort of dance. Sperm transfer is indirect but achieved in different ways. In the Meinertellidae the male deposits a packet of sperm on a silken thread or stalk attached to the ground during courtship. The male may manoeuvre the female into a position such that she can pick it up. In Machilidae, sperm transfer is one step nearer to being direct in that the male actually places sperm on the female's ovipositor, while in others the male deposits numerous small droplets of sperm on a silk thread, which he holds stretched. The female picks up the sperm by means of her ovipositor. Female bristletails lay batches of 10–15 soft, round eggs in cracks and crevices. Good egg-laying sites may contain +100 eggs.

Despite being descendants of the first insect lineages, the evolutionary origins of archaeognathans are particularly challenging to pinpoint, particularly due to their sparse fossil record. Putative Archaeognatha have been found as far back as the Late Devonian (more than 380 mya). Confirmed archaeognathans have been described from exceptionally preserved fossils from the Early Triassic (240 mya), although the origin of the order is undoubtedly much older.

Key reading

Adis, J. and Sturm, H. (1987). On the natural history and ecology of Meinertellidae. (Archaeognatha, Insecta) from dry land and inundation forests of central Amazonia. *Amazoniana* **10**(2): 197–218.

Blanke, A., Machida, R., Szucsich, N.U., Wilde, F., and Misof, B. (2014). Mandibles with two joints evolved much earlier in the history of insects: dicondyly is a synapomorphy of bristletails, silverfish and winged insects. *Systematic Entomology* **40**: 357–64.

Edwards, J. S. (1992). Adhesive function of coxal vesicles during ecdysis in *Petrobius brevistylis* Carpenter (Archaeognatha: Machilidae). *International Journal of Insect Morphology and Embryology* **21**: 369–71.

Mendes, L. F. (2018). Biodiversity of the Thysanurans (Microcoryphia and Zygentoma). In: *Insect biodiversity: science and society, volume* **II**. Foottit, R. G. and Adler, P. H. (eds), pp. 153–98. John Wiley & Sons Ltd., Hoboken.

Montagna, M., Haug, J., Strada, L., Haug, C., Felber, M. and Tintori, A. (2017). Central nervous system and muscular bundles preserved in a 240-million-year-old giant bristletail (Archaeognatha: Machilidae). *Scientific Reports* **7**: 46016.

Sturm, H. (1986). Aspects of the mating behavior in the Machiloidea (Archaeognatha, Insecta). *Braunschweiger Naturkundliche Schriften* **2**(3): 507–18.

Zygentoma

(silverfish and firebrats—alternative name: **Thysanura**)

Common name	Silverfish, firebrat	**Distribution**	Worldwide but predominantly tropical
Derivation	Gk. *zyg*-bridge; *entoma*-insect	**Number of families**	4
Size	Body length 1.4–20 mm	**Known world species**	600 (0.05%)
Metamorphosis	Ametabolous		

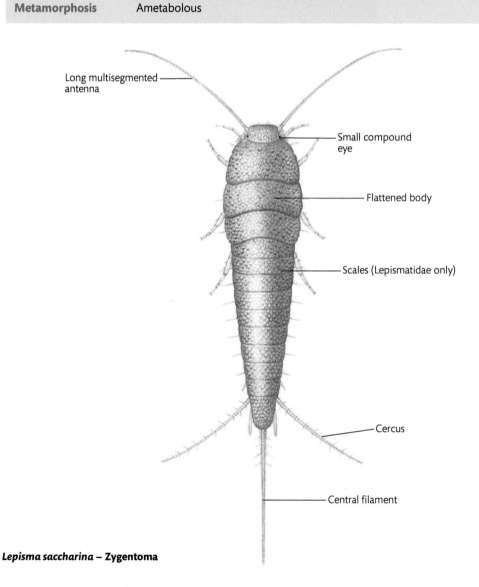

Long multisegmented antenna

Small compound eye

Flattened body

Scales (Lepismatidae only)

Cercus

Central filament

Lepisma saccharina – **Zygentoma**

Essential Entomology. Second Edition. George C. McGavin and Leonidas-Romanos Davranoglou, Oxford University Press.
© George C. McGavin and Leonidas-Romanos Davranoglou (2022). DOI: 10.1093/oso/9780192843111.001.0001

Key features

- ancestrally wingless
- elongate, flattened body, often with silvery scales
- fast running but do not jump
- scavengers
- in soil, litter, burrows, trees, ant nests, sometimes in buildings

> The preference of the firebrat (*Thermobia domestica*) for high temperatures and starchy foods led it to become a common resident of bakeries.

© Paul Brock

Many people will recognize these insects as the small, shiny, silver creatures that scuttle rapidly for shelter when disturbed after dark in a larder or bathroom, and some species damage books and papers and can be minor pests in kitchens. They are, however, quite diverse in shape and appearance and occupy a greater number of habitats than buildings. They can be elongate or oval, flattened or slightly convex, and may have body scales. Like the bristletails, the head has a pair of **long, multi-segmented antennae** but **the eyes, when present, are small and never touch each other on top of the head**. There may be **up to three ocelli**, although some species have none. The maxillary palps are shorter than those of the Archaeognatha with only five or six segments and the **downwards-facing jaws**, although still of a primitive design, **have two points of articulation with the head and act in the transverse plane. Accessory walking appendages or styles may be present on abdominal segments 2–9 but usually on fewer segments (7–9).** Pairs of water absorbing, eversible vesicles usually occur on abdominal segments 2–7, although in some species they are lacking altogether. The end of the abdomen carries **three terminal filaments** but unlike the Archaeognatha, they are all **more or less the same length**.

Zygentoma are mostly nocturnal and omnivorous scavengers and can be found free-living in litter or under bark, underground, in caves, in the burrows of certain mammals, or in association with ant and termite nests. *Atelura formicaria*, a species found in association with ants, is able to follow the pheromone trails laid down by foraging workers of *Lasius* species and it has been suggested that this behaviour is a dispersal mechanism. Some species can survive in habitats with little free water by being able to absorb atmospheric water though their rectum.

Species of the cosmopolitan family Lepismatidae are widespread in many habitats, including litter and in trees. Some lepismatids are almost exclusively found in human habitations. One of these, the silverfish, *Lepisma saccharina*, prefers damp microhabitats such as those found in cellars and bathrooms, while the firebrat, *Thermobia domestica*, can be found living in the warm conditions of kitchens and bakeries. Feeding studies on the firebrat have shown that gut cellulases are produced by the silverfish itself and not by symbiotic micro-organisms as in the guts of more advanced insects like the termites. The domestic species feed on a wide range of starchy materials, such as spilled flour, starched damp textiles and clothing, as well as silk, book bindings, wallpaper, and photographs.

Courtship may consist of a simple, synchronized *pas de deux*, while one species reproduces parthenogenetically, that is, without fertilization by a male. Just before mating males secrete silk threads, usually between two surfaces, onto which they deposit sperm droplets or packets. Receptive females take hold of the sperm using their ovipositor and draw it inside. Individuals may live for three or four years and, as in bristletails, moulting continues until death.

Due to their small size and elusive nature, silverfish are remarkably scarce in the fossil record. As a consequence, their deep evolutionary origins remain obscure. As with Archaeognatha, reports exist of tentative, difficult-to-identify fossils of Zygentoma from the Devonian. Confirmed relatives of modern representatives have been found in Cretaceous amber (100–110 mya), indicating that the current families were well established long ago.

Key reading

Christian, E. (1994). *Atelura formicaria* (Zygentoma) follows the pheromone trail of *Lasius niger* (Formicidae). *Zoologischer Anzeiger* **232**: 213–216.

Mendes, L. F. (2018). Biodiversity of the Thysanurans (Microcoryphia and Zygentoma). In: *Insect biodiversity: science and society, volume II*. Foottit, R. G. and Adler, P. H. (eds.), pp. 153–98. John Wiley & Sons Ltd., Hoboken.

Mendes, L. F. and Poinar, G. (2008). A new fossil silverfish (Zygentoma: Insecta) in Mesozoic Burmese amber. *European Journal of Soil Biology* **44**: 491–4.

Picchi, V. D. (1972). Parthenogenic reproduction in the silverfish *Nicoletia meinerti* (Thysanura). *Journal of the New York Entomological Society* **80**(1): 2–4.

Sturm, H. (1988). The mating behaviour of *Thermobia domestica* (Packard) (Lepismatidae, Zygentoma, Insecta). *Braunschweiger Naturkundliche Schriften* **2**(4): 693–712.

Sturm, H. and Bach de Roca, C. (1993). On the systematics of the Archaeognatha (Insecta). *Entomologia Generalis* **18**: 55–90.

Treves, D. S. and Martin, M. M. (1994). Cellulose digestion in primitive hexapods: Effect of ingested antibiotics on gut microbial populations and gut cellulase levels in the firebrat *Thermobia domestica* (Zygentoma, Lepismatidae). *Journal of Chemical Ecology* **20**: 2003–20

The winged insects

Insects with wings had evolved by the middle of the Carboniferous and their winged descendants make up 99.9% of all extant insect species. Wings now as then allow rapid escape from enemies, efficient mate-finding and dispersal to new feeding, mating, or egg-laying sites. Nevertheless, wings have been lost or reduced in many taxa for various reasons. The most common cause of secondary wing loss is due to the development of a specialized lifestyle. In particular, external parasites like fleas and lice are always wingless. Species in other orders that have parasitic relationships with other species, either living in the nests of termites or ants or feeding on vertebrate blood, are also often wingless or short-winged. Habitat stability over long periods, such as those found in caves, soil, and certain montane environments is another cause of secondary winglessness. In these cases, dispersal is no longer essential to the survival of the species, so not having functional wings and associated muscles is likely to increase fecundity and thereby select for winglessness. The cost of having wings and thus the number of flightless insect species is also greater in habitats where extremes of cold, hot, or wind prevail.

The infraclass Pterygota is divided into two unequal divisions. The mayflies (Ephemeroptera), comprising 0.3% of all insect species, and the dragonflies and damselflies (Odonata), comprising 0.5% of all insect species, are unable to fold their wings back along the body. This condition is known as palaeopterous (from Greek palaeos = old, ancient; pteron = wing). This characteristic wing morphology has given mayflies and dragonflies the name Palaeoptera, which assumes that the two orders are sister groups. This classification remains debated to this day, as the two orders are dissimilar in other regards, especially the occurrence in mayflies of a winged sub-adult stage. Some studies confirm the validity of Palaeoptera, whereas others suggest that both orders represent their own distinct division. The last, and by far the largest, subdivision of the Pterygota includes 99% of all insect species. This division is known as the Neoptera (from Greek neos = new: pteron = wing) and contains 24 orders.

Essential Entomology. Second Edition. George C. McGavin and Leonidas-Romanos Davranoglou, Oxford University Press.
© George C. McGavin and Leonidas-Romanos Davranoglou (2022). DOI: 10.1093/oso/9780192843111.001.0001

Division **Palaeoptera**

Ephemeroptera
(mayflies)

Common name	Mayflies	Metamorphosis	Incomplete (egg, nymph, adult)
Derivation	Gk. *ephemeros*–lasting a day; *pteron*–a wing	Distribution	Worldwide but especially temperate regions
Size	Body length 2–50 mm. Wingspan up to 50 mm	Number of families	42
		Known world species	3,666 (0.32%)

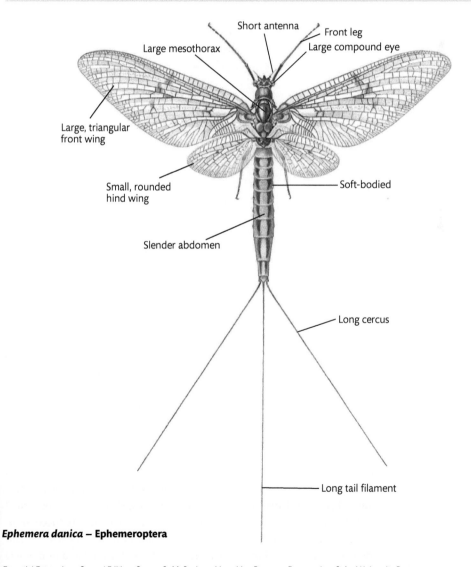

Ephemera danica – **Ephemeroptera**

Essential Entomology. Second Edition. George C. McGavin and Leonidas-Romanos Davranoglou, Oxford University Press.
© George C. McGavin and Leonidas-Romanos Davranoglou (2022). DOI: 10.1093/oso/9780192843111.001.0001

Key features

- aquatic nymphs
- separate winged subimago and adult stages
- two pairs of wings held vertically over body at rest
- adults short-lived and non-feeding
- mating swarms often seen over water
- essential component of freshwater food chains

© George McGavin

❯ Mayflies are unique in that they go through two moults once they have reached the adult stage. *Ephemera danica*, UK.

The Ephemeroptera, together with the Odonata, are the oldest groups of winged insects on Earth today. Ephemeroptera are unique in having a pre-adult winged stage called the subimago—they are the only insects that moult after having developed functional wings. This habit was probably much more common in several extinct Carboniferous and Permian orders, where immature stages had wing-like structures and moulted them throughout their lives. In two extant mayfly families, the Polymitarcyidae and the Palingeniidae, the role of reproductive female has been taken over by the subimago.

Mayflies are **soft-bodied** and relatively delicate insects with nearly **cylindrical bodies**, longish legs, and **two pairs of wings that cannot be folded back along the body**. The head has a pair of **short antennae**, a pair of **large, compound eyes**, and **three ocelli**. Adult mayflies, which are solely concerned with the purpose of reproduction, do not feed and have **very reduced, nonfunctional mouthparts**. The lifecycle is dominated by the nymphal stages and adults live for a very short time, mostly much less than a day, and in a few species, only a matter of minutes. All adults are fully winged, and the front pair of wings are large and triangular, while the hind wings are smaller and more rounded. The end of the abdomen bears **a pair of elongate cerci** and, usually, a single, **long tail filament** between them. The order used to be divided into two suborders: the Schistonota (sometimes called split-back mayflies) and the Pannota (fused-back mayflies). Although this system is no longer used, the terms Pannota–Schistonota are found in most studies (for historical reasons), and it is useful to know them. It is easy to tell mature nymphs of the two suborders apart. Schistonotan nymphs have their wing pads free along the midline whereas, in pannotan nymphs, the wing pads are fused along the midline of the body. However, the higher-level classification and phylogeny of mayflies is complex and highly debated and will likely undergo considerable changes as more representatives of the order are studied with molecular methods.

The aquatic nymphs moult many (10–50) times and can take up to three years to reach adulthood. Mayflies can be found in aquatic habitats of all kinds, from ditches, ponds, and lakes to slow-flowing rivers or fast-flowing streams. Some families, for example, the Baetidae, are found at higher altitudes and latitudes than any others. Different mayfly species are characteristic of different sorts of habitats. Some species are found in still waters, some crawl on gravely bottoms, while others live in burrows that they excavate in mud and silt. Habitat choice is often reflected in nymphal body shape. For instance, nymphs adapted for life in fast-flowing streams are streamlined, flattened, and may have special adhesive gill holdfast organs.

Widespread mayfly families

Family	Nymphal type: feeding habits	Examples of genera
Baetidae*	Active swimmers: saprophagous or herbivorous	*Baetis, Centroptilum, Cloeon, Pseudocloeon*
Ephemeridae*	Burrowers: saprophagous	*Ephemera, Hexagenia*
Heptageniidae*	Poor swimmers or sedentary: saprophagous	*Ecdyonurus, Epeorus, Heptagenia, Rhithrogena*
Leptophlebiidae*	Bottom crawlers: saprophagous or omnivorous	*Habrophlebia, Leptophlebia, Paraleptophlebia, Traverella*
Caenidae†	Crawlers: omnivorous	*Brachycerus, Caenis*
Ephemerellidae†	Crawlers: saprophagous, omnivorous, carnivorous	*Attenella, Ephemerella, Serratella*

* Schistonota; † Pannota

Mayfly nymphs have a closed tracheal system and instead breathe through lateral abdominal gills; they also possess a pair of feathery cerci and a terminal filament at the tip of the abdomen. They have chewing mouthparts and feed on a wide range of submerged plant and animal matter. Most species are detritivorous or herbivorous, but a few are carnivorous. Nymphs of *Ametropis neavei* (Ametropodidae) employ a unique method of suspension feeding:

the front legs, which have modified setae, are used to create a vortex inside specially made pits in sandy sediments. The vortex concentrates algae and fine organic particles suspended in the water at the bottom of the pit, where they are eaten. Vortices are also used as digging tools by other mayfly larvae. By adopting a particular body posture, the larva of *Pseudiron centralis* (Pseudironidae) is able to change the current flow locally to produce spiralling eddy currents which excavate the sediment. As the sand grains are washed away, small invertebrate animals become exposed and are swiftly consumed by the larva.

Commensal relationships have been shown to exist between the larvae of some chironomid midges and mayfly nymphs. The midge larvae, which feed on minute particles of detritus, cling to the mayfly nymph's gill filaments and may move to other locations such as the legs and thorax as they get older. The cerci of mayfly nymphs are used to detect the presence of foraging predators, such as stoneflies, and help them to take avoiding action, although, oddly, in laboratory experiments the removal of cerci in some species has not led to an increase in the chance of being eaten by stonefly nymphs and other predators. There is clear evidence that mayfly nymphs can perceive chemical cues from predators and will take evasive or defensive action when the physical presence of a predator is felt. Mayfly nymphs tend to moult during the day when predators like stonefly nymphs are least active.

When the nymphs are fully grown, they rise to the surface of the water where they moult into the dull-winged subimago, which then flutters to nearby vegetation. The nymphs of large species may crawl to the water's edge and moult to the subimago while clinging to plant stems or marginal stones. The final moult to the shiny-winged adult form can take place in less than an hour or after several days.

When conditions are right, large, sometimes massive, swarms of adults appear that have emerged simultaneously. In rare instances these have caused accidents, the bodies of dead and dying adults piling deep enough on roads to cause vehicles to skid. These mating swarms, which have a characteristic rising and falling flight pattern, take place often at dawn or dusk, over water, in clearings, or near obvious features of the landscape such as trees or bushes called swarm markers. Males of many species are 'bifocal'—they have their eyes divided into upper and lower portions; the upper portion of the eye is sometimes raised on short stalks. Enlarged facets in the upper region help the males see potential mates above them in a swarm. Males are able to recognize swarms of their own species and females of the same species, who recognize the same sorts of makers and enter the swarms to mate. In some species males jostle each other within the swarm but do not interfere with other males when they are actually copulating. There is evidence that males with bigger wings have more success at mating and this might be due to them being able to jostle more effectively. Jostling does not occur in all species but there is some evidence that larger wing size is attractive to females. Males use their long front legs to hang on their chosen partner, engaging their tarsi into special regions on the sides of the female's thorax. Uniquely among the insects, male mayflies have paired reproductive organs which are inserted into paired genital openings in the female. When locked together, copulating pairs gradually sink out of the mating swarm. Mated females either lay hundreds or thousands of eggs by dropping them singly, in small groups, or all at once in packets into the water. Some species release their eggs as they dip the end of their abdomen under the water's surface. The eggs of many species have thread-like extensions that unravel on contact with water and serve to keep the egg from being washed away. A few species retain the eggs within their bodies and give birth to live nymphs. Up to half of all ephemeropteran species may be parthenogenetic, either obligately, facultatively, or accidentally—perhaps as an insurance strategy for times when males have died too early—that is, before they met and mated with the females.

Mayflies are an extremely important component of freshwater food chains and are eaten by most predatory species, both vertebrate and invertebrate. As they make up a major part of the diet of trout and many other fish, it is not surprising that adult and nymphal mayflies are extensively used by anglers as models for tying fishing flies. In the world of fly fishing, subimagos are called duns and the adults are called spinners and mayfly 'artificials' have names like the Greendrake, the Lake Olive, Lunn's Particular, and Greenwell's Glory. Many different elements come together in fly fishing. A well-tied artificial is important but knowing the behaviour of the fish, as well as careful observation of what they are eating, is essential to success.

Many species are intolerant of pollution and are thus useful as water quality indicators. In many countries of the northern hemisphere the decrease in water pH due to acid rain and the reduction in dissolved oxygen due to pollution of various kinds has led to a marked reduction in the abundance of mayfly nymphs, which, in turn has caused a decline in the number of freshwater fish.

Mayflies have an unusual sort of economic impact in the damage they occasionally inflict on the paint work of new cars. Car manufacturers hold large numbers of new cars together in floodlit pounds often located on low-value, marginal land near wet areas or reservoirs. During hot summer months, the surface temperature of car roofs can be very high indeed. Egg-laden females, mistaking a large area of cars for the surface of water, alight to lay eggs and are instantaneously fried. The stricken insects roll over on their backs and the boiling contents of their guts spill out and seep under the outstretched wings, etching a neat insect shape in the paint.

Mayflies have a rich fossil record, with unambiguous representatives dating to the Permian (300 mya). Some fossils from the Carboniferous may represent early mayfly lineages, or close relatives that have since gone extinct. Modern mayflies probably appeared sometime in the Late Triassic–Early Jurassic (200 mya).

Key reading

Alba-Tercedor, J. and Sanchez-Ortega, A. (1991). *Overview and strategies of Ephemeroptera and Plecoptera*. Sandhill Crane Press, Gainesville.

Barber-James, H. M., Gattolliat, J.-L., Sartori, M., and Hubbard, M. D. (2008). Global diversity of mayflies (Ephemeroptera, Insecta) in freshwater. *Hydrobiologia* **595**: 339–50.

Brittain, J. E. (1982) Biology of mayflies. *Annual Review of Entomology* **27**: 119–47.

Campbell, I. C. (ed) (1990). *Mayflies and stoneflies: life histories and biology.* Series Entomologica series 44. Kluwer Academic Publishers, Dordrecht.

Edmunds, G. F. Jr. and McCafferty, W. P. (1988). The mayfly subimago. *Annual Review of Entomology* **33**: 509–29.

Harker, J. (1992). Swarm behaviour and mate competition in Mayflies (Ephemeroptera). *Journal of Zoology* **228**(4): 571–87.

Hubbard, M. D. (1990). *Mayflies of the world. A catalog of the family and genus group taxa (Insecta: Ephemeroptera).* Flora and fauna handbook series 8. Sandhill Crane Press, Gainesville.

Liegeois, M., Sartori, M., and Schwander, T. (2021). Extremely widespread parthenogenesis and a trade-off between alternative forms of reproduction in mayflies (Ephemeroptera). *Journal of Heredity* **112**: 45–57.

Macan, T. T. (1982). *The study of stoneflies, mayflies and caddisflies.* Amateur Entomologist's Society, London.

Misof, B., Liu, S., Meusemann, K. Peters, R. S., Donath, A., Mayer, C., Frandsen, P. B., et al. (2014). Phylogenomics resolves the timing and pattern of insect evolution. *Science* **346**: 7639–7767.

Ruffieux, L., Sartori, M., and L'Eplattenier, G. (1996). Palmen body: a reliable structure to estimate the number of instars in *Siphlonurus aestivalis* (Eaton) (Ephemeroptera: Siphlonuridae). *International Journal of Insect Morphology and Embryology* **25**: 341–4.

Simpson, S. J. and McGavin, G. C. (1996). *The right fly: An angler's guide to identifying and matching natural insects.* Aurum Press, London.

Soluk, D. A. and Craig, D. A. (1988) Vortex feeding from pits in the sand: A unique method of suspension feeding used by a stream invertebrate. *Limnology and Oceanography* **33**(4, pt 1): 638–45.

Soluk, D. A. and Craig, D. A. (1990). Digging with a vortex: Flow manipulation facilitates prey capture by a predatory stream mayfly. *Limnology and Oceanography* **35**(5): 1201–6.

Odonata
(dragonflies and damselflies)

Common name	Dragonflies and Damselflies	Metamorphosis	Incomplete (egg, nymph, adult)
Derivation	Gk. *Odon*–tooth	Distribution	Worldwide
Size	Wingspan 18–200 mm. Body length up to 150 mm	Number of families	30
		Known world species	5,952 (0.53%)

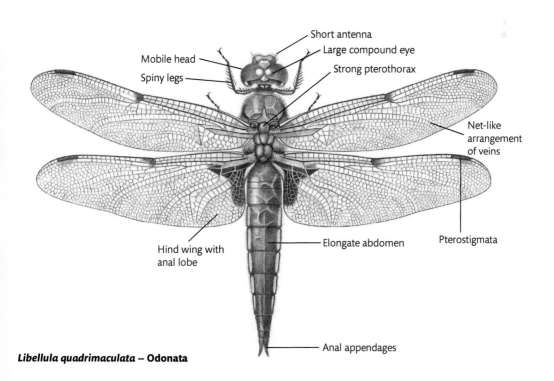

Mobile head

Spiny legs

Short antenna

Large compound eye

Strong pterothorax

Net-like arrangement of veins

Hind wing with anal lobe

Elongate abdomen

Pterostigmata

Anal appendages

Libellula quadrimaculata – Odonata

Essential Entomology. Second Edition. George C. McGavin and Leonidas-Romanos Davranoglou, Oxford University Press.

Key features

- aquatic nymphs, with labium modified into a prey-catching 'mask'
- adults with elongate bodies and two pairs of long, similarly sized wings
- often brightly coloured or metallic
- distinctive fast, hovering, or darting flight
- usually seen near or over water

❯ Dragonflies are among the most efficient aerial predators to have ever evolved.

❯ Many damselfly males possess conspicuous wing patterns which are thought to be attractive to females or used to ward off competing males. *Euphaea fraseri*, India.

Superlative flying skills with speeds over 50 kilometres per hour make these colourful insects instantly recognizable and enduringly popular with artists and poets. In many cultures, dragonflies have had mythological associations. Devil's darning needles, a common name for dragonflies, reflects people's fears, while in South America many believe that certain large, graceful damselflies (Pseudostigmatidae) are the spirits of recently dead humans. In several countries, dragonflies are considered a great delicacy, and are caught, threaded on skewers, and lightly grilled.

Although closely tied to fresh water by virtue of their aquatic nymphs, adults can fly many miles away in search of prey. Despite the fact that they are easily recognizable and distinctive insects, it is estimated that at least a quarter of species worldwide remains undiscovered. The order is split into two major suborders: the dragonflies (suborder: Epiprocta) and the damselflies (suborder: Zygoptera).

In general, the head, which is large and very mobile, **has biting, downwards-pointing mouthparts, short, hair-like antennae** composed of fewer than eight segments, **very large compound eyes**, and **three ocelli**. Dragonflies have round heads and very large eyes while damselflies have broader heads with widely separated eyes. The large eyes give virtually all-round vision and (as expected of these highly active, aerial hunters) they are able to resolve distant objects better than any other insect.

The enlarged thorax is packed with flight muscles to power the front and hind wings, which beat out of phase with each other, producing the whirring noise characteristic of larger species. The sound is produced by the back edge of the front wings brushing past the front edge of the hind wings at midstroke. Odonates can control the frequency and amplitude of their wing beat as well as the phase and angle of attack of the front and hind wings separately, and, as a result, can hover and fly upwards, sideways, and backwards at will. The basalar and subalar muscles that produce the downstroke or power stroke are attached directly to sclerites at the bases of the wings, just outside the hinge point. Contraction of these two sets of muscles can produce a downstroke either with the leading edge down (basalar muscle contracted) or a downstroke with the leading edge up (subalar muscle contracted). The upstroke for each wing is produced by two indirect dorso-ventral muscles (one anterior and one posterior) running from the top of the thorax (the tergum) to the bottom (the sternum) of the segment bearing the wing. Contraction of these muscles pull the tergum down and, as a result, the wing pivots upwards. The same as for the downstroke, the insect can control the angle of the wing on the upstroke by varying the power produced by the dorso-ventral muscles. Contraction of the anterior dorso-ventral muscle will raise the wing with the leading edge down, whereas contraction of the posterior dorso-ventral muscle will raise the wing with the leading edge up. Pairs of wings beat together (i.e., the two front or hind wings are either both up or both down at the same time) but by controlling what each set of flight muscles does, odonates can do different things on opposite sides enabling them to turn around within their own body length. It's a bit like rowing with one oar pulling forwards and the other pulling backwards at the same time to make a tight turn. Dragonflies and damselflies also have unusually low wing loading (mass to wing area), which contributes considerably to their aerobatic capabilities and explains why catching dragonflies can be a tiring occupation. The odonate arrangement of the flight musculature is unique among insects, and it probably arose 300 mya, when their much larger relatives, the Meganisoptera, with wingspans of a little under a metre, were wheeling and gliding through the Coal Measure swamp forests. Species such as aeshnids (hawkers), with long abdomens and their centre of mass well below and behind the wing bases, tend to be more stable and less manoeuvrable than those with short squat abdomens such as libellulids (darters and chasers), whose shape makes them less stable and thus more responsive in the air. Think of the difference between a passenger plane and a fighter plane.

Most species are active during the day and need to warm up their flight muscles to a certain temperature before taking off. Many species bask outstretched but, if they get too hot, they can adopt an obelisk position, pointing their abdomen towards the sun to minimize surface area. Large dragonflies are able to lower their thoracic temperatures when flying in very hot conditions by gliding for longer periods and by increasing haemolymph circulation to their abdomen, which acts as a radiator. While the larger species might only have to worry about vertebrate predators fast enough to catch them, smaller species can fall prey to other aerial predators, such as robber flies (Asilidae).

Apart from the size and shape of the head, adults of the two suborders can be distinguished easily in the field. Resting dragonflies tend to hold their wings sideways from the body at rest, and their hind wings are broader near their bases than the front wings. In damselflies, the wings are held together over the abdomen at rest, and both hind and front wings are of similar shape and narrow at their bases. As a result of the backwards slant of the thoracic segments, it may appear, especially in some of the damselflies, that the wings have been folded back along the body; however, they are in fact held directly above the thorax. Another effect of thoracic rotation has been to bring the legs forward. The three pairs of spiny legs are therefore in an ideal position to act as an in-flight prey capturing net or basket. The shorter, front pair allow in-flight feeding. Although adults have wide-ranging tastes, some species of social wasp are avoided. Dragonflies are generally active hunters, cruising to snatch prey in the air, from vegetation, or the ground, while the more delicate damselflies tend to sit and wait for suitably sized prey to appear before attacking.

The bright colours of odonates, which are involved in mate and rival recognition, very often fade quickly after death. Many species show **pruinescent (powdery) colouration** due to the morphological and biochemical structure of their cuticular waxes thought to be involved in intraspecific communication. The abdomens of males of some tropical species can change colour in response to a drop in ambient temperature. At 30–35°C the abdomen may be bright red or blue, but as the temperature falls and the insects become less active, pigment granules in the epidermis become clumped near the surface, causing a darkening to grey-black. It has been suggested that this might make them less easily seen by enemies.

Courtship is very variable within the order, being at one extreme non-existent—males simply seizing females as they pass. At the other end of the spectrum, courtship might be quite elaborate, with males luring females into a territory by means of special displays. Reproduction in these species can be energetically costly. In many species the mass of adults rises and may even double as they mature. In males this is due to additional flight muscles needed to hold territories, and in females it is due to the growth of ovaries. The size of territory a male holds depends on the species, the population density, and often on the size of the male himself—the larger ones being territory holders—while the smaller ones are consigned to inferior areas. Even so, these satellite males will try to sneak in to mate with a female, so territory holders have to defend their patch vigorously by constant patrolling, hovering, and the engaging in the occasional aerial skirmish. The whole exhausting business is worthwhile, as holding a territory ensures more matings.

Dragonflies and damselflies are unique among modern insects in that the male has secondary sexual organs located ventrally at the front end of the abdomen. Males transfer sperm from the primary genital organs on the ninth abdominal segment to the secondary sperm storage organ located on the underside of the second or third abdominal segment by curling their abdomen round. Once the secondary storage organ is loaded, the males are ready to copulate. The end of the male abdomen is equipped with clasping organs, which are used to hold the female by the head in dragonflies or, on the prothorax behind the head, in damselflies.

The pair, locked together in tandem, may mate straight away or after a nuptial flight. This male grip is maintained during mating, which usually takes place on a perch in the middle of the male's territory or in the air. Unresponsive females try to repel males by raising their front legs over their thorax. The need for protection is witnessed by the fact that the heads of many females are physically damaged by the force exerted by the male's abdominal claspers. To copulate, the female bends the tip of her abdomen round to join with the secondary genitalia of the male. In this posture, called the wheel position, sperm is transferred via the penis to the female's internal storage organ, the spermatheca. Females may mate several times, especially if they want to lay eggs in prime areas controlled by resident males. As in many other insects, male dragonflies and damselflies need to ensure that they will father the eggs laid, especially if they are making an investment in time and energy. Sperm competition is an effective means of ensuring paternity. In most damselflies the hard penis, located on the second abdominal segment, is first used to scrape out the sperm of other males from the female's spermatheca. The removal of sperm may take up most of the copulation time and even though the male may not be able to remove all of the previous male's sperm, it has been shown that last male sperm precedence is high. In many cases, the last male to mate before oviposition takes place will father more than 90% of the offspring. In species of damselfly that mate only once, the males do not have a sperm-removing apparatus, but instead copulate for periods as long as several hours, effectively excluding rivals. Recent work has shown that sperm is

♂

♀

∧ Wheel mating position in damselflies. The male grasps the female behind the head, and she curls her abdomen forward to couple her terminal genitalia with his, located at the base of his abdomen.

transferred either as free motile sperm in a featureless seminal fluid or aggregated into special structures called spermatodesms. Spermatodesms are found in species who do not defend a territory and do not guard the females, which tend to move away after copulation to lay their eggs. Free sperm is found in species that hold territories and guard their mates, who lay eggs immediately after mating. In the latter situation males have predictable access to females. Several ideas have been put forward to explain the function of spermatodesms. The structures may protect the sperm in some way, increase sperm life, or provide some benefit in terms of nutrients to the female.

Female dragonflies mostly lay their eggs directly into water or attach them to plants underwater, but species belonging to the family Aeshnidae and all damselflies lay eggs in slits they cut into marginal or submerged plant stems using an ovipositor. Males may retain in tandem or guard the females closely as they lay their eggs.

The aquatic nymphs (referred in old works as 'naiads') are highly predacious and equipped with a specialized prehensile modification of the labium for rapid prey capture. The labium, or facial mask, bears hooked or toothed labial palps and is hinged and can be shot forward from under the head in a matter of milliseconds to seize prey which, for larger nymphs, may include frogs and even small fish.

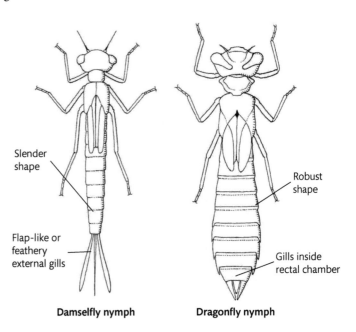

Slender shape

Robust shape

Flap-like or feathery external gills

Gills inside rectal chamber

❯ Nymphs.

Damselfly nymph   **Dragonfly nymph**

Odonate larvae respond to a range of mechanical, chemical, and visual cues that warn them of the presence of predators such as fish, water bugs, and beetles. In experiments, damselfly larvae responded to water from tanks that contained predatory fish that had been feeding on damselflies or small, sympatric fish, but not to water from tanks where the fish had been feeding on exotic prey such as mealworms.

Damselfly nymphs have a closed tracheal system with no spiracles. Instead they obtain their oxygen from the water by means of three, feathery or flap-like external gills located at the end of the abdomen. Known as caudal lamellae, they are used for swimming and evidence shows that damselfly nymphs use them to escape from dragonfly nymphs and similar predators. In dragonfly nymphs, the gills are internal and are located in the rectal chamber. Water is drawn in and forced out to provide a respiratory current over the gills. If required, water can be forced out of the rectal chamber at high speed to provide a jet-propelled escape. Some damselfly nymphs also pump water in an out of the rectum for a great deal of the time but for different reasons. The rate and stroke volume increases in the nymphs of some species if they are stressed or short of oxygen. It has also been shown that the pumping motions are concerned with the exchange of ions in the epithelium of the hindgut, in particular to maintain a good flow of water over the rectal pads, where special chloride cells are responsible for the uptake of inorganic ions. Constant movement of water around the caudal lamellae will also maximize gaseous exchange.

The habitats favoured by nymphs of different families ranges from water-filled tree holes and temporary pools of water to streams, rivers, and lakes. Some make burrows in swamps and others are able to hunt for prey after dark out of water, among marginal vegetation. The nymphs of several families have left the aquatic realm, and instead live in moist terrestrial environments, such as leaf litter, mosses, and logs.

Dragonfly nymphs will readily engage in cannibalism, particularly if there is a shortage of food and nymphs of different instars are present. The removal of younger instars promotes the synchronization of hatching and population development. Development is very much dependent on temperature as well as food supply, and in cool temperate regions nymphs may take several years before they crawl out of the water onto stones or up vegetation, to emerge as adults. They do not mate right away; instead, they spend some time feeding and maturing, gaining strength and body mass for the rigours of reproduction. Tropical species may have two or even three generations a year.

Although dragonflies sometimes capture and eat beneficial insects such as honey bees, they are not pests and may rather have a useful role in the control of natural insect populations. In some tropical countries, dragonfly nymphs have been kept in drinking water storage tanks to control the numbers of larval *Aedes aegypti* mosquitoes far more safely and economically than pesticides.

Odonata are very well represented in the fossil record, due to their large size, their dispersal capabilities, and their association with aquatic habitats, which facilitate fossilization. Extinct groups related to Odonata, such as the Meganisoptera and the Geroptera, represent the oldest known flying insects, likely originating in the Late Carboniferous (300 mya). Like other very old groups such as the mayflies, we are uncertain when the ancestors of modern Odonata appeared, although the Late Permian or Early Triassic (around 250 mya) are likely.

Key reading

Bybee, S. M., Ogden, T. H., Branham, M. A., and Whiting, M. F. (2008), Molecules, morphology and fossils: a comprehensive approach to odonate phylogeny and the evolution of the odonate wing. *Cladistics* **24**(4): 477–514.

Corbet, P. S. (1980). Biology of Odonata. *Annual Review of Entomology* **25**: 189–217.

Corbet, P. S., Longfield, C., and Moore, N. W. (1985). *Dragonflies*. Collins, London.

Dijkstra, K.-D. B., Bechly, G., Bybee, Dow, R. A., Dumont, H. J., Fleck, G., Garrison, R. W., Hämäläinen, M., Kalkman, V. J., Karube, H., May, M. L., Orr, A. G., Paulson, D. R., Rehn, A. C., Theischinger, G., Trueman, J. W. H., Van Tol, J., Von Ellenrieder, N., and Ware, J. (2013). The classification and diversity of dragonflies and damselflies (Odonata). *Zootaxa* **3703**(1): 36–45.

Fincke, O. M., Waage, J. K., and Koenig, W. D. (1997). Natural and sexual selection components of odonate mating patterns. In: Choe, J. C. and Crespi, B. J. (eds), *The evolution of mating systems in insects and arachnids*, pp. 58–74. Cambridge University Press, Cambridge.

Grimaldi, D. A. and Engel, M. S. (2005). *Evolution of the insects*. Cambridge University Press, Cambridge.

Johnson, D. (1991). Behavioural ecology of larval dragonflies and damselflies. *Trends in Ecology and Evolution* **6**(1): 8–13.

Miller, P. L (1994). The responses of rectal pumping in some zygopteran larvae (Odonata) to oxygen and ion availability. *Journal of Insect Physiology* **40**(4): 333–9.

Polcyn, D. M. (1994). Thermoregulation during summer activity in Mojave Desert dragonflies (Odonata: Anisoptera). *Functional Ecology* **8**: 441–9.

Rehn, A. C. (2003). Phylogenetic analysis of higher-level relationships of Odonata. *Systematic Entomology* **28**(2): 181–240.

Siva-Jothy, M. T. (1997). Odonate ejaculate structures. *Odonatologica* **26**(4): 415–37.

Theischinger, G. and Hawking, J. (2006). *The complete field guide to dragonflies of Australia*. CSIRO Publishing, Collingwood.

Waage, J. K. (1979). Dual function of the damselfly penis: sperm removal and transfer. *Science* **203**: 916–18.

Waage, J. K. (1984). Sperm competition and the evolution of odonate mating systems. In: *Sperm competition and the evolution of animal mating systems*. Smith R. L. (ed), pp. 251–90. Academic Press, Cambridge, MA.

Waage, J. K. (1986). Evidence for widespread sperm displacement ability in Zygoptera (Odonata) and the means for predicting its presence. *Biological Journal of the Linnean Society* **28**(3): 285–300.

Watson, J. A. L. (1982). A truly terrestrial dragonfly larva from Australia (Odonata: Corduliidae). *Australian Journal of Entomology* **21**: 309–11.

Division **Neoptera**

....

The arrangement of moveable sclerites at the base of the wings allows neopterans to fold their wings out of the way back along the abdomen when not in use. This ability has allowed species in these orders to colonize all kinds of specialized terrestrial microhabitats unavailable to palaeopterous adult insects. The Neoptera is split into two unequal divisions: the Exopterygota, where the immature stages are similar to the adults (incomplete metamorphosis, or hemimetaboly), and the Endopterygota, where the immature stages are very different from the adults and there is an intermediate pupal stage (complete metamorphosis, or holometaboly). The evolution of a pupal stage took place around 350 mya and was a major contributor to the success of insects, of which more than 80% belong to the Endopterygota. Several theories attempt to explain the advantages conferred by holometaboly, proposing factors such as protection from climatic instability, the exploitation of different resources between immatures and adults, the allocation of tissue growth to the larval stage, and tissue differentiation to the pupal stage, among others. It is likely that several of these factors acted in conjunction. We know that the earliest insects did not undergo metamorphosis (ametaboly), much like silverfish and bristletails today. The difference between hemimetabolous and holometabolous development might, at first, seem striking. In fact, some theories proposed that the holometabolan larva is a prolonged embryonic stage, and that the pupa corresponds to the immature stages of hemimetabolan insects. However, the latest developmental and endocrinological studies suggest that the immatures of both Exopterygota and Endopterygota in fact represent the same developmental stage, and that the holometabolan pupa corresponds to the last instar(s) of hemimetabolan insects. Future studies will undoubtedly further elucidate the origins and mechanisms of insect developmental modes.

Key reading

Belles, X. (2020). The evolution of metamorphosis. In: *Insect metamorphosis. From natural history to regulation of development and evolution*. Belles, X. (ed), pp. 251–72. Academic Press, Cambridge, MA.

Jindra, M. (2019). Where did the pupa come from? The timing of juvenile hormone signalling supports homology between stages of hemimetabolous and holometabolous insects. *Philosophical Transactions of the Royal Society B* **374**: 20190064.

Rainford, J. L., Hofreiter, M., Nicholson, D. B., and Mayhew, P. J. (2014). Phylogenetic distribution of extant richness suggests metamorphosis is a key innovation driving diversification in insects. *PLoS One* **9**(10): e109085.

Rédei, D. and Štys, P. (2016). Larva, nymph and naiad—for accuracy's sake. *Systematic Entomology* **41**: 505–10.

Rolff J., Johnston, P. R., and Reynolds, S. (2019). Complete metamorphosis of insects. *Philosophical Transactions of the Royal Society B* **374**: 20190063.

Essential Entomology. Second Edition. George C. McGavin and Leonidas-Romanos Davranoglou, Oxford University Press.
© George C. McGavin and Leonidas-Romanos Davranoglou (2022). DOI: 10.1093/oso/9780192843111.001.0001

Superorder **Exopterygota** (= Hemimetabola)

In these orders, the young stages are called nymphs and, in most cases, they feed on the same foods as the adults. Their wings develop on the outside of the body (exopterygote) and metamorphosis is simple or incomplete (hemimetabolous). An exception can be found in thrips (Thysanoptera) and many scale insects (Hemiptera: Coccoidea), which undergo a non-feeding, pupa-like stage during their development (neometabolous).

Polyneoptera

The Polyneoptera are one of the major lineages of Neoptera, and comprise earwigs (Dermaptera), angel insects (Zoraptera), grasshoppers, crickets, and katydids (Orthoptera), stick and leaf insects (Phasmatodea), webspinners (Embioptera), stoneflies (Plecoptera), cockroaches and termites (Blattodea), mantids (Mantodea), ice crawlers (Grylloblattodea), and the newly discovered heel walkers (Mantophasmatodea). We describe these orders in a sequence that reflects the taxonomic groupings proposed by the latest phylogenetic studies.

Haplocercata

This name (meaning simple-cerci in Greek) refers to a group of two closely related orders: earwigs (Dermaptera) and the elusive angel insects (Zoraptera), which are characterized by simple, one-segmented cerci (albeit of remarkably different morphology and function). The evolutionary relationships of either order had remained enigmatic until recently, when morphological and molecular studies established their place in the insect tree of life.

Dermaptera
(earwigs)

Common name	Earwigs	Distribution	Worldwide, predominantly tropical regions
Derivation	Gk. *Derma*–skin, hide; *pteron*–a wing		
Size	Body length 3–85 mm	Number of families	12
Metamorphosis	Incomplete (egg, nymph, adult)	Known world species	1,978 (0.17%)

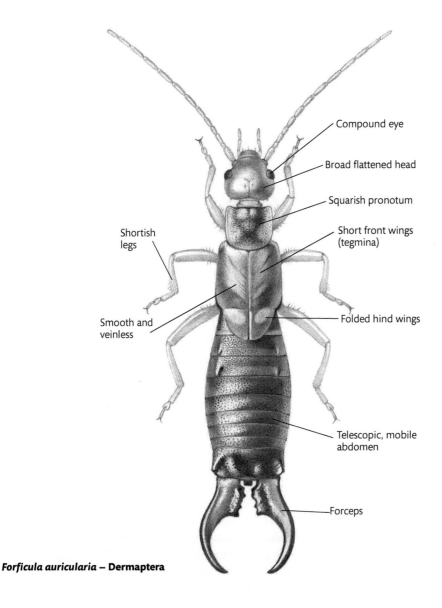

Compound eye

Broad flattened head

Squarish pronotum

Short front wings (tegmina)

Shortish legs

Folded hind wings

Smooth and veinless

Telescopic, mobile abdomen

Forceps

Forficula auricularia – **Dermaptera**

Essential Entomology. Second Edition. George C. McGavin and Leonidas-Romanos Davranoglou, Oxford University Press.
© George C. McGavin and Leonidas-Romanos Davranoglou (2022). DOI: 10.1093/oso/9780192843111.001.0001

Key features

- elongate and slightly flattened
- abdomen telescopic with distinctive terminal, forceps-like cerci
- first pair of wings short and hardened
- second pair of wings fan-shaped and foldable
- prefer confined, humid microhabitats such as soil, litter, or under bark

> A pair of *Forficula dentata* browsing for nectar. Note that the male has larger cerci than the female. UK.

© Paul Brock

Mostly drab, nocturnal and generally reluctant to fly, it is possible to distinguish the vast majority of earwigs from all other insects by their **distinctive abdominal forceps.** The eyes may large, small, or absent, the antennae are **long, thread-like, and multi-segmented,** and there are **no ocelli.**

The order is divided into one extant suborder, the Neodermaptera, and two extinct suborders. The most peculiar members of Dermaptera belong to the families Arixeniidae (five species) and Hemimeridae (11 species), which are blind, wingless, and their slender, unsegmented cerci are much less well developed.

Arixeniid earwigs are stout, fairly hairy species found in south Asia, where they live epizoically (i.e., non-parasitically on the body surface of an animal or plant) on the fur or roosts of two species of molossid bat, feeding on their skin fragments and excreta. The Hemimeridae are flattened, cockroach-like earwigs found only in Africa, where all but one species live epizoically on the fur of giant rats belonging to the genus *Cricetomys*. A single species has only been found in the nests of the pouched rat *Beamys major*. It is thought that these epizoic species live mutualistically on their hosts, where they feed on the tough outer layers of the skin and a species of fungus that, left unchecked, would lead to hair loss in the host. The females of both families give birth to live nymphs. The eggs do not have a food supply of their own and develop inside the mother's reproductive tract, where they obtain nutrients directly from a special tissue similar to a mammalian placenta. The morphological differences of these two families from all other earwigs led some researchers in the past to propose that they might even represent a different order. Molecular studies, however, have shown that both Arixeniidae and Hemimeridae definitely belong to Dermaptera, and that their modified morphology and epizoic habits evolved independently from each other, from more generalized earwigs. It is possible that both arixeniids and hemimerids are in fact modified members of other earwig families and will no longer be considered families of their own in the near future.

A characteristic feature of the order (with the exception of the epizoic species already mentioned), which enables these insects to enter small crevices in soil, bark, and stones, is the **small toughened or leathery, veinless, front wings,** which cover and protect the much larger, semicircular hind wings. The hind wings have a greatly expanded fanlike area, which possesses a sequence of alternating long and short 'ribs'. Each long rib has a bending point that acts as a hinge, which allows the wings to be folded like a fan and then folded again along two transverse folds to enable them to fit under the front wings. In earwigs, the mechanism that folds the hind wings and packs them under the fore wings in an origami-like manner is so effective that it serves as an invaluable model in the design of biologically inspired technology. This mechanism is very ancient, having been found in extinct earwig relatives from the Permian (280 mya), and is thought to have contributed to the success of this order, as it allows them to inhabit tight spaces but also retain their flight capacity—something that few other polyneopteran insects can do. Another peculiarity of earwigs are their terminal forceps, which are usually straight in females and curved in males and are used in a variety of ways, such as weapons for defence, prey handling, courtship displays, and to fold the fan-like hind wings under the toughened front wings. The flexible and telescopic abdominal segments allow earwigs to use their forceps in all directions. In those cases studied, individuals dominant in male–male interactions are able to gain exclusive access to females. Defence in the adults of some species relies on the production of a repellent secretion from abdominal glands. Earwigs like to be in confined spaces (rarely ears) and, when resting, will ensure that as much of their body as possible is in contact with the substrate. There is some disagreement about the origin of the common name. It either refers to earwigs entering ears or possibly to the ear-shape of the hind wing.

Earwigs have simple chewing mouthparts that face forwards. They are omnivorous, some species preferring predominantly a plant diet, while others are more predacious. Some species can be serious pests of flowers and crops, while others have been shown to be of benefit in eating small insect pests in fruit trees. Earwigs can be parasitized by tachinid flies such as *Triarthria setipennis*, which is an endoparasitoid of the European earwig *Forficula auricularia*. The female flies lay their eggs near a suitable host and the first instar larvae climb on to the earwig and enter the body by penetrating the intersegmental membranes. This fly was introduced to North America early in the twenty-first century in an attempt to control the accidentally introduced European earwig.

Copulation takes place end to end, and females may retain the spermatophore for months before the sperm it contains is used to fertilize the eggs. Females typically lay their eggs in tunnels they dig in the soil and show a high degree of maternal care. Sometime the males assist in tunnelling but take no further part in raising the young. Females groom their eggs, licking and turning them to remove fungal spores and they will carry their eggs up and down the tunnel to keep them in an even ambient temperature. Females also guard their eggs from predators. Maternal care continues for some time after the eggs hatch and the mother will feed the young nymphs by bringing food into the nest or by regurgitation of part of her own meal. Eventually the nymphs have to disperse, for as they grow older, their mother stops guarding them and starts regarding them as a potential meal. Earwigs moult up to five times and, apart from getting bigger and gaining additional antennal segments at each moult, they look similar to their parents.

The earliest earwig relatives appear in the Late Permian–Early Triassic (about 250 mya) and differed in some features from modern Dermaptera. Their cerci were long and segmented, as is the general condition in most other hemimetabolans, and their forewings were veined. The ancestors of modern earwigs started making their appearance sometime between the Late Jurassic–Early Cretaceous (150–100 mya), perhaps earlier.

Key reading

Deiters, J., Kowalczyk, W., and Seidl, T. (2016). Simultaneous optimisation of earwig hindwings for flight and folding. *Biology Open* **5**(5): 638–44.

Engel, M. S. and Haas, F. (2007). Family-group names for earwigs (Dermaptera). *American Museum Novitates* **3567**(1): 1–20.

Grimaldi, D. A. and Engel, M. S. (2005). *Evolution of the insects*. Cambridge University Press, Cambridge.

Jarvis, K. J., Haas, F., and Whiting, M. F. (2004). Phylogeny of earwigs (Insecta: Dermaptera) based on molecular and morphological evidence: reconsidering the classification of Dermaptera. *Systematic Entomology* **30**(3): 442–53.

Sato, K., Pérez-de la Fuente, R., Arimoto, K., Seong, Y., Aonuma, H., Niiyama, R., and You, Z. (2020). Earwig fan designing: biomimetic and evolutionary biology applications. *Proceedings of the National Academy of Sciences of the United States of America* **117**(30): 17622–6.

Kuhlmann, U. (1995). Biology of *Triarthria setipennis* (Fallen) (Diptera: Tachinidae), a native parasitoid of the European earwig, *Forficula auricularia* L. (Dermaptera: Forficulidae), in Europe. *Canadian Entomologist* **127**(4): 507–17.

Moore, A. J. and Wilson, P. (1993). The evolution of sexually dimorphic earwig forceps: social interactions among adults of the toothed earwig, Vostox apicedentatus. *Behavioural Ecology* **4**: 40–8.

Naegle, M. A., Mugleston, J. D., Bybee, S. M., and Whiting, M. F. (2016). Reassessing the phylogenetic position of the epizoic earwigs (Insecta: Dermaptera). *Molecular Phylogenetics and Evolution* **100**: 382–90.

Nakata, S. and Maa, T. C. (1974). A review of the parasitic earwigs. *Pacific Insects* **16**: 307–74.

Popham, E. J. (1965). A key to the Dermaptera subfamilies. *Entomologist* **98**: 126–36.

Simmons, L. W. and Tomkins, J. L. (1996). Sexual selection and the allometry of earwig forceps. *Evolutionary Ecology* **10**(1): 97–104.

Terry, M. D. and Whiting, M. F. (2005). Mantophasmatodea and phylogeny of the lower neopterous insects. *Cladistics* **21**(3): 240–57.

Zoraptera
(angel insects)

Common name	Angel insects, ground lice	Distribution	Primarily tropical, some species in subtropical and temperate regions
Derivation	Gk. *zoros*–pure; *a+pteron*–wingless		
Size	Body length 1.5–3 mm	Number of families	2 (Spiralizoridae, Zorotypidae)
Metamorphosis	Incomplete (egg, nymph, adult)	Known world species	44 (0.004%)

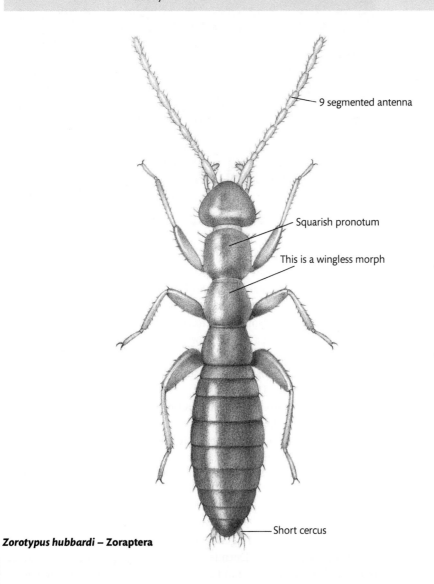

9 segmented antenna

Squarish pronotum

This is a wingless morph

Short cercus

Zorotypus hubbardi – Zoraptera

Essential Entomology. Second Edition. George C. McGavin and Leonidas-Romanos Davranoglou, Oxford University Press.
© George C. McGavin and Leonidas-Romanos Davranoglou (2022). DOI: 10.1093/oso/9780192843111.001.0001

Key features

- very small, dimorphic (coming into two forms)
- wingless form is eyeless, with a termite-like appearance
- winged form with compound eyes, ocelli, and two-segmented tarsi
- short, one-segmented cerci
- extremely complex male genitalia
- rarely encountered, mostly associated with rotting wood

© Piotr Naskrecki

> Zorapterans, such as this *Latinozoros* sp. from Costa Rica, are among the most elusive of all insects, with only a handful of known species.

These small, delicate insects, whose wingless forms resemble minuscule termites, are among the least known insect orders. Their elusive habits in rotting wood and leaf litter, combined with their small size, make them difficult to collect, and as a result they have been poorly studied. Furthermore, their extreme morphological uniformity has puzzled scientists regarding their classification for more than a century. Zoraptera were initially placed close to the Paraneoptera (thrips, true bugs, bark lice, and lice), but subsequent studies showed that they are in fact most closely related to earwigs (Dermaptera) in the Polyneoptera. The possession of **short, one-segmented abdominal cerci** and the generalized type of **downward-pointing mouthparts** betray their polyneopteran affinities. Zoraptera is an ancient group, with a few fossils known from Burmese amber (about 100 mya). Molecular clock estimates suggest an origin in the Jurassic (200 mya), before the breakup of the supercontinent Pangaea, which explains the worldwide distribution of Zoraptera (although they have not yet been discovered in Australia).

Zorapterans are sub-social, forming aggregations of up to 120 individuals under bark or in piles of wood dust and leaf litter, where they eat fungal threads, spores, springtails, and occasionally engage in cannibalism. Zoraptera appear to be dependent on these aggregations, as individuals removed from the group soon perish. It is possible that the grooming behaviour that these insects exhibit spreads essential antimicrobial substances or perhaps endosymbiotic organisms, which are crucial for their survival.

Adults are dimorphic, being either blind, pale, and wingless (resembling the nymphs), or darkly pigmented with eyes, ocelli, and two pairs of pale, sparsely veined wings. Winged morphs disperse to new locations and the wings are then shed. This order was first discovered in the early twentieth century, and all known species were initially contained in the genus *Zorotypus* in the single family Zorotypidae. Subsequent morphological and molecular analyses split *Zorotypus* into nine genera belonging to two families—Spiralizoridae and Zorotypidae—while many more species are likely to be discovered in the near future.

Although species from both families are morphologically extremely similar to each other, each family shows striking differences in male genitalia and in the choice of reproductive strategy. In Spiralizoridae, male genitalia are symmetrical, containing a very long, coiled intromittent organ. During mating, the intromittent organ uncoils inside the female reproductive tract, where it serves to remove or destroy sperm deposited by previous males. Insemination is **internal** in this family. The family Zorotypidae, whose male genitalia are asymmetrical and lack a coiled intromittent organ, follows an entirely different reproductive strategy, which is unique not only among insects, but Hexapoda as a whole. In some zorotypid species, the male deposits several spermatophores **externally** on the abdominal tip of the female, each containing **a single giant sperm**, which is as long as the male's entire body length. The spermatophore and large sperm likely function as physical obstacles that prevent the females from copulating with more than one male. Therefore, males from both families go at great lengths at ensuring the paternity of their offspring, but they achieve this in remarkably different ways.

Both families of Zoraptera also exhibit elaborate pre-copulatory courtship behaviours, and males of many species give females gifts of secretions from their cephalic gland to induce females to accept them as mates. It seems that females can judge the quality of the gift and will terminate a copulation before sperm is transferred if the male is inferior. Females mate two or three times in succession with suitable males and evidence shows that by adopting this strategy, they can lay more eggs and get more cephalic secretions than if they mated with different males. In some species, males are larger than females and wrestle, often using hind-leg kicking and head butting to gain matings. Mating success appears to be determined by age and not necessarily by size, and dominant males may get three quarters of all matings.

Zorapterans have been increasingly studied in the last few years, and there is much to learn from these elusive yet fascinating insects.

Key reading

Choe, J. C. (1995). Courtship feeding and repeated mating in *Zorotypus barberi* (Insecta: Zoraptera). *Animal Behaviour* **49**: 1511–20.

Choe, J. C. (1994). Sexual selection and mating in system in *Zorotypus gurneyi* Choe (Insecta: Zoraptera): II determinants and dynamics of dominance. *Behavioural Ecology and Sociobiology* **34**: 233–7.

Choe, J. C. (1994). Sexual selection and mating system in *Zorotypus gurneyi* (Insecta: Zoraptera): I dominance hierarchy and mating success. *Behavioural Ecology and Sociobiology* **34**: 87–93.

Choe, J. C. (1997). The evolution of mating systems in Zoraptera: mating variations and sexual conflicts. In: *The evolution of mating systems in insects and arachnids*. Choe, J. C. and Crespi, B. J. (eds), pp 130–45. Cambridge University Press, Cambridge.

Dallai, R., Mercati, D., Gottardo, M., Machida, R., Mashimo, Y., Matsumura, Y., and Beutel, R. G. (2014). Giant spermatozoa and a huge spermatheca: A case of coevolution of male and female reproductive organs in the ground louse *Zorotypus impolitus* (Insecta, Zoraptera). *Arthropod Structure & Development* **43**(2): 135–51.

Dallai, R., Mercati, D., Gottardo, M., Machida, R., Mashimo, Y., Matsumura, Y., and Beutel, R. G. (2013) Divergent mating patterns and a unique mode of external sperm transfer in Zoraptera: an enigmatic group of pterygote insects. *Naturwissenschaften* **100**: 581–94.

Dallai, R., Mercati, D., Gottardo, M., Machida, R., Mashimo, Y., and Beutel, R. G. (2011). The male reproductive system of *Zorotypus caudelli* Karny (Zoraptera): Sperm structure and spermiogenesis. *Arthropod Structure & Development* **40**(6): 531–47.

Engel, M. S. (2007). The Zorotypidae of Fiji (Zoraptera). *Bishop Museum Occasional Papers* **91**: 33–8.

Kočárek, P., Horká, I., and Kundrata, R. (2020). Molecular phylogeny and infraordinal classification of Zoraptera (Insecta). *Insects* **11**: 51.

Matsumura, Y., Beutel, R. G., Rafael, J. A., Yao, I., Câmara, J. T., Lima, S. P., and Yoshizawa, K. (2020). The evolution of Zoraptera. *Systematic Entomology* **45**: 349–64.

Valentine, B. D. (1986). Grooming behavior in Embioptera and Zoraptera. *Ohio Journal of Science* **86**(4): 150–2.

Wipfler, B., Letsch, H., Frandsen, P. B., Kapli, P., Mayer, C., Bartel, D., Buckley, T. R., Donath, A., Edgerly-Rooks, J. S., Fujita, M., Liu, S., Machida, R., Mashimo, Y., Misof, B., Niehuis, O., Peters, R. S., Petersen, M., Podsiadlowski, L., Schütte, K., Shimizu, S., Uchifune, T., Wilbrandt, J., Yan, E., Zhou, X., and Simon, S. (2019). Evolutionary history of Polyneoptera and its implications for our understanding of early winged insects. *Proceedings of the National Academy of Sciences of the USA* **116**: 3024–9.

Plecoptera
(stoneflies)

Common name	Stoneflies	Metamorphosis	Incomplete (egg, nymph, adult)
Derivation	Gk. *plectos*–plaited, woven; *pteron*–a wing	Distribution	Worldwide, but mainly cool temperate regions
Size	Body length 3–48 mm. Maximum wing span about 100 mm	Number of families	16
		Known world species	3,743 (0.33%)

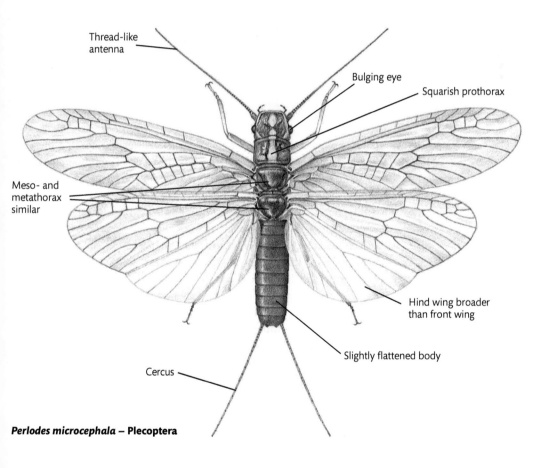

Thread-like antenna

Bulging eye

Squarish prothorax

Meso- and metathorax similar

Hind wing broader than front wing

Slightly flattened body

Cercus

Perlodes microcephala – **Plecoptera**

Essential Entomology. Second Edition. George C. McGavin and Leonidas-Romanos Davranoglou, Oxford University Press.
© George C. McGavin and Leonidas-Romanos Davranoglou (2022). DOI: 10.1093/oso/9780192843111.001.0001

Key features

- aquatic nymphs
- weak flying insects near water
- adults typically rest with wings held flat or rolled around body
- important in aquatic food chains

> Stoneflies use vibrational signals to attract each other and mate. *Nemoura* sp., UK.

© Anne Riley

Stoneflies are soft or leathery looking, generally slender insects with soft, slightly flattened bodies. Although they have **two pairs of membranous wings,** they are not good fliers, seldom travel far from water, and spend a great deal of time resting on marginal rocks and vegetation. The front wings are much narrower and slightly longer than the hind wings, which are folded in pleats beneath. Both pairs of wings may have **complex patterns of cross veins.** The head has **bulging eyes, two or three ocelli,** and **thread-like, multi-segmented antennae.** The mouthparts are weakly developed or non-functional and many species, which live less than a fortnight as adults, do not feed at all. In other species, adults feed by scraping algae, lichen, and similar material from rocks and other substrates. The legs are quite long, robust, and appear widely separated. **The elongate abdomen** may be cylindrical or slightly flattened and has a **pair of single- or multi-segmented cerci** at its posterior end. The order is divided into two suborders: the Arctoperlaria (12 families) and the Antarctoperlaria (four families). With the exception of one family, the Notonemouridae, all Arctoperlaria are found in the northern hemisphere. All families in the Antarctoperlaria, (Austroperlidae, Eustheniidae, Diamphipnoidae, and Gripopterygidae) are found in the southern hemisphere.

Males attract females by rubbing, tapping, or drumming a species-specific signal on the substrate to which the females respond with acoustic signals of their own. The noises are produced by a hardened finger or hammer-like structure on the underside of the abdomen being struck or rubbed against the ground, although in a few species belonging to the family Chloroperlidae, no contact is made, and the sound is produced by tremulation—body vibrations transmitted to the substrate via the legs and tarsi. Theories suggest that this method of signalling may have evolved as it is less likely to be intercepted by acoustically searching predators. In most cases, male and female duetting occurs and in some, parts of the female's call may overlap with that of the male. Duetting has been shown to reduce the time spent by the male searching for the stationary responding female. Male stoneflies typically follow a pattern of drumming and directed searching until the source of the vibrations (the female) is found. Females will move to a different location if the males take too long to find them, which may have the double advantage of reducing the risk of predation and selecting a fit mate. Once the male has located her, mating takes place almost immediately on the ground or on plants. Male stonefly calls vary from series of single beats to quite complex songs with distinct phases varying in beat interval, beat number, and other characteristics, whereas female responses are generally simple.

Females, who can lay many hundreds or thousands of eggs, deposit slimy egg masses into the water. The eggs may be spherical and adhesive, flattened, spindle-shaped, with filamentous projections or other devices to ensure that they stick to objects or lodge in crevices. The aquatic nymphs are quite broad and flattened with noticeable developing wing pads and a pair of long terminal, abdominal tails. They have a closed tracheal system and obtain their oxygen by straightforward diffusion through the surface of the body or by means of a variety of branched or feathery gill tufts located on different parts of the body. Some species have gill tufts at the end of their abdomens, while others have gills on the neck and thorax. Remnants of thoracic and abdominal gill tufts may still be visible in the adult stages. Cold, fast-flowing, stony- or gravel-bottomed streams are preferred, and many species are very intolerant of pollution. Stonefly nymphs are mostly herbivorous but some families (Perlidae, Chloroperlidae, and Perlodidae) contain omnivorous and predacious species. Nymphs pass through anything up to thirty moults and may take from one up to four years before moving away from the water to emerge as adults, often during the hours of darkness. The nymphs of predatory stoneflies (Perlodidae) have been shown to be able to respond to the swimming patterns of their prey and can distinguish between prey and non-prey species of mayfly nymphs. Females may mate with several males and mate guarding to prevent access by rival males has been recorded for some species. Males may guard a newly moulted female until she is ready to mate, often before her cuticle has fully hardened.

Plecopterans are generally adapted to cold conditions and in the northern hemisphere, many species, such as those belonging to the Capniidae, may become adults, mate, and die during the cold months of Winter.

Stonefly nymphs are not defenceless. It has been shown that species such as *Pteronarcys dorsata* (Pteronarcyidae) have complex defensive repertoires. They are able recognize different types of predator and respond either by feigning death if the predators are fish, or by backing away and using reflex bleeding if the predators are tactile hunters such as crayfish. Some adult pteronarcyids also show reflex-bleeding if attacked and, in some species, the haemolymph is forced out a distance of many centimetres, accompanied by a distinct popping noise. Nymphs are able to learn paths to shelters like particular groups of stones that provide a refuge from predatory fish.

Plecopterans are a very important component in aquatic food chains and, like other aquatic insect groups such as mayflies and caddisflies, are imitated in patterns of fishing fly used by anglers. As stone flies are usually very intolerant of low water oxygen, they are useful as aquatic bio-indicators. The large nymphs of the California salmonfly (*Pteronarcys californica*) may have been eaten by native Americans of the Achomawi, Modoc, and Wintun nations of northern California.

The origins of Plecoptera are lost in the abyss of time. Possible plecopteran relatives are known from the Permian (300–250 mya), although the ancestors of the modern representatives likely appeared much later, around the Jurassic (200 mya), based on phylogenomic and biogeographical data.

Key reading

Alba-Tercedor, J. and Sanchez-Ortega, A. (1991). *Overview and strategies of Ephemeroptera and Plecoptera*. Sandhill Crane Press, Gainesville.

Campbell, I. C. (ed) (1990). *Mayflies and stoneflies: life histories and biology*. Entomologica series 44. Kluwer Academic Publishers, Dordrecht.

Du Bois, C. A. (1935). *Wintu ethnography*, **36**. University of California Press, Berkeley.

Illies, J. (1965). Phylogeny and zoogeography of the Plecoptera. *Annual Review of Entomology* **10**: 117–40.

Landolt, P. and Sartor M. (eds) (1997). *Ephemeroptera & Plecoptera: biology, ecology, systematics*. MTL—Mauron, Tingley & Lachat SA, Fribourg.

Marden, J. H. and Kramer, M. G. (1994). Surface-skimming stoneflies: a possible intermediate in insect flight evolution. *Science* **266**(5184): 427–30.

Macan, T. T. (1982). *The study of stoneflies, mayflies and caddisflies*. Amateur Entomologist's Society, London.

Moore, K. A. and Williams, D. D. (1990). Novel strategies in the complex defensive repertoire of a stonefly (*Pteronarcys dorsata*) nymph. *Oikos* **57**(1): 49–56.

Pekarsky, B. L. and Wilcox, R. (1989). Stonefly nymphs use hydrodynamic cues to discriminate between prey. *Oecologia* **79**(2): 265–70.

Szczytko, S. W. and Stewart, K. W. (1979). Drumming behaviour of four western Nearctic *Isoperla* (Plecoptera) species. *Annals of the Entomological Society of America* **72**: 771–86.

Stewart, K. W. and Maketon, M. (1991). Structures used by Nearctic stoneflies (Plecoptera) for drumming, and their relationship to behavioral pattern diversity. *Aquatic Insects* **13**(1): 33–53.

Stewart, K. W. (1994). Theoretical considerations of mate finding and other adult behaviors of Plecoptera. *Aquatic Insects* **16**: 95–104.

Stewart, K. W. (1997). Vibrational communication in insects: epitome in the language of stoneflies. *American Entomologist* **43**(2): 81–91.

Sutton, M. Q. (1985). The California salmon fly as a food source in northeastern California. *Journal of California and Great Basin Anthropology* **7**(2): 176–82.

Ward, J. V. (1994). Ecology of Alpine streams. *Freshwater Biology* **32**(2): 277–94.

Zwick, P. (2000). Phylogenetic system and zoogeography of the Plecoptera. *Annual Review of Entomology* **45**: 709–46

Orthoptera
(grasshoppers and crickets)

Common name	Grasshoppers, crickets, and their relatives	**Metamorphosis**	Incomplete (egg, nymph, adult)
Derivation	Gk. *orthos*–straight; *pteron*–a wing	**Distribution**	Worldwide but mostly in warm regions
Size	Body length 2–200 mm	**Number of families**	40
		Known world species	23,434 (2.07%)

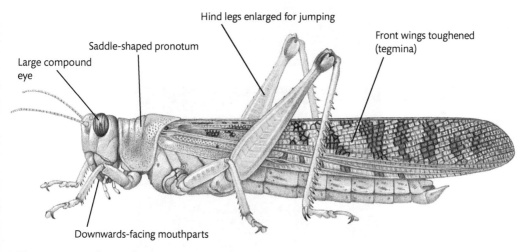

Saddle-shaped pronotum

Hind legs enlarged for jumping

Front wings toughened (tegmina)

Large compound eye

Downwards-facing mouthparts

Schistocerca gregaria – **Orthoptera**

Essential Entomology. Second Edition. George C. McGavin and Leonidas-Romanos Davranoglou, Oxford University Press.

Key features

- hind legs usually much larger and longer than other legs and used for jumping.
- many species make sounds using hind legs and/or front wings
- characteristic 'grasshopper' shape with pronotum extended down at each side
- several species can be serious crop pests

❯ A spermatophore transferred by the male to a female katyidid, serves as a nutritious meal. Sabah, Malaysia.

❯ A female katydid ovipositing inside a branch. *Onomarchus* sp., Singapore.

© George McGavin

> Also known as dune crickets, schizodactylids often construct deep tunnels in the sandy riverine habitats. *Schizodactylus* sp., Northern India.

Springing and singing might describe the species in this, the largest order of hemimetabolous insects that have chewing mouthparts. Of course, a few cannot jump and quite a few do not sing but very many of the ones commonly encountered do both things quite well. Orthopterans can be found in just about every conceivable terrestrial habitat, even in caves or burrowing in soil, and some live permanently in association with ant nests. The range of the order is divided into two suborders: the Ensifera and the Caelifera. These taxa share several common characteristics. Both have conspicuous eyes and may have ocelli. They all possess **downward-pointing chewing mouthparts, an enlarged, saddle or shield-shaped pronotum, toughened, narrow front wings (or tegmina) with larger fan-folded hind wings (when present), and large hind legs modified for jumping. The abdomen has a pair of short, terminal cerci.**

The Ensifera, comprising the crickets, bush-crickets, and katydids, have long or very long, multi-segmented antennae, and are mainly nocturnal and solitary (Table 2.1). Most species are cryptically coloured brown or

Table 2.1 **Principal ensiferan families**

Family	Habits	Examples of genera
Tettigoniidae (katydids) (~ 7,100 spp.)	Mostly in foliage. Herbivorous, saprophagous, predacious. Some pests	*Acanthodis, Anabrus, Ephippiger, Orchelimum, Tettigonia*
Gryllidae and related families (true crickets) (~ 4,900 spp.)	Mainly ground-living. Omnivorous	*Acheta, Gryllodes, Gryllus, Oecanthus*
Rhaphidophoridae (cave crickets) (~ 660 spp.)	Many in caves, burrows. Saprophagous, predacious	*Ceuthophilus, Dolichopoda, Tachycines, Troglophilus*
Gryllotalpidae (mole crickets) (~ 110 spp.)	Burrowing. Omnivorous. Some are pests	*Gryllotalpa, Scapteriscus*

green to mimic dead or living leaves. For example, the new world katydid, *Markia hystrix*, is a remarkable mimic of the lichens on which it lives. In female ensiferans, the ovipositor is always prominent and sword-, sickle-, or stiletto-shaped. Many species are herbivorous, eating mostly plants other than grasses, but some are partly or wholly predacious. The Caelifera, comprising grasshoppers and locusts, have short antennae and the females never have prominent ovipositors. They are generally ground-living, diurnal, grass-feeders, and many are cryptically coloured to blend in with the ground, or they are warningly coloured to advertise their ability to produce unpalatable or repellent secretions from special glands. Some species regurgitate their gut contents as a defensive strategy. The brightest colours and striking patterns of black, red, yellow, and orange are seen in the chemically defended species belonging to the small families Pyrgomorphidae and Romaleidae. Cryptic species may have bright colour patches on their hindwings to distract predators as they jump up, fly, and then drop to the ground out of sight. Members of several families (e.g., Chorotypidae, Trigonopterygidae) take crypsis to the extreme, with their entire body shape and colouration imitating that of a leaf.

Nymphal development goes through 4–11 instars. As in the Odonata, wing pad orientation changes in development. In young nymphs the front edge of the wing pads is lateral and ventrally directed. In that last instar the orientation changes so that the front edges of the wing pads are directed dorsally, and the hind wings overlap the front wings. The normal orientation is resumed in the adult stage (front edges of the wings directed ventrally).

The ability to jump, seen in the majority of species, explains an older name for the order—Saltatoria. Jumping can be part of an escape from potential danger or to enable the insect to take to the air. The strong hind legs are also used to kick out at attackers or rivals. Some groups of orthopterans, who have evolved habits such as burrowing or digging, cannot jump very well. Although orthopteran legs vary greatly in size, the mechanically ingenious jumping mechanism of the hind leg is essentially the same. The tibia is controlled by a large extensor muscle, which occupies most of the volume of the femur, and a much smaller flexor muscle. Energy, produced by contraction of the large extensor muscle, is stored prior to the jump. The energy is stored in two cuticular springs at the knee joint (semilunar processes) and in the slightly elastic extensor apodeme (= tendon). The catch, which keeps the tibiae from extending too soon, is formed by a pocket in the flexor apodeme, which sits over a lump at the end of the femur. When the grasshopper has contracted its extensor muscle fully and has stored as much energy as it can, it releases the tension on the flexor muscle, which causes the catch to disengage and the full force of the jump to be developed. The system allows the power of the muscles to be amplified. In the case of the adult Desert Locust, the amplification is about ten times, allowing a long jump of up to 80 cm.

Singing, usually by males, is another well-known behavioural characteristic of the order. The songs, and the manner in which they are produced, are diverse. The songs are generally designed to attract females but can also be involved in territoriality or signalling aggression or alarm. As a very rough guide grasshoppers (Acrididae) generally sing during the day, katydids (Tettigoniidae) sing at night, and crickets (Gryllidae) sing during the day or night.

Male ensiferans typically produce songs by stridulating, that is, rubbing parts of their front wings against each other. In gryllids, for example, the underside of both wings has a vein near the base which is modified as a file with between 35 and 300 teeth. The stiffened edge of the other wing forms a scraper or plectrum. As the right wing overlaps the left wing, the file on the right wing and the scraper on the left wing rasp against each other and set up vibrations that pass through the wings and excite a special clear harp or mirror region on each wing. The resonant frequency of these harp regions closely matches the song frequency. When singing, crickets raise their wings at an angle to allow the sound to be radiated more effectively. A pulse of sound is made on the closing stroke of the front wings and the songs vary in the number and character of the pulses produced (4–200 per second) and the way the pulses are grouped. Although the stridulatory structures on the front wings of crickets, bush-crickets, and katydids are superficially very similar, remarkably, they have evolved independently at least four times—suggesting that sound communication also evolved several times. Both winged and wingless crickets can make vibrations that pass through the substrate (e.g., bark, soil, but not air) by tremulating (shaking their body) or by drumming their abdomen or feet against a solid surface. Substrate vibrations are considered to be the original mode of signalling in Ensifera, with sound appearing later in their evolution.

Cricket songs can be very musical, and frequencies vary from 1500 Hz–10 kHz. Katydids tend to be noisier and can employ frequencies from 20–100 kHz, but the details of sound production differ in the two groups. Several species of tree crickets (*Oecanthus* spp.) increase the volume of their song by positioning their wings across a hole they chew in a large leaf, which acts as a baffle. An even more incredible example of acoustic engineering is seen in mole crickets (Gryllotalpidae). Mole crickets are remarkable for their incredible morphological resemblance to mammalian moles. The body is very robust, cylindrical, and covered with very short, velvety hairs. The legs are short and strong, and the front legs are particularly modified for digging, being much broadened and armed with stout teeth. The eyes are small, and the ovipositor is very short or vestigial. But they are also famous for their ability to construct elaborate singing burrows to make their songs more audible. The burrow has a bulb region tuned to resonate at the song's carrier frequency (3.4 kHz) and twin, exponentially flared horns to amplify and carry the song to the surface. The burrow is very effective and at a metre above the burrow the peak sound level can be 92 decibels, which is enough to carry nearly a mile in quiet conditions.

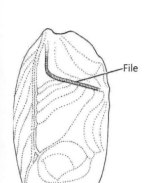

∧ The underside of the front wing of a gryllid showing the stridulatory file.

∧ The unique acoustic burrow of the more cricket.

❯ *Tropidacris cristata* **is a large grasshopper from Central and South America with a wingspan of up to 18 cm. Costa Rica.**

Once a female has been attracted, the two mate above ground. Eggs are laid in an underground chamber and full development through the ten or so nymphal stages may take up to two years.

Caeliferan species make sounds in a variety of ways, ranging from hind-wing snapping and palp and/or mandible rubbing to the more familiar rubbing of the inner face of the hind femur against the edge of the tegmina. The inside of each femur has a row of tiny teeth or pegs that set up vibrations in the front wings. As in ensiferan singers, the tone of the song is dependent on the spacing of the teeth and the speed at which the leg is moved (Table 2.2).

As for the mechanisms of sound production, the tympanal or hearing organs in Orthoptera take two distinct forms. In the Caelifera, a tympanum is situated on either side of the first abdominal segment. Each organ consists of a thin tympanic membrane located over an air sac. Attached to the underside of the membrane is a chordotonal organ of varying complexity. A more complicated system is found in the Ensifera, where a single or a pair of tympanal

Table 2.2 Principal caeliferan families

Family	Habits	Examples of genera
Acrididae (grasshoppers, locusts) (~ 6,700 spp.)	On ground and low plants. Herbivorous. Many serious pests	*Acrida, Chortoicetes, Locusta, Melanoplus, Nomadacris, Schistocerca*
Tetrigidae (pygmy locusts) (~ 1,823 spp.)	Ground-living, woods and wet areas. Saprophagous, herbivorous	*Discotettix, Paratettix, Tetrix, Xerophyllum*
Pyrgomorphidae (Bush hoppers, Bush locusts) (~ 500 spp.)	Ground-living. Bushes, shrubs, some in trees and on grasses. Some local pests	*Atractomorpha, Petasida, Phymateus, Pyrgomorpha*

organs are situated on the front tibiae close to the knee joint. The leg around the tympana is slightly swollen and typically has two longitudinal slits that each open to a tympanic chamber. Where there is a pair of tympanic membranes, they are unequal in size and the trachea behind them is divided by a rigid membrane. Vibration of the tympanic membranes is picked up by a number of chordotonal organs, the most important of which is the called the crista acustica. The crista acustica comprises a row of sensory cells attached to the anterior tympanal membrane. In gryllids the trachea associated with the tympana are part of the normal tracheal system but the most complex form of orthopteran hearing, which is found in tettigoniids, relies on a specialized acoustic trachea that runs up the middle of each front leg and is not part of the normal tracheal system. The acoustic tracheae on each side are connected and joined to specialized acoustic spiracles in the prothorax. If hearing is to be of use in locating a distant sound source, it must be directional. The problem common to insects is that their small size makes differences in intensity or time of arrival of sound at two different tympanal organs very small indeed. The problem is overcome by the introduction, mechanically or acoustically, of a time delay between the two ears. In tettigoniids the main input of sound to the system comes from the acoustic spiracles on the thorax; it then travels down and is amplified by the tapering acoustic tracheae before it reaches the tympanal organs. The effects of these sounds are modified by sound vibrations picked up directly through the slits' opening to the outside from the tympanic chambers. The precise mechanism of cricket hearing is, however, exceedingly complex and not yet fully understood.

Courtship in the Orthoptera varies enormously from very simple affairs where males do not engage in any sort of foreplay, to long and very complex 'wooing' involving precise movements and mutual contact of many parts of the body accompanied, at the right time, by different types of songs. Humans may find the singing of crickets attractive and in several countries, they are kept as pets for their songs. But the intended audience are just as choosy. It has been shown that the quality of sound a male cricket makes is very important to his chances of a successful mating. Smaller males with asymmetrical harps produce less pure tones or a higher frequency than do larger males with larger symmetrical harps. Not surprisingly, females prefer males who sing better, louder, and longer.

Being the last male to mate with a female is very advantageous. The way in which sperm is used—a last-in first-out basis—makes it likely that the last mate will father most of the eggs. In the gryllid, *Truljalia hibinonis*, males are able to remove more than 87% of the sperm of a previous mating after they have copulated. The rival's sperm is not wasted but eaten. Simply hanging on tightly to the female's back until she lays her eggs and kicking out at any other male who comes near is a commoner strategy. Where females are in short supply, lone males will make strenuous attempts to take over. Fighting between male crickets is common and has even been popular in some countries as a gambling sport.

Ensiferans typically mate with the smaller male beneath or below the female and the resultant fertilized eggs are often laid singly or in small batches in plant tissues. Males transfer sperm in spermatophore consisting of an ampulla containing the sperm and a nuptial gift in the form of a spermatophylax, which the female eats after mating. The consumption of the spermatophylax enables the female to produce more offspring. Males who have the chance of lots of matings provide smaller individual gifts than those males who mate with few females. This suggests that the gift might have to do with guaranteeing paternity when the father's investment is high. But some other research has shown that spermatophylax size simply correlates with sperm number and ejaculate volume, suggesting a simple sperm protective function. A larger spermatophylax will take the female longer to eat so, by the time she has finished, the sperm will be transferred. There are probably two types of gift, depending on species: one, a low-value, easy-to-produce spermatophylax, which is protective; the other, a large, highly nutrient-rich spermatophylax, which also increases the fecundity of the females. In some gryllids the spermatophylax provided by the male is a bit of a confidence trick. True, it does contain a significant amount of amino acids—enough to persuade the female to eat it but not so much that the male is depleted. She might actually use up more energy eating it than she gets back.

The mating of behaviour of North American sagebrush crickets belonging to the genus *Cyphoderris* is remarkable for it involves males offering portions of their body tissue as nuptial gifts to the female. In *Cyphoderris strepitans*, once the female has mounted the male's back, she begins to eat his hind wings. These are especially fleshy, and the female also drinks the haemolymph that issues from the wounds she makes. Males, rather than wishing to terminate the cannibalism, make it easy for the female by raising their wings into a convenient position and make sure that she stays put by grasping the end of her abdomen very tightly with a special hooked clamping device while they transfer their spermatophore. It has been suggested that this device may allow males to force females to copulate either because the male has an insufficient amount of hind wing to keep her interest or, more likely, it has evolved to prevent females from taking advantage of free meals. In species of another genus, *Bradyporus*, the female simply bites their partner's back and drinks haemolymph.

An interesting reversal of the normal sexual roles in seen in the Mormon cricket, *Anabrus simplex*, and other similar situations are known. In poorer habitats it is only well-nourished males who choose a mate. These males have a large, nutritious nuptial gift, which can be one-third of their total body weight and their love song will attract several females who compete for access. The female who succeeds will mount the male's back but might still not be chosen. The male is able to tell how heavy she is (and therefore how many eggs she might be carrying), and if not satisfied, he may choose another mate. The reversal of normal sexual role is seen when paternal investment is high and can be induced experimentally in some bush cricket species.

In the Caelifera, mating takes place with the male on top of the female and the eggs (up to 200) are normally laid in the soil, protected by an egg pod made of a special protective foam produced by the female. Caeliferan ovipositors are short and squat, composed of pairs of valvulae that expand and grip the side of the egg-laying burrow to pull the abdomen down, in some cases, to more than double its length.

Many acridid species are agricultural pests, capable of causing immense crop losses in tropical, subtropical, and even in temperate regions. In North America, several species in the spur-throated grasshopper genus *Melanoplus* can be serious pests. The most notorious acridid species, however, are commonly referred to as 'locusts'. Locust is the generic common name given to fewer than ten species in the genera *Schistocera*, *Locusta*, *Locustana*, *Nomadacris*, *Anacridium*, and *Chortoicetes*. The defining feature of these species is that they can exist in solitary or gregarious forms. As solitary individuals, even large populations do little lasting damage, but as gregarious swarms they can devastate vast swathes of vegetation. The Desert locust, *Schistocerca gregaria*, probably one of the most damaging insects in the world, has been a pest since agriculture first developed and is mentioned as the eighth of the ten biblical plagues of Egypt. When conditions are right, and rain brings about the growth of lush vegetation, the locusts are able to feed and lay more eggs than usual. More nymphs hatch but now, rather than shy away from each other, they mass together in growing numbers. They have become gregarized. These gregarious black, yellow, and orange-striped hoppers (as they are called) look very different from the pale green solitary nymphs. Once locust nymphs become aggregated in the habitat and reach a critical local population density, they suddenly shift from uncoordinated movement to collective mass marching. This striking behavioural transition has been explained by computer modelling and is a classic case of an emergent phenomenon, arising because of local interactions among individuals. The hoppers feed and march until they become adult. The result can be massive swarms of adults, covering as much as 1000 km². Although they are good fliers, they tend to move in the direction of the prevailing winds. When a food source is located, the swarm will descend, strip it bare, and move on. As each locust in a single swarm, consisting of up to 50 billion individuals (5×10^{10}), can eat its own weight of food in a day (1.5–2.0 grams), it takes but a few moments to calculate the amount of food consumed; somewhere between 75,000–100,000 tons.

Today, locust swarms can be tracked by satellite and the use of insecticides and biopesticides at the right time and in the right place can do much to control outbreaks of certain species. Several studies using modelling and biochemical analysis have revealed the mechanisms of gregarization. If food is scarce, a female will do better to lay eggs destined to become solitary individuals, which are well camouflaged (green or brown in colour) and avoid one another. However, as population density rises when food is plentiful, it is more advantageous for locusts to band together and become gregarious. In the Desert locust, this transition in behaviour occurs rapidly upon crowding and is induced by touch-sensitive hairs on the hind legs, which, when stimulated by contacting another locust, elicit the release of serotonin within the CNS, which sets in train behavioural gregarization and ultimately leads to other changes in colour and shape.

Female locusts use their own experience of being crowded to set the developmental trajectory of their unhatched embryos. The more recently the

mother was crowded before laying her eggs in the soil, the higher the proportion of gregarious hatchlings that emerge. This 'epigenetic' effect is mediated
by a chemical added by the female to the foam surrounding her eggs.

As large and easily cultured insects, locusts have also proved themselves to be incredibly useful as model systems for the investigation of
neurophysiology. For instance, they were once thought to be gluttons but
now turn out to be gourmets. Studies on locust feeding behaviour have shown
them to be capable of regulating their food intake to a high degree. Furthermore, their selection of macronutrients in response to their own changing
nutritional needs and a variable nutritional environment has led to the development of a geometric framework for considering feeding and nutrition
in insects and other animals, including vertebrates. Indeed, insights gained
from studying locust appetites has revolutionized our understanding of human obesity and its causes—in particular, the role of a powerful appetite for
protein.

Orthoptera are remarkably well-represented in the fossil record, with the
origins of the order going back to the Late Permian (around 250 mya). Extinct
relatives of Orthoptera appeared even earlier, and some occupied ecological
niches that are now filled by other orders. For example, the Chresmodidae
possessed very long legs and may have walked on water, much like modern
water striders (Hemiptera) do.

Key reading

Anstey, M. L., Rogers, S. M., Ott, S. R., Burrows, M., and Simpson, S. J. (2009). Serotonin
    mediates behavioral gregarization underlying swarm formation in desert locusts. *Science*
    323: 627–30.
Bailey, W. J. and Rentz, D. C. F. (eds) (1990). *The Tettigoniidae: biology, systematics and
    evolution*. Crawford House Press, Bathurst.
Bennet-Clark, H. C. (1975). The energetics of the jump of the locust, Schistocerca gregaria.
    Journal of Experimental Biology **63**: 53–83.
Bennet-Clark, H. C. (1987). The tuned singing burrow of mole crickets. *Journal of Experi-
    mental Biology* **128**: 383–409.
Bennet-Clark, H. C. (1989). Songs and the physics of sound production. In: *Cricket be-
    haviour and neurobiology*. Huber, F., Moore, T.E. and Loher, W. (eds), pp. 227–61.
    Cornell University Press, Ithaca.
Bennet-Clark, H. C. (1990). Jumping in Orthoptera. In: *Biology of grasshoppers*. Chapman,
    R. F. and Joern, A. (eds), pp. 173–203. John Wiley and Sons, New York.
Bennet-Clark, H. C. (1998). Size and scale effects as constraints in insect sound communi-
    cation. *Philosophical Transactions of the Royal Society of London B* **353**: 407–19.
Bidau, C. J. (2014). Patterns in Orthoptera biodiversity. I. adaptations in ecological and
    evolutionary contexts. *Journal of Insect Biodiversity* 2(20): 1–39.
Buhl, J., Sumpter, D. J. T., Couzin, I. D., Hale, J. J., Despland, E., Miller, E. R., and Simpson,
    S.J. (2006). From disorder to order in marching locusts. *Science* **312**: 1402–6.
Chapman, R. F. and Joern, A. (eds) (1990). *Biology of grasshoppers*. John Wiley and Sons,
    New York.
Desutter-Grandcolas, L., Jacquelin, L., Hugel, S., Boistel, R., Garrouste, R., Henrotay, M.,
    Warren, B. H., Chintauan-Marquier, I. C., Nel, P., Grancolas, P., and Nel, A. (2017). 3-D
    imaging reveals four extraordinary cases of convergent evolution of acoustic communi-
    cation in crickets and allies (Insecta). *Scientific Reports* **7**: 7099.
Grimaldi, D. A. and Engel, M. S. (2005). *Evolution of the insects*. Cambridge University Press,
    Cambridge.

Gage, A. R. and Barnard, C. J. (1996). Male crickets increase sperm number in relation to competition and female size. *Behavioural Ecology and Sociobiology* **38**(5): 349–53.

Gwynne, D. T. (1984). Sexual selection and sexual differences in Mormon crickets (Orthoptera: Tettigoniidae), Anabrus simplex. *Evolution* **38**: 1011–22.

Gwynne, D. T. and Simmons, L. W. (1990). Experimental reversal of courtship roles in an insect. *Nature* **346**(6280): 172–4.

Gwynne, D. T. (1997). The evolution of edible 'sperm sacs' and other forms of courtship feeding in crickets, katydids and their kin (Orthoptera: Ensifera). In: *The evolution of mating systems in insects and arachnids.* Choe, J. C. and Crespi, B. J. (eds), pp. 110–29. Cambridge University Press, Cambridge.

Huber, F., Moore, T. E., and Loher, W. (1990). *Cricket behaviour and neurobiology.* Cornell University Press, Ithaca.

McCaffery, A. R., Simpson, S. J., Islam, M. S., and Roessingh, P. (1998). A gregarizing factor present in the egg pod foam of the Desert Locust Schistocerca gregaria. *Journal of Experimental Biology* **201**(3): 347–63.

Miller, G. A., Islam, M. S., Claridge, D. W., Dodgson, T., and Simpson, S. J. (2008). Swarm formation in the desert locust (*Schistocerca gregaria*): isolation and NMR analysis of the primary maternal gregarizing agent. *Journal of Experimental Biology* **211**: 370–6.

Preston-Mafham, K. (1990). *Grasshoppers and mantids of the world.* Blandford Press, London.

Raubenheimer, D. and Simpson, S. J. (2020). *Eat like the animals: what nature teaches us about the science of healthy eating.* Houghton Mifflin Harcourt: New York.

Sakaluk, S. K., Bangert, P. J., Eggert, A. C., Gack, C., and Swanson, L. V. (1995). The gin trap as a device facilitating coercive mating in sagebrush crickets. *Proceedings of the Royal Society of London B* **261**(1360): 65–71.

Simmons, L. W. (1995). Male bushcrickets tailor spermatophores in relation to their remating intervals. *Functional Ecology* **9**(6): 881–6.

Simmons, L. W. and Ritchie, M. G. (1996). Symmetry in the songs of crickets. *Proceedings of the Royal Society of London B* **263**(1375): 1305–11.

Simpson, S. J., McCaffery, A. R., and Hägele, B. (1999). A behavioural analysis of phase change in the desert locust. *Biological Reviews* **74**: 461–80

Simpson, S. J. and Raubenheimer, D. (2000). The hungry locust. *Advances in the Study of Behaviour* **29**: 1–44.

Stritih, N. and Čokl, A. (2013). Mating behaviour and vibratory signalling in non-hearing cave crickets reflect primitive communication of Ensifera. *PLoS One* **7**(10): e47646.

Sword, G. A., LeCoq, M., and Simpson, S. J. (2010). Phase polyphenism and preventative locust management. *Journal of Insect Physiology* **56**: 949–57.

Vahed, K. and Gilbert, F. S. (1996). Differences across taxa in nuptial gift size correlate with differences in sperm number and ejaculate volume in bushcrickets (Orthoptera: Tettigoniidae). *Proceedings of the Royal Society of London B* **263**(1374): 1255–63.

Weddell, N. (1994). Dual function of the bushcricket spermatophore. *Proceedings of the Royal Society of London B* **258**(1352): 181–5.

Ingrisch, S. (2011). 'Order Orthoptera Oliver, 1789' in Zhang, Z.-Q. (ed.) *Animal biodiversity: An outline of higher-level classification and survey of taxonomic richness. Zootaxa* **3148**: 195–7.

Zuk, M. and Simmons, L. W. (1997). Reproductive strategies of the crickets (Orthoptera: Gryllidae). In: *The evolution of mating systems in insects and arachnids.* Choe, J. C. and Crespi, B. J. (eds), pp. 89–109. Cambridge University Press, Cambridge.

Dictyoptera

Cockroaches, mantids, and their extinct relatives are united under the super-order called the Dictyoptera because they share morphological features to do with wing venation, internal musculature, the structure of the mouthparts, the male genitalia, and the production of an ootheca. Termites, who were (until recently) assigned to their own distinct order, have now been shown to be highly modified cockroaches, based on an overwhelming amount of molecular and morphological evidence.

Essential Entomology. Second Edition. George C. McGavin and Leonidas-Romanos Davranoglou, Oxford University Press.
© George C. McGavin and Leonidas-Romanos Davranoglou (2022). DOI: 10.1093/oso/9780192843111.001.0001

Blattodea
(cockroaches and termites–alternative name: **Blattaria**)

Common name	Cockroaches and termites	**Distribution**	Worldwide
Derivation	L. *blatta*–cockroach	**Number of families**	19
Size	Body length 3–100 mm	**Known world species**	7,314 (0.65%)
Metamorphosis	Incomplete (egg, nymph, adult)		

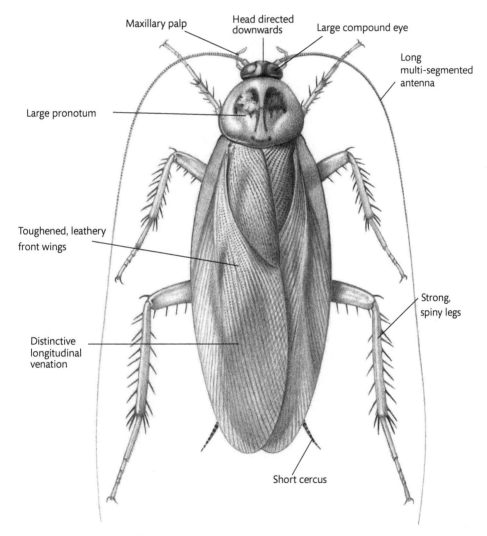

Periplaneta americana – **Dictyoptera**

Essential Entomology. Second Edition. George C. McGavin and Leonidas-Romanos Davranoglou, Oxford University Press.
© George C. McGavin and Leonidas-Romanos Davranoglou (2022). DOI: 10.1093/oso/9780192843111.001.0001

Key features

- flattened, broadly oval and leathery-bodied insects (cockroaches)
- front wings toughened as protective 'tegmina' to cover the membranous hind wings
- eggs typically laid in a toughened case
- median ocellus lost
- obligate mutualism with endosymbiont bacteria involved in nitrogen recycling
- soft-bodied, social insects with different castes (termites)
- in soil, litter, wood, or in association with habitations

❯ A female Epilamprinae cockroach guarding her nymphs, which are hiding under her body and wings. Singapore.

❯ Female *Perisphaerus* (Blaberidae) cockroaches have evolved a body type similar to that of a pill millipede and will roll into a ball when disturbed. Their nymphs feed on milk-like secretions from specialized pouches under their mother's thorax. Singapore.

Cockroaches and termites are of immense ecological importance in tropical and subtropical ecosystems across the globe, while their behaviour and physiology have revolutionized our understanding on the evolution of sociality and mutualism in animals. Cockroaches and termites will be described here separately, due to their highly divergent morphology and habits.

Cockroaches

Cockroaches are fairly uniform in general appearance, being leathery-bodied, and mostly sombre-coloured insects with an oval outline. Some tropical day-active species are brightly coloured to advertise distastefulness to potential predators. The body, when viewed from the side, is flattened, especially in species living under bark and stones, and this enables them to squeeze through very tight spaces. The **downward-directed head,** bearing generalized **chewing and biting jaws,** is shielded and largely concealed from above by the **large pronotum.** The compound eyes are well developed, there are **two ocelli-like spots,** and the **long, thread-like antennae** are multi-segmented. There are **two pairs of wings,** although many species are wingless. The front pair of wings, called **tegmina,** are toughened and overlap at rest to cover the larger **fan-shaped, membranous hind wings.** Wing venation is mainly longitudinal in orientation with numerous pale cross veins in the hind wing. Cockroaches are generally fast runners, and many species fly infrequently. The abdomen, which is normally completely concealed by the folded tegmina, bears a pair of sensitive, **single- or multi-segmented cerci** at the posterior end. A unique species that departs from this general body plan is the leaproach (genus *Saltoblattella*), which has enormous hind legs that allow it to move around by jumping, much like a grasshopper. Although fossils of extinct cockroach-like 'roachoids' of ambiguous affinities were among the most numerous neopteran insects in the hot, humid coal forests of the upper Carboniferous around 300 mya, the origins of modern cockroaches only go back to the Cretaceous (139–66 mya), which is considerably younger compared to other insect groups (e.g., beetles and wasps).

Cockroaches play an important role in the nitrogen cycle of soil and litter communities in tropical and subtropical regions, as the vast majority of species are free-living scavengers and omnivores that eat a wide range of dead or decaying organic materials, including bird droppings and bat guano, and a few are able to digest wood. Some species damage growing plants, and cannibalism and predation also occur. Many cockroaches live in association with other species and can be found in wasp, ant, or termite colonies, or in the nests of birds, caves with bats, or the lairs and burrows of mammals, while several species are semi-aquatic. Cockroaches are an important source of food for many species, both invertebrate and vertebrate, and are hosts for numerous parasitic species. Some species of the family Blatellidae are the main pollinators of certain woody climbing plants in the lowland rain forests of Sarawak. The flowers produce odours of rotting wood or fungi to attract the cockroaches, which feed on a secretion produced by the stigma and on the pollen itself.

In spite of their ecological diversity, cockroaches have been inadequately studied, as most of the extensive literature concerning these ubiquitous insects deals with effects of pest species and their control. The bad press regarding the unpleasant habits of certain species is misguided, as in fact, only 0.005% (or about 40 cockroach species) are considered pests because of their close association with humans, of which 20 have a significant impact. Synanthropic species have become widely distributed around the world due to extensive commercial trade and are associated with warm conditions and poor hygiene and sanitation. The main problem is that these species are attracted to both filth and foodstuffs, and, as a result, may carry a huge diversity of pathogenic organisms on their feet and other body parts. The list of diseases caused by mechanically transmitted bacterial species includes dysentery, diarrhoea, and gastroenteritis. They also act as intermediate hosts for a number of helminths that are pathogenic to vertebrates and are able to harbour and excrete dangerous viruses and protozoans. When they feed, they regurgitate partly digested food and leave behind their faeces and a characteristic offensive odour. The exposure, especially of children, to high levels of cockroach allergens in house dust can produce serious health problems such as allergies, dermatitis and eczema, and asthma. Insecticides of many types are routinely used to control pest species, but the development of resistance has generated interest in integrated pest management using biological control agents, such as hymenopteran parasitoids and fungal pathogens. The most common method of control remains the use of toxic sugar baits. However, in a fascinating example of natural selection, populations of the German cockroach, *Blattela germanica*, have evolved an aversion for sugar baits, greatly reducing the traps' effectiveness. The cockroaches have achieved this by having mutations that modified their taste buds, which make them taste sugary foods as highly bitter, therefore avoiding them altogether.

Principal cockroach families

Family	Distribution	Examples of genera
Blaberidae	Tropical	*Blaberus, Gromphadorhina, Leucophaea, Nauphoeta*
Blattellidae	Worldwide	*Blattella, Parcoblatta*
Pseudophyllodromiidae	Mostly Tropical	*Saltoblattella, Supella*
Blattidae	Tropical	*Periplaneta*
Cryptocercidae	N. America and East Asia	*Cryptocercus*
Nocticolidae	Africa, Asia, Australia	*Nocticola, Spelaeoblatta*
Corydiidae	Worldwide	*Arenivaga, Polyphaga, Tivia*

In terms of their general habits, cockroaches are very sensitive to air- and substrate-borne vibration, and their normal defence is to run and hide, but some species have warning colouration to signal non-palatability to predators, and several tropical species mimic the colour patterns of distasteful beetles, such as lycids, ladybirds, and blister beetles. Many cockroaches, especially those in the blattid subfamily Polyzosterinae, produce noxious-smelling repellent chemicals from ventrally located intersegmental or other glands. These compounds can produce skin rashes and temporary blindness, and a number of species, such as the Florida cockroach, *Eurycotis floridana*, are able to spray secretions over a distance of many centimetres.

Sexual behaviour ranges from relatively simple to quite complex repertoires of behaviour involving sequences of distinctive body movements. In general, females produce sexual pheromones to which males respond to over relatively long distances. Males sometimes also produce sexual pheromones and, when they meet, there is usually much antennal contact and a variety of body-rocking or trembling movements. The males of many species have special dorsal abdominal glands that produce aphrodisiac secretions to encourage the female. By raising their wings and adopting a particular posture, the male entices the female to mount his back partially or completely. In this position the female can lick the tergal gland secretions and the male can engage his genitalia with hers. The female then dismounts and turns to face backwards to complete the coupling.

In some species, such as members of the blaberid subfamily, Oxyhaloinae, courtship involves the production of sounds. Males of *Nauphoeta cinerea* stridulate by rubbing a ridged portion of the costal veins of the front wing against similar fine ridges on the hind corners of the pronotum. The Madagascan Hissing cockroach, *Gromphadorhina portentosa*, a large, wingless species that lives under forest litter, has an unusual mating behaviour. The males, which are larger than the females and can be up to 10 cm long, have enlarged thoracic humps with which they batter rivals in head-to-head contests. The fights are accompanied by a loud hissing produced when air is forcibly squeezed out of the second pair of abdominal spiracles, and are won by the biggest, noisiest males. The hissing noises also act as an aphrodisiac to the females, who will refuse to couple with silent partners.

Sperm is transferred in a spermatophore, and, after mating, female cockroaches lay up to forty eggs in two rows surrounded by a tough, protective egg case or ootheca. Some cockroaches are oviparous and stick their egg cases to the ground and cover them or carry them about partly projecting from the end of the abdomen. In all species of the family Blaberidae and a few in the Blattellidae, females are ovoviviparous. The egg case, which is much softer, thinner, and more flexible than in oviparous species, is fully extruded from the end of the abdomen and then rotated by 90° so that the long axes of the eggs within are horizontal. The ootheca is then drawn back inside and brooded within the body of the female until the eggs hatch.

Ovoviviparity differs from viviparity in that the young do not receive any nutrients from the mother while retained inside her body. Viviparity, the most derived reproductive strategy, is seen in the blaberid species *Diploptera punctata*, where females produce only half a dozen eggs, which are stored inside their abdomen in a 'uterus'. The developing embryos are supplied with a unique form of cockroach 'milk', which flows through the walls of the uterine wall. This milk is rich in fats and proteins and has even been considered as a potential superfood for humans. Female cockroaches in many species are known to provide varying degrees of maternal care after their young hatch. In the simplest cases, young nymphs may crowd under their mother's body until their cuticle becomes sclerotized. In some species, the young may crawl into the space beneath the tegmina of the female to find refuge. Females of the species *Thorax porcellana* take this approach one step further. Once her young are born, they hide underneath her wings, on her abdomen. She then starts making compressing motions with her abdomen, which squeeze a nutritious fluid out of microscopical pores from her abdominal membranes, on which the nymphs feed. The first instar nymphs are entirely dependent on their mother for nutrition and protection and perish rapidly if they are removed from her. Perhaps the most peculiar type of maternal care is provided by the females of *Perisphaerus*, also known as pill cockroaches. Males have a typical cockroach appearance, but females are wingless and look remarkably like pill millipedes, and they can roll themselves into a ball in a similar manner. Their nymphs are born blind and depend on their mother of survival. They cling on the underside of her abdomen, where they insert their elongated head in a lock and key manner in small holes at the base of their mother's legs, which are thought to provide the nymphs with nourishment, much like suckling mammals. In the presence of predators, the mother is also able to roll herself into a defensive ball with her nymphs still attached, in order to ensure their protection.

One of the best-known species, the so-called American cockroach, *Periplaneta americana* (Blattidae), originally spread from Africa on ships and is now cosmopolitan. Due to its relatively large size and ease of culturing, it has been used extensively as a laboratory animal for studies ranging from insecticide testing to research on the function of the nervous system. The German cockroach, *Blattella germanica*, is probably the most significant single pest species found in domestic and commercial buildings in many parts of the world, and considerable efforts are directed towards its control. A single female, which can live for six months, is capable of producing around 20,000 offspring.

The family Cryptocercidae contains a single genus *Cryptocercus* with a handful of unusual species found in North America and East Asia. Cryptocercids are sub-social species that live in family groups inside tunnels eaten out of dead trees. A pair of adults and their developing nymphs stay together in a family group, and it may take the parents several years to raise a complete brood, feeding the young nymphs and protecting them from enemies. They are able to eat wood because they have symbiotic protozoans in their guts to digest the cellulose. At each moult the nymphs have to re-acquire their symbionts by eating the excrement of their parents or other nymphs. The protozoans belong to the same genus as those found in early termites (Mastotermes: Mastotermitidae); these termites also share with all cockroaches (except the family Nocticolidae) intracellular bacterial endosymbionts, which are passed from mother to offspring and are essential for their survival. Both types of microbial symbionts provided the first indication that they might have been inherited in cockroaches and termites from a common ancestor—termites were until recently placed in their own order (Isoptera) and were thought to be related to cockroaches. The latest molecular studies have now shown that termites are in fact a lineage of highly modified cockroaches, and their closest living relatives are cryptocercids, suggesting that wood feeding and complex sociality evolved in their common ancestor.

Termites

Termites are undoubtedly the most successful cockroach lineage, and one of the dominant organisms on the planet. There can be no more impressive a sight in the African savannahs than the spectacularly large mounds built by termites in the genus *Macrotermes* (Termitidae). These towering, multi-vented chimneys, which can be more than six metres high, provide essential air conditioning for the nests below ground level, without which the termites might overheat. An intricate system of tunnels and vents throughout the structure of the nest maintains optimum temperature and gas levels in the brood chambers and gardens, where they cultivate a special fungus to eat. Other equally impressive termite mounds can be seen in northern Australia, where wedge-shaped nests more than three metres tall are made by the magnetic or compass termite, *Amitermes meridionalis* (Termitidae). The long axis of the nest always runs roughly north–south, so that the flat surface of the nest is orientated to act as a solar collector that warms the nest in the morning and evening. When the sun is at its highest during the heat of the day, the small surface presented by the top of the nest prevents overheating. Termite mounds are so effective in keeping cool in spite of the scorching heat that their structure inspired architects to design buildings in Africa and Australia that use the same passive cooling mechanism, instead of using air-conditioning.

Confined to tropical and subtropical regions, between 45–50° north and south of the Equator, termites have an immense impact in many habitats. They may consume up to one-third of the annual production of dead wood, leaves, and grass, and can be present in huge numbers, sometimes more than 8,000/m^2 (800/ft^2). Although individually small, their total biomass per unit area can vary from 1–5 g/m^2 to 22 g/m^2, which is double that of the biggest herds of African hoofed grazing mammals. In Amazonian *terra firma* forests, termites constitute up to 10% of the biomass of soil fauna, but more than 80% of the biomass of dead wood fauna.

Termites are sometimes called 'white ants' because their workers are pale-bodied, and because of similarities in their social organization. Termites live in permanent social colonies and are **polymorphic**, that is, they have a number of distinct **castes**. Termites evolved their social structures much earlier than ants, but they are, nevertheless, truly social, as they feed, groom, and protect each other, and offspring of one generation help their parents raise the next generation.

In general, termites are **pale, soft-bodied, and wingless** with shortish antennae composed of fewer than 32 segments. The rear end of the abdomen has a **pair of short cerci,** just like other cockroaches. A colony, which may vary in size from a small nest with a couple of dozen individuals, to a massive, architecturally complex, air-conditioned mound containing millions of termites, contains four different castes: primary reproductives, supplementary reproductives, soldiers, and workers. Primary reproductives (the kings

❯ **polymorphic**
having more than two forms

❯ **caste**
any group of individuals in a colony of social insects that is structurally or behaviourally different from individuals in other groups

and queens) are large, darkly hardened individuals whose sole job is to reproduce. Primary reproductives were once winged termites that left to found the
new colony, and which shed their **two pairs of wings** after a short dispersal or
nuptial flight. Colonies normally have a single queen and a few reproductive
males. Supplementary reproductives are less hardened than kings and queens
and these will become reproductive if the king or the queen unexpectedly die.
Soldiers are sterile males and females with heavily sclerotized and modified
pear-shaped heads.

> Termites are among the most
abundant and ecologically
important organisms on the
planet. Northern India.

© George McGavin

Soldiers have the primary function of defending the colony against predators,
which usually come in the form of termite-hunting ants. Although soldier
termites are usually blind, they are variously armed with large jaws that slice
ants, or they possess the ability to eject sticky secretions from a tube-shaped
head gland that entraps and kills ant invaders. The ongoing arms race between ants and termites have led some species of the genus *Pericapritermes* to
develop remarkable ways to defend themselves. The soldiers have peculiarly
shaped, twisted asymmetric mandibles. When the soldiers are attacked by an
ant, their mandibles strike against each other at a speed of 132.4 m/s^2, in what
is the fastest movement in the animal kingdom. When the mandibles touch
the ant, the force generated by this ultra-fast impact is so large (4,600 times
the termite's body weight), that it can dismember ants at a single blow, with
nearly 100% success.

   Considering this combative strength, worker termites are rather unim-
pressive in their appearance, and they resemble nymphs. However, their
role is essential in maintaining the colony, and they are the most numerous caste, outnumbering soldiers by 50 to 1. Workers, like soldiers, can be
of both sexes (in contrast to the social Hymenoptera, where workers are all
females). Normally blind with simple, chewing mouthparts, workers build
and repair the nest, forage, feed nymphs, and groom other colony members.

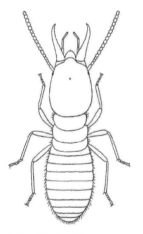

⌃ A soldier termite
(Rhinotermitidae).

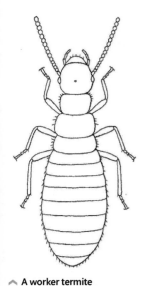

A worker termite (Rhinotermitidae).

❯ **trophallaxis**
the exchange of food between colony members in social insects

In the Kalotermitidae, there is no true worker caste, and the jobs are done instead by older nymphs.

The founding of new colonies happens at certain times of the year with the production of large numbers of winged reproductives that usually emerge around sunset. After a short nuptial flight, they settle on the ground, shed their wings, and females adopt a characteristic stance with their abdomen pointed up in the air. In this pose they release sexual pheromones to attract males. A pair will then move off to find a good place nearby in which to burrow. They make a small chamber for themselves in soil or wood, seal it, and mate. The job of rearing the first brood falls to the reproductives, thereafter all tasks are taken over by workers or nymphs. Soldiers may not be produced until the colony grows larger, and new reproductive individuals may not appear for many years.

Although some termites eat a range of plant matter, including seeds, grasses, and leaves, the majority eat dead or decaying wood.

All termites digest cellulose, an ability dependent on symbiotic interactions with bacteria, protozoa, and fungi. Six families make up the so-called lower termites. In these taxa, the gut contains symbiotic flagellate protozoans to enable cellulose digestion, and they are passed on by **trophallaxis** via oral or anal feeding, just like in their relatives, the cryptocercid cockroaches.

In addition to cellulose-digesting symbionts, all termites (except the fungus-growing Macrotermitinae) have nitrogen-fixing gut bacteria.

Members of the Termitidae, comprising some 70% of all species, digest cellulose, that is pre-digested either by fungus or by the action of anaerobic bacteria. The family is made up of four distinct subfamilies. The Apicotermitinae, or soldier-less termites, are humus-feeding species with cultures of bacteria in their hindgut to ferment plant material. In species belonging to the genus *Globitermes*, workers act as soldiers and are able to self-destruct by bursting their abdomens and covering their enemies (usually ants) with slimy, sticky gut contents. The Termitinae, or subterranean termites, also have bacterial cultures in their hindgut to ferment plant material, as do the Nasutitermitinae (snouted termites). This latter subfamily has strange snouted soldiers with reduced jaws. Their heads contain well-developed frontal glands, which produce sticky and noxious chemicals that are ejected at predators. The Macrotermitinae (fungus-growing termites) construct a permanent mud-covered network of tunnels to reach their food. The fungus gardens are spongy, greyish brown combs made of their own faeces. The fungus, belonging to the genus *Termitomyces*, is only found inside termite nests, and it breaks down the faeces and termites that eat it. Fungal hyphae are mixed with regurgitated food and fed to nymphs. Fungus-growing species are very important in decomposition processes in seasonal habitats in Asia and Africa, because, by cultivating the fungus, the termites are able to survive through the dry season.

Termite families

Family	Distribution	Habits
Mastotermitidae* (Darwin termite)	1 species in northern Australia (*Mastotermes darwiniensis*)*	Pest of wooden structures and crops
Kalotermitidae (dry wood termites)	> 460 species, widespread	Prefer dead wood, timber pests
Hodotermitidae (harvester termites)	21 species in Africa and Asia	Grass-harvesters, pasture pests
Archotermopsidae (dampwood termites)	> 6 Species in North America and Asia	Prefer decaying wood, timber pests
Serritermitidae (saw-toothed termites)	3 species in South America	Live in mound walls of another termite, food unknown
Stolotermitidae	10 species in Australia, New Zealand and South America	Prefer decaying wood, occasional timber pests
Stylotermitidae	> 50 species in Indian sub-continent and Southeast Asia	Inhabit living trees, on which they feed
Rhinotermitidae (subterranean termites)	> 300 species, widespread	Prefer decaying wood, crop and timber pests
Termitidae (higher termites)	> 2,070 species, widespread	Food variable (grass, wood, humus, etc.), many pests

* The Darwin termite shares many characteristics with woodroaches (Cryptocercidae), especially in that they possess similar protozoan symbionts that they transfer in faecal food to the young.

> **pseudergate**
a caste in lower termites made up from immatures who take the role of workers

Lower termites have flexible development pathways and there is no true worker caste; this role is taken by a **pseudergate** caste made up of nymphs. Moulting in lower termites is complicated and, depending on colony structure and the presence of certain pheromones, can produce no change (a station-ary moult), can produce an individual belonging to a different caste (i.e., a pseudergate moults to become a reproductive), or can produce a reversion to an earlier caste. To make it worse, instars can also be added or missed out entirely.

Termitids have a more elaborate and rigid caste system than other 'lower' termites and, in some species, the **physogastric** queens can grow to 1,000% of their original size. The abdomen, which is grossly swollen with fat body and ovaries, can produce several thousand eggs each day. Generally, after eggs hatch there are two developmental pathways. One leads, through a se-ries of nymphal stages, to individuals that develop wing buds, and eventually become reproductives. The other path, known as the sterile pathway, leads through a series of moults to the production of sterile workers and soldiers. This pathway can be complicated in some species with the production of two separate subpathways, one leading to the production of small workers and small soldiers and the other to the production of large soldiers and workers.

Another way that termites are essential components in ecosystems is that they are food for a great many other species. Many ant species specialize on termites as their main prey. Another, less-aggressive approach is followed by species of ants that live inside termitaria and defend the termites from attack by other ants. The relationship is almost certainly mutualistic as the ants receive shelter (and perhaps food) in return for their services as bodyguards. Representing such an abundant source of food as they do, it is not at all surprising that termites have many specialized vertebrate predators around the world. Many bird species eat termites, but also many mammals, such as the aardvark, aardwolf, pangolins, armadillos, anteaters, sloth bears, and echidnas. Chimpanzees spend more than one-quarter of their foraging time all year round using sticks and branches from particular tree species as tools to extract termites from their mounds. Two types of tool are used: a sharp-ended stick the width of a thick pencil and up to 50 cm long to make holes in the walls, and a thinner, flexible probe to act as a lure. The end of the probe is made brush-like, and it is pushed into the hole where the soldier termites seize it. The chimpanzee withdraws the probe and quickly eats the insects. Humans, very close relatives of chimpanzees, harvest termites as well, especially when huge numbers of reproductives leave the nest to mate and disperse. They are a rich food source, providing more than twice as much energy, twice as much iron and phosphorus, and four times as much calcium per 100 g as lean beef. Unsurprisingly, in parts of Africa, Australia, Asia, and the Americas, where termites are abundant, all stages are readily consumed by locals and indigenous communities, while their mounds can form important family-owned resources.

Despite the fact that they are an essential component in tropical terrestrial food chains and primary nutrient recyclers, and they are considered the functional equivalent of earthworms in temperate regions, the termites are (somewhat unfairly) notorious, as many species are destructive pests and attack structural timbers, wooden buildings, furniture, wooden artefacts, books, and crops in many areas of the world. In parts of Africa, there may be 800 termite mounds per hectare, which can tie up a huge amount of soil, and can encourage erosion and subsidence. Recent studies using carbon-14 dating techniques show that some South African termite nests or termitaria may have been in existence for 4,000 years. Termites cause huge commercial damage to pasture grasses and crops both above and underground (sugarcane, potatoes, yams). Termite damage is a major problem in tropical forestry and explosives and persistent insecticides are routinely used to destroy and poison colonies before areas are planted. The use of resistant tree varieties will probably be most cost-effective in the long run. Biological control has not been widely successful so far, but some products containing bacteria that are genetically engineered to express the *Bacillus thuringiensis* endotoxin are useful against subterranean termites (Rhinotermitidae), and a number of fungal pathogens have been used to control termites that damage crops. Another unusual feature of termites is that, as result of bacterial fermentation in their hindguts, vast quantities of methane (a greenhouse gas) are released by the termite's rear ends to the atmosphere (similarly to ruminants such as farmed cattle), which represent 1–3% of the world's methane emissions. However, unlike cattle and rice paddies, which release harmful amounts of methane, before they reach the atmosphere more than half of the termites' methane emissions are broken down and recycled into the soil by methane-consuming bacteria in the mound's walls. Other termite gut symbionts instead release hydrogen in the atmosphere, as they can convert 1 g of wood into 10 l of hydrogen, which is harmless for the planet. This process is so effective that scientists are trying to harness the powers of these symbionts in order to massively produce hydrogen as an ecologically friendly alternative source of energy. It is evident that cockroaches and termites are much more than the annoying pests that most of us think they are, and that, without them, major ecosystems on the planet would collapse.

Extinct cockroach relatives are among the most abundant insect groups in the fossil record, especially during the Carboniferous (350–300 mya). These 'roachoids' had a similar body plan with modern cockroaches, but many had important differences, such as a long ovipositor, which is absent in their extant relatives. Some roachoids even had beak-like mouthparts to perforate plant tissue, while certain basal cockroaches looked like beetles or praying mantises. The ancestors of modern cockroaches evolved much later, around the Jurassic (200 mya). Termites evolved from a *Cryptocercus*-like ancestor in the Late Jurassic (about 150 mya) and are unique among insects in having no extinct families, which attests to the endurance of this group.

Key reading

Barth, R. H. (1968). The mating behaviour of *Gromphadorhina portentosa* (Schaum) (Blattaria, Blaberoidea, Blaberidae, Oxyhaloinae): an anomalous pattern for a cockroach. *Psyche* **75**(2): 124–31.

Bell, J. W., Roth, L. M., and Nalepa, C. A. (2007). *Cockroaches: ecology, behaviour and natural history*. Johns Hopkins University Press, Baltimore.

Bhoopathy, S. (1998). Incidence of parental care in the cockroach *Thorax porcellana* (Saravas) (Blaberidae: Blattaria). *Current Science* **74**(3): 248–51.

Bohn, H., Picker, M., Klass, K.-D., and Colville, J. F. (2010). A jumping cockroach from South Africa, *Saltoblattella montistabularis*, gen. nov., spec. nov. (Blattodea: Blattellidae). *Arthropod Systematics and Phylogeny* **68**(1): 53–69.

Grimaldi, D. A. and Engel, M. S. (2005). *Evolution of the insects*. Cambridge University Press, Cambridge.

Evangelista, D. A., Wipfler, B., Béthoux, O., Donath, A., Fujita, M., Kohli., M. K., Legendre, F., Liu, S., Machida, R., Misof, B., Peters, R. S., Podsiadlowski, L., Rust, J., Schuette, K., Tollenaar, W., Ware, J. L., Wappler, T., Zhou, X., Meusemann, K., and Simon, S. (2019). An integrative phylogenomic approach illuminates the evolutionary history of cockroaches and termites (Blattodea). *Proceedings of the Royal Society B* **286**(1895): 20182076.

Grandcolas, P. (1994). Phylogenetic systematics of the subfamily Polyphaginae, with the assignment of *Cryptocercus* Scudder to this taxon (Blattaria, Blaberoidea, Polyphagidae). *Systematic Entomology* **19**: 145–58.

Howse, P. E. (1970). *Termites: a study in social behaviour*. Hutchinson University Library, London.

Jacklyn, P. M. (1992). 'Magnetic' termite mound surfaces are orientated to suit wind and shade conditions. *Oecologia* **91**: 385–95.

Krishna, K., Grimaldi, D. A., Krishna, V., and Engel, M.S. (2013). Treatise on the Isoptera of the world: introduction. *Bulletin of the American Museum of Natural History* **2013**(377): 1–200.

Krishna, K. and Weesner, F. M. (eds) (1970). *Biology of termites* (2 Vols). Academic Press, London.

Kuan, K.-C., Chiu, C.-I., Shih, M.-S., Chi, K.-J., and Li, H.-F. (2020). Termite's twisted mandible presents fast, powerful, and precise strikes. *Scientific Reports* **10**: 9462.

Lo, N., Bandi, C., Watanabe, H., Nalepa, C., and Beninati, T. (2003). Evidence for coclado-genesis between diverse dictyopteran lineages and their intracellular endosymbionts. *Molecular Biology and Evolution* **20**(6): 907–13.

Lo, N., Tokuda, G., Watanabe, H., Rose, H., Slaytor, M., Maekawa, K., Bandi, C., and Noda, H. (2000). Evidence from multiple gene sequences indicates that termites evolved from wood-feeding cockroaches. *Current Biology* **10**(13): 801–4.

Martius, C. (1994). Diversity and ecology of termites in Amazonian forests. *Pedobiologia* **38**: 407–28.

Nalepa, C. A. (1984). Colony composition, protozoan transfer and some life history characteristics of the woodroach, *Cryptocercus punctulatus* Scudder (Dictyoptera: Cryptocercidae). *Behavioural Ecology and Sociobiology* **14**: 273–9.

Nalepa, C. A. (1988). Reproduction in the woodroach *Cryptocercus punctulatus* Scudder (Dictyoptera: Cryptocercidae): mating, oviposition, and hatch. *Annals of the Entomological Society of America* **81**(4): 637–41.

Nauer, P. A., Hutley, L. B., and Arndt, S. K. (2018). Termite mounds mitigate half of termite methane emissions. *Proceedings of the National Academy of Sciences* **115**(52): 13306–11.

Pearce, M. J. (1997). *Termites: biology and pest management*. CAB International, Wallingford.

Roth, L. M. (1970). Evolution and taxonomic significance of reproduction in Blattaria. *Annual Review of Entomology* **15**: 75–96.

Roth, L. M. (1981). The mother-offspring relationship of some blaberid cockroaches (Dictyoptera: Blattaria: Blaberidae). *Proceedings of the Entomological Society of Washington* **83**: 390–8.

Roth, L. M. and Alsop, D. W. (1978). Toxins of Blattaria. In: *Arthropod venoms*. Handbook of experimental pharmacology series **48**. Bettini, S. (ed), pp. 465–87. Springer-Verlag, Berlin/Heidelberg.

Roth, L. M. and Willis, E. R. (1960). *The biotic associations of cockroaches*. Smithsonian miscellaneous collections series **141**. Smithsonian Institution, Washington, DC.

Schal, C., Gautier, J. Y., and Bell, W. J. (1984). Behavioural ecology of cockroaches. *Biological Reviews* **59**(2): 209–54.

Schal, C. and Hamilton, R. L. (1990). Integrated suppression of synanthropic cockroaches. *Annual Review of Entomology* **35**: 521–51.

Sreng, L. (1993). Cockroach mating behaviors, sex pheromones and abdominal glands (Dictyoptera: Blaberidae). *Journal of Insect Behaviour* **6**(6): 715–35.

Wada-Katsumata, A., Silverman, J., and Schal, C. (2013). Changes in taste neurons support the emergence of an adaptive behavior in cockroaches. *Science* **340**(6135): 972–5.

Watson, J. A. L. and Abbey H. M. (1985). Seasonal cycles in *Nasutitermes exitiosus* (Hill) Isoptera: Termitidae. *Sociobiology* **10**: 73–92.

Wendelken, P. W. and Barth, R. H. (1987). The evolution of courtship phenomena in neotropical cockroaches of the genus *Blaberus* and related genera. *Ethology* **27**(Suppl): 98.

Wilde, de, J. and Beetsma, J. (1982). Caste development in social insects. *Advances in Insect Physiology* **16**:167–246.

Williford, A., Stay, B., and Bhattacharya, D. (2004). Evolution of a novel function: nutritive milk in the viviparous cockroach, *Diploptera punctata*. *Evolution & Development* **6**(2): 67–77.

Mantodea
(praying mantids)

Common name	Mantids	**Distribution**	Predominantly tropical and subtropical regions
Derivation	Gk. *mantis*–diviner, prophet	**Number of families**	15
Size	Body length 8–20 mm	**Known world species**	2,500 (0.22%)
Metamorphosis	Incomplete (egg, nymph, adult)		

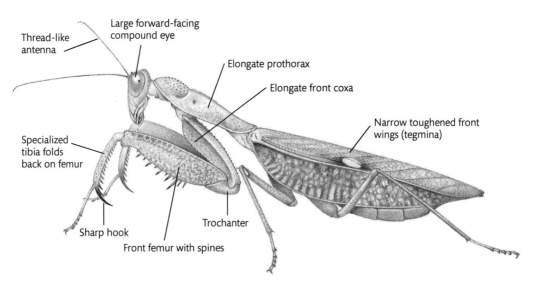

Thread-like antenna

Large forward-facing compound eye

Elongate prothorax

Elongate front coxa

Narrow toughened front wings (tegmina)

Specialized tibia folds back on femur

Sharp hook

Front femur with spines

Trochanter

Sphodromantis centralis – Mantodea

Essential Entomology. Second Edition. George C. McGavin and Leonidas-Romanos Davranoglou, Oxford University Press.
© George C. McGavin and Leonidas-Romanos Davranoglou (2022). DOI: 10.1093/oso/9780192843111.001.0001

Key features

- distinctive highly mobile, triangular head with large, forward-facing compound eyes
- front pair of legs characteristically modified for prey capture
- asymmetric male genitalia
- eggs laid in papery, foam-like, or cellophane-like egg cases

❯ A *Liturgusa* sp. mantis from Ecuador, feeding on a cricket.

© Anne Riley

❯ As their principal prey-capturing appendages, the mantis's sickle-shaped forelegs require regular cleaning. Brazil.

© Anne Riley

› An exquisitely camouflaged
female *Empusa pennata*
from France.

© Paul Brock

The common name 'praying' mantid comes from the way in which these insects rest with their front legs held up and together as if in prayer. An equally
appropriate homonym would be 'preying' mantid, because the features that
separate these insects from their close relatives the cockroaches are mainly
concerned with their highly specialized predatory lifestyle. Mantids are masters of ambush and rely on being difficult to see. They keep very still or sway
gently to mimic the movement of the foliage, and their colour patterns are
usually sombre greens and browns to blend in perfectly with the background.
Some tropical species may be very brightly coloured and ornamented with
strange cuticular outgrowths and texturing to mimic the flowers upon which
they rest. Certain species take this to the extreme by having a body shape,
colouration, and texture that mimics that of dead or live leaves, sticks, and
grasses. Others, such as *Vespamantoidea wherlyi*, instead want to be seen—
they mimic the form and colouration of potentially dangerous wasps, so that
predators confuse them with the latter and leave them alone.

Mantids have distinctive triangular and very mobile heads with a pair of
large compound eyes placed laterally and facing forward to give true binocular vision. Binocular triangulation to calculate the distance of prey objects is
known in many vertebrates, but has also been proved conclusively in mantids,
showing that even the smallest brains can undertake complex sensory tasks.
Other predacious insects, such as tiger beetles and dragonfly larvae, might
also use triangulation when hunting. In addition to the eyes, there may be

three ocelli. The first segment of the thorax is very long indeed and bears the **specialized raptorial, front legs.**

The coxa of the front legs is also very elongate, and the femur is enlarged and variously equipped with rows of sharp spines and teeth. The tibia, which is also spined or toothed, folds back on the inner face of the femur like a jackknife.

The middle and hind pair of legs are of normal construction and appear to arise much further back down the body. The **front wings are narrow and toughened** while the **fan-folded, hind wings** are much larger, membranous, and have a more obvious net-like arrangement of veins, although some species have short wings or may be wingless. Some mantids can be incredibly stick insect-like, but the shape of the head and the raptorial front legs are characteristic.

Mantids, which are active during the day and some by night, are predacious on a wide range of insects, spiders, and other arthropods, which they ambush or stalk. Larger species have even been recorded catching and eating vertebrates such as salamanders, mice, lizards, and even hummingbirds. Using triangulation, mantids are able to calculate the exact distance, speed, and direction of their victim, and can snatch insects out of the air as they fly past. The speed and elegance of their movements makes mantids popular insect pets and has even inspired a style of karate. The strike, which takes place in two distinct phases, lasts less than 100 ms. In the initial phase, the tibiae are fully extended in readiness for the second phase, which takes the form of a rapid sweeping action. The femora are quickly extended, and, at the same time, the tibiae are flexed around the prey. It has been shown that some of the spines projecting from the ventral surface of the femora are moveable and, when stimulated, elicit a tibial flexion reflex, which helps the insect maintain a strong grip on struggling prey. The size of prey that mantids can cope with depends on the species, but, generally, if prey can kick or jump, the mantid will not attack anything much larger than half its own body weight. Portions of the prey containing poisonous secretions are left uneaten. However, mantises of the genus *Carrikerella* follow a different hunting approach. The spines of their raptorial fore tibiae are forwardly oriented and barbed—much like a spear or a harpoon. While most species use the spines of their raptorial forelegs to simply restrain their prey, *Carrikerella* use their foretibial spines to impale microscopic arthropods that they fish out of small crevices that most mantises would not be able to access. This allows *Carrikerella* to hunt insects of multiple size categories, which makes them more effective predators.

Many species have been shown to be cannibalistic, and mating for the smaller male can be a dangerous business if he attempts to mate with a nonreceptive female. Females will signal their receptivity by emitting pheromones and by changes in their behaviour. Nevertheless, males are very cautious to keep out of the way of the front legs of females, and normally approach and jump on their prospective mate from behind, keeping low and out of the way of the female's front legs. Males do get eaten, even while copulating, and although this happens often in captivity, the extent to which it occurs in the

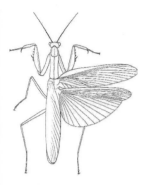

The mobile head of the mantid (*Mantis religiosa*) supported by a very elongate prothorax, which also bears the raptorial front legs.

Mantid front legs are well equipped with spines and teeth to impale and hold prey.

wild is unclear. Female-biased sex ratios seen in nature might be due to sexual cannibalism but could also be due to higher predation rates on males, the fact that females live longer, or simply that more female offspring are produced. However, it matters little to the female if her suitor loses his head. Due to reflexes in the CNS, the half-devoured male will continue to copulate for some time ensuring the transfer of his sperm. Eggs are laid within a protective papery or foam-like ootheca made from abdominal gland secretions. This egg case contains anything from a dozen to a few hundred eggs, depending on the species.

In some cases, a balloon of a cellophane-like secretion surrounds an inner ootheca. The ootheca is fixed to twigs or rock surfaces and, in some species, is guarded against predators by the female. Several species of parasitic wasps with appropriately long ovipositors seek out mantid egg cases and will parasitize every egg within. In guarding species, maternal care may even continue for a few days after the young have hatched. On hatching, the very small nymphs moult again before they start feeding; they become adult after ten or twelve moults.

Mantids not only have superb eyesight for hunting, but also they have an additional remarkable sense to avoid being hunted by bats. An ultrasonic-detecting 'cyclopean' ear, responsive to frequencies of between 25 and 45 kHz, is located in a deep groove on the ventral midline of their bodies between their metathoracic legs. Pressed against the insides of the groove walls are three pairs of tracheal air sacs, the largest pair of which have a neural connection. Ultrasound triggers immediate evasive behaviour. Rapid extension of the front legs and a flick up of the abdomen causes a rapid stall and forces the insect into a roll or a steep dive. They can recover control from such an extreme manoeuvre, but if the pursuing bat is too close, they will crash-land without injury. Predation pressure by bats to flying mantises must be considerable, as cyclopean ears are present in many mantises and may have evolved independently multiple times. They are frequently best developed in only the sex that flies the most (usually the male) while short-winged flightless species and females usually have reduced or no auditory capability. However, cyclopean ears are maintained in certain flightless species and may be used in communication, prey detection, or the avoidance of different predators, although we currently have no concrete proof. Recently, a species of mantid has been shown to have additional, equally sensitive hearing at lower frequencies (2–4 kHz), but the reason for this is not yet known. The sensitivity of the mantid system is similar to bat-detecting devices found in crickets and lacewings, but less sensitive than the tympanal hearing organs of moths.

Mantids also need to defend themselves against birds, reptiles, and mammals. They can run, fly, and hide, but also use threat postures, holding their front legs out sideways to look bigger and displaying bright markings on the inside of the femora. No idle threat, a strike from a mantid can easily draw blood or damage an eye, and a predator might be persuaded to seek an easier meal elsewhere. Like stick insects, mantids will shed a leg if trapped, but **autotomy** rarely involves the front legs, without which the insect would eventually die of hunger.

⌃ A typical mantid egg case, which may contain a few hundred developing eggs.

❭ autotomy
the casting off of a part of the body, for example, when under threat

As to the origins of mantids, their earliest fossils date back to the Cretaceous, although they started diversifying considerably only from the Tertiary onwards. The Alienopteridae, a remarkable family discovered in Cretaceous Myanmar amber, is thought to represent a lineage that is either ancestral to Mantodea, or its sister group. Alienopterids are unique in that they possess characteristics present in both Mantodea and cockroaches, further reaffirming the sister-group relationship between the two orders.

Key reading

Bai, M., Beutel, R.G., Klass, K.-D., Zhang, W., Yang, X., and Wipfler, B. (2016). Alienoptera—A new insect order in the roach—mantodean twilight zone. *Gondwana Research* **39**: 317–26.

Grimaldi, D. (2003). A revision of Cretaceous mantises and their relationships, including new taxa (Insecta: Dictyoptera: Mantodea). *American Museum Novitates* **3412**: 1–47.

Johns, P. M. and Maxwell, M. R. (1997). Sexual cannibalism: who benefits? *Trends in Ecology and Evolution* **12**: 127–8.

Reitze, M. and Nentwig, W. (1991). Comparative investigations into the feeding ecology of six Mantodea species. *Oecologia* **86**: 568–74.

Rivera, J. and Callohuari, Y. (2020). A new species of praying mantis from Peru reveals impaling as a novel hunting strategy in Mantodea (Thespidae: Thespini). *Neotropical Entomology* **49**: 234–49.

Rosner, R., von Hadeln, J., Tarawneh, G., and Read, J. C. A. (2019). A neuronal correlate of insect stereopsis. *Nature Communications* **10**: 2845.

Roussel, S. (1983). Binocular stereopsis in an insect. *Nature* **302**: 821–2.

Yager, D. D. (1990). Sexual dimorphism of auditory function and structure in praying mantises (Mantodea: Dictyoptera). *Journal of Zoology* **221**: 517–38.

Yager, D. D. (1996). Serially homologous ears perform frequency fractionation in the praying mantis, Creobroter (Mantodea, Hymenopodidae). *Journal of Comparative Physiology A, Sensory, Neural and Behavioural Physiology* **178**: 463–75.

Yager, D. D. and Hoy, R. R. (1986). The cyclopean ear: a new sense for the praying mantis. *Science* **231**: 727–9.

Wieland, F. (2013). *The phylogenetic system of Mantodea (Insect: Dictyoptera): species, phylogeny and evolution.* Universitätsverlag Göttingen: Göttingen.

Wieland, F. & Svenson, G. J. (2018). Biodiversity of Mantodea. In: *Insect biodiversity: science and society*, **II**. Foottit, R. G. and Adler, P. H. (eds), pp. 389–416. John Wiley and Sons, Hoboken.

Xenonomia

This is one of the most obscure clades of insects, which comprises two wingless orders—the ice crawlers (Grylloblattodea), and the heel walkers (Mantophasmatodea), which are the most recently described order of insects. Some classifications unite both groups under the single order Notoptera, although the latest molecular and morphological studies validate their status as separate, deeply divergent orders.

Essential Entomology. Second Edition. George C. McGavin and Leonidas-Romanos Davranoglou, Oxford University Press.
© George C. McGavin and Leonidas-Romanos Davranoglou (2022). DOI: 10.1093/oso/9780192843111.001.0001

Grylloblattodea
(ice crawlers–alternative name: **Notoptera**)

Common name	Ice crawlers or rock crawlers	Metamorphosis	Incomplete (egg, nymph, adult)
Derivation	Gk. combination of cricket (gryllus) and cockroach (blatta)	Distribution	Cooler regions of North America and north-east Asia
Size	Body length 14–43 mm	Number of families	1 (Grylloblattidae)
		Known world species	33 (0.003%)

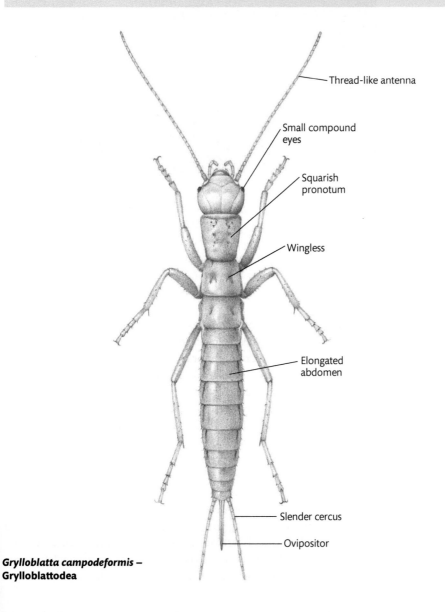

Thread-like antenna

Small compound eyes

Squarish pronotum

Wingless

Elongated abdomen

Slender cercus

Ovipositor

***Grylloblatta campodeformis* –
Grylloblattodea**

Essential Entomology. Second Edition. George C. McGavin and Leonidas-Romanos Davranoglou, Oxford University Press.
© George C. McGavin and Leonidas-Romanos Davranoglou (2022). DOI: 10.1093/oso/9780192843111.001.0001

Key features

- eyes reduced or absent, ocelli absent
- wingless
- eversible vesicles on first abdominal segment
- confined to certain cold, high-altitude regions

❯ Grylloblattodea, one of the smallest insect orders, can only survive in very damp and cold environments. This species, *Galloisiana nipponensis* lives underground, in the wet and cool forests of Japan.

© Piotr Naskrecki

These slender, wingless, slightly hairy, grasshopper-like insects were first discovered in the Canadian Rockies in 1913. They were originally thought of as being a small 'primitive' family of the Orthoptera, but a number of earwig-, stonefly-, cockroach-, and even silverfish-like features made it inevitable that they should have an order all of their own. For a long time, there was much disagreement on exactly where these 'living fossils' fit in relation to the other polyneopteran orders.

The head has **small, compound eyes,** although these are sometimes absent; **slender, thread-like antennae**; and **simple chewing, forward-facing mouthparts.** There are **no ocelli.** The **pronotum is square** and the abdomen is cylindrical with a pair of **slender, multi-segmented cerci.**

Species in only five genera are currently known: *Galloisiana* recorded from China, Japan, and South Korea; *Grylloblattina* from South Korea; *Grylloblatta* from western parts of North America and Canada; *Grylloblatella* from China and southern Siberia; and *Namkungia* from South Korea.

Grylloblattodea can be found in a range of different habitats and elevations, although all are characterized by a cool climate with high humidity. South Korean species of *Galloisiana* have been found in limestone caves, whereas, in Japan, the commonest species *Galloisiana nipponensis* is a mountain dweller and lives among rocks, under stones, and in decaying wood in subalpine, deciduous forests up to 2000 m. The dependence of Grylloblattodea on cool and glacial habitats means that most populations are fragmented and could be severely affected by climate change.

Not much is known about the behaviours of Grylloblattodea. Mating, which has been observed in some *Galloisiana* species, involves some preliminary tactile courtship and antennal fencing. The male then grips the female's prothorax with his mandibles and mounts her back. He then twists his abdomen round to engage his genitalia and partially slides off sideways (invariably to his mate's right-hand side). The two may remain in this position for up to four hours. The female may not lay her eggs for another ten weeks or more. It is thought that there may be eight nymphal instars, and complete nymphal development may take up to five or six years. As nymphs age they become darker coloured and add segments to their antennae at each moult. The adults typically live for fewer than two years.

With a few exceptions, most species of *Grylloblatta* are extremophiles adapted to life in the cold conditions of mountainous regions, where they can be found moving about on the surface of snow and ice, particularly after dark. During the day they can be found in rock crevices or under snow. Nymphs and adults are primarily scavengers, their preferred food being dead or torpid insects and other small arthropods that they find. Wind-borne insects, particularly aphids, that end up as fallout on the surface of snow in mountainous regions are the most important source of food for rock crawlers. There is some evidence that they feed on a range of organic matter, including moss and, perhaps, some plant material, and, in laboratory cultures, they are cannibalistic. In the summer months, they retreat underground to avoid the desiccating conditions. Females use a short ovipositor to lay relatively large, dark eggs in moss or soil and these may enter diapause for a year. Species are distinguished on the structure of the male genitalia. Life history details of most species are not well known. Due to their reliance for low temperatures, climate change may pose a severe challenge for the survival of ice crawlers across the globe.

The origins of Grylloblattodea are very obscure, as they are very scarce in the fossil record. Currently available fossils from the Permian and the Jurassic may represent groups that are only distantly related to modern ice crawlers and may not be that informative on their evolution. However, morphological data and phylogenomics overwhelmingly support a sister-group relationship to the Mantophasmatodea, placing the origin of both orders to the Jurassic (around 170 mya).

Key reading

Jarvis, K. J. and Whiting, M. F. (2006). Phylogeny and biogeography of ice crawlers (Insecta: Grylloblattodea) based on six molecular loci: designating conservation status for Grylloblattodea species. *Molecular Phylogenetics and Evolution* **41**(1): 222–37.

Kamp, J. W. (1979). Taxonomy, distribution and zoogeographic evolution of *Grylloblatta* in Canada (Insecta: Notoptera.) *Canadian Entomologist* **111**: 27–38.

Kim, B.-W. and Lee, W. (2007). A new species of the genus *Galloisiana* (Grylloblattodea, Grylloblattidae) from Korea. *Zoological Science* **24**: 733–45.

Schoville, S. D., Simon, S., Bai, M., Beethem, Z., Dudko, R. Y., Eberhard, M. J. B., Frandsen, P. B., Küpper, S. C., Machida, R., Verheij, M., Willadsen, P. C., Zhou, X., and Wipfler, B. (2021). Comparative transcriptomics of ice-crawlers demonstrates cold specialization constrains niche evolution in a relict lineage. *Evolutionary Applications* **14**(2): 360–82.

Wipfler, B., Bai, M., Schoville, S., Dallai, R., Uchifune, T., Machida, R., Cui, Y., and Beutel, R. G. (2014). Ice crawlers (Grylloblattodea)—the history of the investigation of a highly unusual group of insects. *Journal of Insect Biodiversity* **2**(2): 1–25.

Mantophasmatodea

(heel walkers—alternative name: **Notoptera**)

Common name	Heel walkers or gladiators	**Distribution**	Southern and east Africa (Namibia, South Africa, Tanzania)
Derivation	Gk. combination of praying mantis and stick insect (phasma)		
		Number of families	3 (Austrophasmatidae, Mantophasmatidae, Tanzaniophasmatidae)
Size	Body length up to 35 mm		
Metamorphosis	Incomplete (egg, nymph, adult)	**Known world species**	20 (0.0018%)

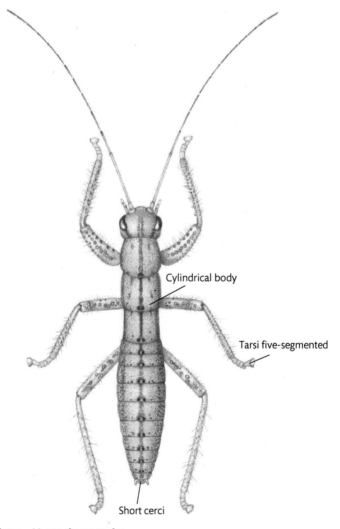

Tyrannophasma gladiator – **Mantophasmatodea**

Essential Entomology. Second Edition. George C. McGavin and Leonidas-Romanos Davranoglou, Oxford University Press.
© George C. McGavin and Leonidas-Romanos Davranoglou (2022). DOI: 10.1093/oso/9780192843111.001.0001

Key features

- wingless
- mantis-like predatory head, ocelli absent
- tarsal pads distinctly enlarged and held upwards
- drumming organ on male abdomen
- confined to arid and semi-arid regions of Southern and East Africa

❯ Mantophasmatodea are the most recently described extant insect order. They resemble stick insects and mantids, hence the name. They are also called heelwalkers, as their tarsal arolia are always held upwards. *Sclerophasma* sp., South Africa.

© Mike Picker

Heel walkers represent the most recently discovered insect order, only found in 2002. Since their initial discovery, a comprehensive body of literature has described their behaviour, morphology, physiology and evolution, rendering them one of the best-known orders of insects.

The details of their discovery are fascinating. The first known specimen was described from Eocene Baltic amber (48–34 mya) collected in 1992, as belonging to an undetermined order. It was only in 2002, when a series of unidentified museum specimens collected in the 1950s allowed the examination of additional characters that made it clear that these insects are part of a previously unrecognized group. Mantophasmatodea are the sister group to Grylloblattodea, with which they share a common ancestor in the Jurassic (about 170 mya).

Heel walkers bear a superficial resemblance to stick insects and praying mantises. Mantophasmatodeans walk in a characteristic manner, where their **enlarged tarsal pads (arolia) are held upwards** to protect them, with their 'heels' being the only part touching the ground—hence their common name, heel walkers. This is a feature shared with many stick insects (Phasmatodea). All heel walkers are **medium sized** with a **slender body**. Adults are **wingless**, which made some researchers to initially believe that they are nymphs of winged insects. Their head is **triangular, mantis-like, and facing vertically,** and it is equipped with **powerful mandibles** that are used to chew small insects. The antennae are long. All of their legs are fairly long and slender, with the

fore and middle tibiae bearing small spines that assist heel walkers in catching and restraining small insects. The abdomen bears **one-segmented cerci,** and the **male genitalia are asymmetrical.**

Much is known about the habits of Mantophasmatodea. They generally inhabit shrubland and grassland areas in the arid and semi-arid regions in southern and eastern Africa. All life stages are found on grasses, bushes and short trees. They have not been found in forested areas, and seem to occupy habitats at varying altitudes, with particular diversification in the Fynbos biome. All species are well-camouflaged, being green, brown, yellow, and grey, which allows them to blend with their surroundings. Their colour can gradually change if they are placed on a differently coloured environment.

All mantophasmatodeans are generalist predators of small and medium-sized insects, which they typically hunt during the evening and at night-time. They use their armed fore- and mid-legs to grasp their prey, which they quickly devour using sharp mandibles. Their enlarged tarsal pads enable them to hold on to surfaces while their other legs are busy restraining their struggling prey.

When it comes to courtship, heel walkers likely use pheromones to attract mates from long distances. Once in closer range, both sexes communicate and locate each other with low-frequency vibrational signals, which they detect using highly sensitive scolopidial organs on their legs. The vibrations are produced by a distinct drumming behaviour, where the abdomen is rhythmically tapped against the substrate. In males, the tip of their abdomen has a lip-like spine which they use to tap the substrate—much like a drumstick. Like praying mantises, the male may run a risk of being eaten by the female and tries to mount her cautiously and rapidly.

Oviposition takes place on small holes in sandy soil, where the eggs are deposited inside a foam that hardens and incorporates sand particles to form a protective pod. Females lay multiple egg pods, each containing between 10–30 eggs, depending on the species. The egg pods hatch after approximately eight months, with the first rains that mark the end of the dry season. Due to the highly seasonal environments they inhabit, heel walkers are generally short-lived. Once hatched, the nymphs mature in five months, and the adults survive for only a couple more months.

Key reading

Eberhard, M. J. B., Lang, D., Metscher, B., Pass, G., Picker, M. B., and Wolf, H. (2010). Structure and sensory physiology of the leg scolopidial organs in Mantophasmatodea and their role in vibrational communication. *Arthropod Structure & Development* **39**(4): 230–41.

Eberhard, M. J. B. and Picker, M. D. (2008). Vibrational communication in two sympatric species of Mantophasmatodea (Heel walkers). *Journal of Insect Behaviour* **21**: 240.

Huang, D., Nel, A., Zompro, O., and Waller, A. (2008). Mantophasmatodea now in the Jurassic. *Naturwissenschaften* **95**: 947–52.

Klass, K.-D., Zompro, O., Kristensen, N. P., and Adis, J. (2002). Mantophasmatodea: a new insect order with extant members in the Afrotropics. *Science* **296**: 1456–9.

Roth, S., Molina, J., and Predel, R. (2014). Biodiversity, ecology, and behavior of the recently discovered insect order Mantophasmatodea. *Frontiers in Zoology* **11**: 70.

Terry, M. D. and Whiting, M. F. (2005). Mantophasmatodea and phylogeny of the lower neopterous insects. *Cladistics* **21**(3): 240–57.

Wipfler, B., Letsch, H., Frandsen, P. B., Kapli, P., Mayer, C., Bartel, D., Buckley, T. R., Donath, A., Edgerly-Rooks, J. S., Fujita, M., Liu, S., Machida, R., Mashimo, Y., Misof, B, Niehuis, O., Peters, R. S., Petersen, M., Podsiadlowski, L., Schütte, K., Shimizu, S., Uchifune, T., Wilbrandt, J., Yan, E., Zhou, X., and Simon, S. (2019). Evolutionary history of Polyneoptera and its implications for our understanding of early winged insects. *Proceedings of the National Academy of Sciences of the United States of America* **116**: 3024–9.

Wipfler, B., Theska, T., and Predel, R. (2018). Mantophasmatodea from the Richtersveld in South Africa with description of two new genera and species. *ZooKeys* **746**: 137–60.

Eukinolabia

This group includes two morphologically divergent orders: the stick insects (Phasmatodea) and the webspinners (Embioptera). Their close relationship was first proposed based on structures of their mouthparts and their eggs, which was then confirmed by a suite of morphological and phylogenomic studies.

Essential Entomology. Second Edition. George C. McGavin and Leonidas-Romanos Davranoglou, Oxford University Press.
© George C. McGavin and Leonidas-Romanos Davranoglou (2022). DOI: 10.1093/oso/9780192843111.001.0001

Phasmatodea
(stick insects and leaf insects)

Common name	Stick and leaf insects, walking-sticks	Metamorphosis	Incomplete (egg, nymph, adult)
Derivation	Gk. *phasma*-apparition, spectre	Distribution	Mainly tropical and subtropical
Size	Body length up to 360 mm. Mostly 10–100 mm	Number of families	Uncertain
		Known world species	Uncertain

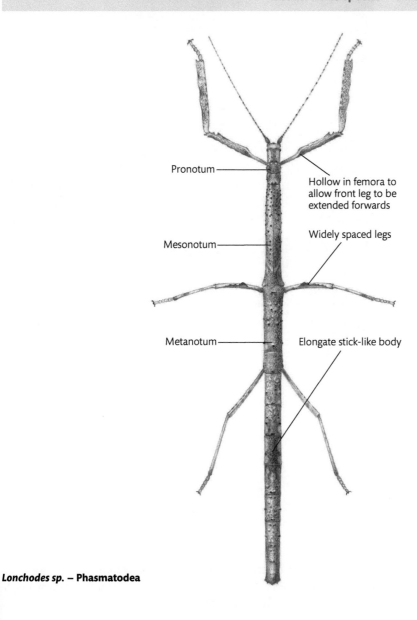

Pronotum

Hollow in femora to allow front leg to be extended forwards

Mesonotum

Widely spaced legs

Metanotum

Elongate stick-like body

Lonchodes sp. – **Phasmatodea**

Essential Entomology. Second Edition. George C. McGavin and Leonidas-Romanos Davranoglou, Oxford University Press.
© George C. McGavin and Leonidas-Romanos Davranoglou (2022). DOI: 10.1093/oso/9780192843111.001.0001

Key features

- slender, stick-like, or broad, leaf-like body with widely spaced legs
- slow moving and herbivorous
- confined to vegetation, which many mimic

❯ Stick and leaf insects are masters of camouflage, such as this *Phyllium* sp., which perfectly imitates the shape, colour, and veins of a leaf.

© Rupert Soskin

❯ Phasmatodea may exhibit marked sexual dimorphism in size, colouration, and structure, as can be seen from these mating *Spinopeplus* sp. from Sumaco, Ecuador.

© Anne Riley

Stick insects are mostly nocturnal, slow moving and herbivorous. Their **characteristic stick-like or leaf-like bodies, the cryptic colouring,** and the way they move make them very difficult to see among foliage and afford them a high degree of protection from predators.

The relationships within stick and leaf insects still remain poorly understood, hence the uncertainty in the number of families—some classification systems propose as few as three, while others up to 30. The evolutionary origins of Phasmatodea are also hotly debated, with some studies suggesting they are among the youngest insect orders dating back to the Late Jurassic–Early Cretaceous (about 150–120 mya), and others proposing a significantly older origin in the Permian–Triassic (more than 250 mya). What is more certain, is that visual predators have driven the remarkable evolution of camouflage in Phasmatodea, and there is strong evidence that most modern stick and leaf insects have diversified their morphology as a result of bird predation.

Some species freeze motionless when disturbed, holding the middle and hind legs tightly along the body and stretching the front legs out. Others sway gently, imitating the movements of the vegetation. The leaf insects are contained in one group, the Phylliinae, of which there are about 50 species confined to southeast Asia, New Guinea, and Australia. Leaf insects are broad and flattened, brown or green, and with fantastic leaf-like expansions on all leg and body segments.

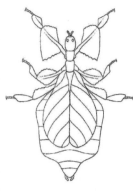

⌃ The characteristic leaf-like bodies of leaf insects make them very difficult to see among foliage and afford them a high degree of protection from predators.

The disguise is further improved by surface texturing, and blotches of colour or mottling, resembling leaf damage and veining on the short, front wings.

The body of stick insects can be very variable from short and smooth to large and very spiny. Many species have thorny or leafy outgrowths on various body parts to break up the outline of the insect and to help it blend in with the background. The head is often characteristically domed and carries quite long, **thread-like antennae, simple chewing mouthparts**, a pair of **small, laterally placed compound eyes**, and, in winged species, ocelli. **The front wings (tegmina) are short, toughened**, and, at rest, slightly overlapped to protect the much larger fan-shaped, membranous hind wings. Many species have very short wings or none at all. The ovipositor of the females is very short and often difficult to see. The abdomen carries a pair of small, terminal cerci.

Males are generally smaller than females, sometimes differently shaped, and have one less moult. Mating in phasmids is, generally, rather a dull affair. The male is presumably attracted by a sexual pheromone emitted by the female, and then simply mounts his mate, engages his genitalia by twisting his abdomen round underneath, and hangs on. Males may remain mounted for many days; in the genus *Timema* (Timematidae), males may spend their entire adult lives being carried about by the much larger female. A bit of excitement is seen in the North American species *Diapheromera veliei* (Phasmatidae), where competing males fight each other, kicking out with spined legs as they try to gain exclusive access to a mature female. Parthenogenesis is common in stick insects, and, in some species, males are very rare or even unknown.

⌃ Stick insect eggs look very seed-like and have a protuberance called a capitulum.

The eggs of stick insects look remarkably like seeds in size, surface texture, colour, and shape, and may be dropped, scattered, stuck to vegetation, or lightly covered with sand or soil. In species where the eggs are dropped to the ground, rather than being glued to vegetation or buried, the eggs have protuberance called a capitulum. The resemblance between these eggs and seeds of higher plants, whose seeds have elaiosomes, is striking and for good reason. Both capitula-bearing eggs and elaiosome-bearing seeds attract ants that take them back to their nests. Here, seeds and eggs are safe from herbivores and parasitic wasps, respectively. Nymphs, which look very much like the adults (especially in wingless species), may go through as few as two or as many as eight instars before becoming mature.

Although the main defensive strategy in stick insects is crypsis and death feigning, some will jump and fall from the foliage if disturbed by predators, such as birds. Others flash bright colours on their hind wings to startle enemies, adopt threatening scorpion-like postures by raising their abdomen, or they produce hissing sounds. A few species possess thoracic glands, which can produce noxious substances to repel ants, and even vertebrate predators, and they may be brightly coloured to advertise this ability. The hind femora of *Eurycantha* males from New Guinea have a very large inner spine and a reflex closing action, which can trap and crush. The spine is so hard and sharp that it is used as a fishing hook. An interesting feature of many stick insects is that, if attacked, they can shed their legs and partially regenerate them during subsequent moults. Autotomy of limbs takes place at a fracture line at the base of the legs, between the femur and the trochanter, and a special diaphragm seals off the wound rapidly to prevent loss of body fluids. Many species, such as *Carausius morosus*, can change colour. Epidermal cells contain granules of pigment, which are able to come together or disperse. When it is hot and sunny, the pigment clumps making the animals pale. When it is cool and overcast, the granules disperse, making the surface appear dark and allowing the insect to absorb more heat. Becoming darker in colour at night is also a good antipredator strategy.

Most species are of no economic significance, but a few species of stick insect can be pests. For example, the Australian spur-legged phasmid *Didymuria violescens* is responsible for defoliation in highland *Eucalyptus* forests. Many species, such as Macleay's Spectre (*Extatosoma triaratum*), Thailand stick insects (*Bacillus* spp.), the Pink-winged stick insect (*Sipyloidea sipylus*), the Jungle Nymph (*Heteropteryx dilatata*), and the Giant Spiny stick insect (*Eurycantha calcarata*), are popular as pets, and all can be easily reared on a diet of fresh bramble leaves.

Key reading

Bässler, U. (1983). *Neural basis of elementary behavior in stick insects.* Studies of brain function series **10**. Springer-Verlag, Berlin.

Bradler, S. (2009). *Die Phylogenie der Stab- und Gespenstschrecken (Insecta: Phasmatodea).* Species, phylogeny, and evolution series **2**. University of Göttingen Press, Göttingen.

Brock, P. D. (1999). *The amazing world of stick and leaf insects.* The amateur entomologist series **26**. A.E.S. Publications.

Brock, P. D. (1991). *Stick-insects of Britain, Europe and the Mediterranean.* Fitzgerald Publishing, London.

Bedford, G. O. (1978). Biology and ecology of the Phasmatodea. *Annual Review of Entomology* **23**: 125–30.

Hughes, L. and Westoby, M. (1992). Capitula on stick insect eggs and elaiosomes on seeds: convergent adaptations for burial by ants. *Functional Ecology* **6**: 642–8.

Robertson, J. A., Bradler, S., and Whiting, M. F. (2018). Evolution of oviposition techniques in stick and leaf insects (Phasmatodea). *Frontiers in Ecology and Evolution* **6**: 216.

Schmitz, J. (1993). Load compensating reactions in the proximal leg joints of stick insects during standing and walking. *Journal of Experimental Biology* **183**: 15–33.

Simon, S., Letsch, H., Bank, S., Buckley, T. R., Donath, A., Liu, S., Machida, R., Meusemann, K., Misof, B., Podsiadlowski, L., Zhou, X., Wipfler, B. and Bradler, S. (2019). Old World and New World Phasmatodea: phylogenomics resolve the evolutionary history of stick and leaf insects. *Frontiers in Ecology and Evolution* **7**: 345.

Sivinski, J. (1978). Intrasexual aggression in the stick insect *Diapheromera veliei* and *D. covilleae* and sexual dimorphism in the Phasmatodea. *Psyche* **85**: 395–403.

Strauß, J., von Bredow, C.-R., von Bredow, Y. M., Stolz, K., Trenczek, T. E., and Lakes-Harlan, R. (2017). Multiple identified neurons and peripheral nerves innervating the prothoracic defense glands in stick insects reveal evolutionary conserved and novel elements of a chemical defense system. *Frontiers in Ecology and Evolution* **5**: 151.

Tihelka, E., Cai, C., Giacomelli, M., Pisani, D., and Donoghue P. C. J. (2020). Integrated phylogenomic and fossil evidence of stick and leaf insects (Phasmatodea) reveal a Permian–Triassic co-origination with insectivores. *Royal Society Open Science* **7**(11): 201689.

Embioptera

(webspinners–alternative name: **Embiidina**)

Common name	Webspinners	**Metamorphosis**	Incomplete (egg, nymph, adult)
Derivation	Gk. *embios*-having life, *pteron*-a wing	**Distribution**	Mainly tropical; some in warm temperate regions
Size	Body length 3–25 mm. Mostly under 12 mm	**Number of families**	13
		Known world species	400 (0.035%)

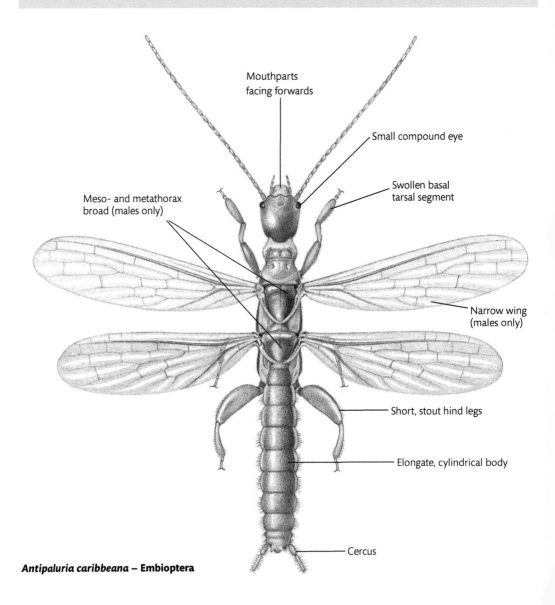

Mouthparts facing forwards

Small compound eye

Swollen basal tarsal segment

Meso- and metathorax broad (males only)

Narrow wing (males only)

Short, stout hind legs

Elongate, cylindrical body

Cercus

Antipaluria caribbeana – Embioptera

Essential Entomology. Second Edition. George C. McGavin and Leonidas-Romanos Davranoglou, Oxford University Press.
© George C. McGavin and Leonidas-Romanos Davranoglou (2022). DOI: 10.1093/oso/9780192843111.001.0001

Key features

- gregarious in silk galleries
- swollen first tarsal segment contains silk glands
- hind femur distinctly swollen

> Embioptera are unique among insects in that they produce silk from their enlarged front tarsi, as can be seen in this female *Embia* sp. from Mozambique.

© Piotr Naskrecki

Web spinners are pale or dark-coloured, gregarious or sub-social insects who get their common name from their ability to make elaborate (and sometimes extensive) silk tunnels and galleries in soil, litter, and on or beneath bark. The tunnels protect the web spinners and their young from enemies, desiccation, and rain. They are narrow-bodied, elongate, and cylindrical or slightly flattened, with **small, kidney-shaped eyes, no ocelli**, antennae with 10–35 segments, and **simple, forward-pointing, biting mouthparts**. Both sexes are equipped with **swollen, muscular hind legs,** which, when threatened, allow them to rapidly move backwards inside their tunnel without having to turn.

The remaining legs are short, and the front legs of both sexes and nymphs are characterized by having swollen basal tarsal segments, which contain numerous silk glands.

The liquid silk produced by each secretory gland passes through a minute tubule to one of about a hundred hollow, hair-like or bristle-like silk ejectors on the ventral surface. The mid-tarsal segments also have silk glands and ejectors but are not swollen. As the front feet are rubbed to and fro against the substrate, silk strands are pulled from the ends of the ejectors to build up a mat of silk. As the colony grows, galleries and tunnels are extended to take in new food sources such as dead plant material, litter, lichens, and mosses. Only the adult females and the nymphs, which they resemble, feed. Males do not feed as adults, and their forward-facing jaws are used to grasp the female during copulation. The only individuals to leave the confines of the colony are mature males who disperse to find mates elsewhere. Females are wingless, but the males usually have two equally sized pairs of long, narrow wings,

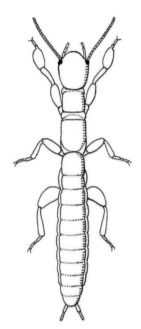

^ A typical adult female web spinner with a swollen tarsal segments containing silk glands.

which are very unusual in that they have certain hollow veins which can be inflated with haemolymph to make them stiff for flight. When the veins are not inflated, the wings can fold forwards to enable the male to run backwards through the galleries (if necessary) without damaging them.

When males locate females, they typically court each other with low frequency substrate vibrations. After mating, eggs are laid in the galleries and the female builds a protective nest over them composed of silk mixed with chewed food and debris. The female will remain nearby and guard the eggs against parasites and other females competing for oviposition sites. Some species have a rich repertoire of territorial vibrational signals that they produce against female competitors and unwanted males, which allows them to communicate over long distances without having to leave their oviposition site. Once the eggs hatch, the female must break open the nest to release her young. In some species, maternal care extends to providing the first instar nymphs with a paste of masticated food. Some species are parthenogenetic.

Despite the protection of their silk tunnels and cryptic habits, web spinners are attacked by ants and parasitic wasps. The very few known species in the bug genus *Embiophila* (Plokiophilidae) live in association with web spinners and suck out the eggs, and dead or dying larvae and adults. The web spinners do not seem to mind their presence; therefore, it is likely that they get something in return. Ten species of wasp in the family Sclerogibbidae are specialist ectoparasitoids of web spinner nymphs. The larvae feed on their hosts through the intersegmental membranes and then pupate in the web spinners' tunnels.

The systematics of Embioptera are still poorly understood, and most of the current families have been shown to be rather artificial. Furthermore, their cryptic habits make them difficult to collect, and they are generally scarce in most entomological collections. It is likely that at least 1,500 species remain undescribed, which outnumber described species by more than 3 times. This highly specialized insect order is in dire need of further study. Fossils of web spinners are also scarce, with the oldest representatives of the order having been found in Jurassic deposits.

Key reading

Büsse, S., Hörnschemeyer, T., Hohu, K., McMillan, D., and Edgerly, J. S. (2015). The spinning apparatus of webspinners—functional-morphology, morphometrics and spinning behaviour. *Scientific Reports* **5**: 9986.

Choe, J. C. (1994). Communal nesting and subsociality in a webspinner, *Anisembia texana* (Insecta: Embiidina: Anisembiidae). *Animal Behaviour* **47**: 971–3.

Dejan, K. A., Fresquez, J. M., Meyer, A. M., and Edgerly, J. S. (2013). Maternal territoriality achieved through shaking and lunging: an investigation of patterns in associated behaviors and substrate vibrations in a colonial embiopteran, Antipaluria urichi. *Journal of Insect Science* **13**: 82.

Edgerly, J. S. (1988). Maternal behaviour of a web-spinner (Order: Embiidina): mother-nymph associations. *Ecological Entomology* **13**: 263–72.

Edgerly, J. S. (1994). Is group living an antipredator defense in a facultatively communal webspinner (Embiidina: Clothodidae)? *Journal of Insect Behaviour* **7**: 135–47.

Edgerly, J. S., Büsse, S., and Hörnschemeyer, T. (2012). Spinning behaviour and morphology of the spinning glands in male and female *Aposthonia ceylonica* Enderlein, 1912 (Embioptera: Oligotomidae). *Zoologischer Anzeiger—A Journal of Comparative Zoology* **251**: 297–306.

Hodson, A. M., Cook, S. E., Edgerly, J. S., and Miller, K. B. (2014). Parthenogenetic and sexual species within the *Haploembia solieri* species complex (Embioptera: Oligotomidae) found in California. *Insect Systematics and Evolution* **45**: 93–113.

Huang, D.-Y. and Nel, A. (2009). Oldest webspinners from the Middle Jurassic of Inner Mongolia, China (Insecta: Embiodea). *Zoological Journal of the Linnean Society* **156**: 889–95.

Miller, K. B., Hayashi, C., Whiting, M. F., Svenson, G. J., and Edgerly, J. S. (2012). The phylogeny and classification of Embioptera (Insecta). *Systematic Entomology* **37**: 550–70.

Ross, E. S. (1970). Biosystematics of the Embioptera. *Annual Review of Entomology* **15**: 157–72.

Ross, E. S. (1987). Studies on the insect order Embiidina: a revision of the family Clothodidae. *Proceedings of the California Academy of Sciences* **45**: 9–34.

Ross, E. S. (2001). EMBIA: contributions to the biosystematics of the insect order Embiidina. Part 3: The Embiidae of the Americas (Order Embiidina). *Occasional Papers of the California Academy of Sciences* **150**: 1–86.

Valentine, B. D. (1986). Grooming behavior in Embioptera and Zoraptera. *Ohio Journal of Science* **86**: 150–2.

Paraneoptera (= Acercaria)

This assemblage of orders comprises barklice and true lice (Psocodea), true bugs (Hemiptera), and thrips (Thysanoptera). The latter two orders are further classified under the category of Condylognatha to highlight the specialization of their mouthparts and their sister-group relationship. The origins of Paraneoptera go back to before the Carboniferous, placing them among the oldest insect groups. Hemiptera are the most diverse hemimetabolan lineage and, together with the remaining paraneopteran orders, constitute 10% of all known insect species.

Essential Entomology. Second Edition. George C. McGavin and Leonidas-Romanos Davranoglou, Oxford University Press.
© George C. McGavin and Leonidas-Romanos Davranoglou (2022). DOI: 10.1093/oso/9780192843111.001.0001

Psocodea
(barklice, booklice, and true lice)

Common name	Barklice, booklice, and true lice	Metamorphosis	Incomplete (egg, nymph, adult)
Derivation	Gk. *psokos*–gnawed	Distribution	Worldwide
Size	Body length 1–10 mm. Mostly under 6 mm	Number of families	61
		Known world species	> 11,270 (0.99%)

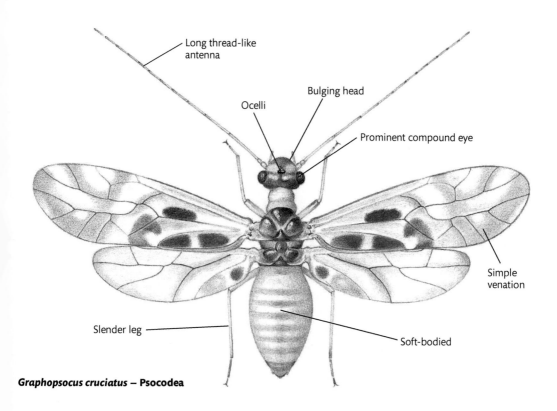

Long thread-like antenna

Ocelli

Bulging head

Prominent compound eye

Simple venation

Slender leg

Soft-bodied

***Graphopsocus cruciatus* – Psocodea**

Essential Entomology. Second Edition. George C. McGavin and Leonidas-Romanos Davranoglou, Oxford University Press.
© George C. McGavin and Leonidas-Romanos Davranoglou (2022). DOI: 10.1093/oso/9780192843111.001.0001

Key features

Bark lice:

- small and generally cryptic
- clypeus distinctly bulging outwards, wings macropterous, brachypterous, or absent
- very common on trees, vegetation, and in litter
- some species are pests of stored products

True lice:

- small, wingless, flattened ectoparasites living permanently on vertebrate hosts
- eyes with two ommatidia, ocelli absent, legs modified for clinging to fur or feathers
- feed on skin debris, secretions, feathers, or blood
- several species are significant vectors of human and animal diseases

> An aggregation of strikingly coloured barklice (Psocidae: *Clematoscenea* sp.) from Singapore.

© Zestin Soh

Barklice and booklice are very common, small, squat, and soft-bodied insects, which, on account of their size, cryptic colouration, and habits, are often overlooked and certainly understudied. True lice, on the other hand, are ectoparasitic, wingless insects found on mammal or bird hosts, and spend all their lives in the microhabitat provided by the skin, fur, hair, or feathers. True lice are better studied as they infest humans and are of significant veterinary and financial importance. Psocodea can be found in a very wide range of terrestrial habitats including caves and the nests of birds, bees, and wasps, but are particularly abundant in litter, soil, and on the bark and foliage of trees and shrubs. Despite the fact that they can make up a significant proportion of the total biomass in tree canopies and are preyed on by ants, spiders, and birds, their ecological role and impact has not been well studied and is little understood. Although true lice bear little resemblance to bark- and booklice

> A male head louse, *Pediculus humanus capitis*, on its natural habitat – a human scalp. Lice use their claw-like tarsi to hold onto hairs.

© Gilles San Martin

> A pair of mating *Ibidoecus* sp. lice on a feather of the Australian white ibis (*Threskiornis molucca*).

© John Gausas, The Academy of Natural Sciences of Drexel University

and were previously classified in a distinct order known as the Phthiraptera, overwhelming morphological and molecular evidence suggest that their systematic position lies deeply within Psocodea. Due to their radically different morphology and life history, we describe true lice separately from the rest of Psocodea.

The Psocodea are subdivided into three main subgroups: the Trogiomorpha, the Troctomorpha (which also includes true lice), and the Psocomorpha (containing the vast majority of the known species). The relationships within and between different psocodean families are poorly understood, and current classification systems are likely to change in the future.

The head, which is large and appears very bulging at the front, carries a pair of **prominent compound eyes, simple chewing mouthparts, long thread-like**

Typical front wing venation in the Psocidae.

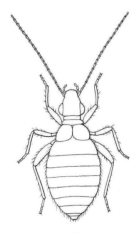

A typical wingless psocodean.

antennae, and, in winged psocodea, **three ocelli.** The antennae may have anything between 10 and 50 segments and are generally half as long as the body or longer. Most species can be classified as epiphytic grazers. The chewing mandibles are asymmetrical, and parts of the maxillae are rod-shaped and used to brace the head against the substrate as the mandibles scrape away at the microflora, which may consist of algae, lichens, and moulds and fungal spores. Some species eat pollen and other plant tissues. The genus *Psilopsocus* (Psilopsocidae) contains some very unusual species, which have wood-boring habits.

When wings are present, they are held roof-like over the body at rest and arise from a thorax that appears slightly humped in side view. The front wings are slightly larger than the hind wings, and both pairs have simple venation.

A few species have scales on their wings, and bodies that can make them look superficially like very small moths. There are many species in which short-winged and wingless forms occur in either sex.

Sexual dimorphism in wing development is typical of certain genera and families, and species living in stable habitats that do not need to disperse as much as those in other habitats tend to have shortened wings. The three pairs of legs are of a normal, walking design, and although some species have expanded hind femora and can jump weakly, most species are simply quick runners. Many species are solitary scavengers, living on vegetation, bark, and leaf litter, and under stones or anywhere that has adequate moisture. Some species exhibit varying degrees of gregariousness; groups of adults and nymphs living and feeding. Members of the tropical family Archipsocidae are very gregarious, and remarkable in that the dense silken sheets produced by their labial silk glands can be so extensive that they cover whole bushes and trees.

Courtship in some species may involve the production of sound from specialized organs at the base of the hind legs. These structures, called Pearman's organs, take the form of a ridged area and tympanum on the inner face of the coxa. In general, the males, following courtship, which may involve wing fluttering and other vibrational signals, transfer sperm to the females in a sperm packet known as a spermatophore. The courtship behaviour of many Psocodea is quite unusual. In the trogiid species *Lepinotus patruelis*, males and females compete for mates and will try to struggle with already copulating couples. Females do not appear to be very fussy about their mates. Males, on the other hand, exhibit a great deal of mate choice, and, interestingly, males on a high-quality diet are choosier in picking mates than males on a low-quality diet. This may have something to do with the investment that the males put into their special sclerotized spermatophore, coupled with the fact that sperm may serve as a source of nutrients for the females. An extreme of this condition has been found in two members of the family Prionoglarididae (*Afrotrogla* and *Neotrogla* spp.), where the sexual roles have been completely reversed. The males possess vagina-like genitalia, while the females have a penis-like organ known as the gynosome. The females compete for access to males, and mate with several individuals. Mating lasts for a remarkably long period (40–70 hours), during which the males transfer to the females voluminous and nutritious spermatophores. It is thought that the nutrient-poor cave environments in which these species live have forced females to mate with as many males as possible (up to 11 spermatophores have been found in some females), as the spermatophores provide them with necessary food resources that would be hard to obtain only by scavenging.

Eggs may be laid singly or in clusters, on vegetation or under bark, where they may be exposed or covered with silk, detritus, or adult faecal material. Females of many species are attracted to fresh conspecific eggs and may add their own eggs to the batch or near to it. Guarding of egg batches by males and females has been observed in some species, but in most species the eggs are not attended. The nymphs of many species stay together in groups and normally pass through six instars before becoming adults, although in some species the number is reduced to five or four.

The females of a single genus (*Archipsocopsis* spp.) give birth to live young, and the reproductive tract may contain more than ten embryos, which obtain nutrition from the walls of the ovarian tubules down which they pass. **Thelytoky** is common, and species in which this occurs tend to have wider distributions than bisexual species.

> **thelytoky**
> parthenogenesis where only
> female offspring are produced

Many psocodean species, such as those belonging to the genera *Lepinotis*, *Liposcelis*, and *Trogium*, are associated with humans and have become very widespread. Pest species are associated with houses and stored products, such as dried fruit, grains, and flour, and some can be pests in museum collections. Outbreaks are often due to poor house-keeping, and a combination of warm and damp conditions. However, non-parasitic Psocodea overall are of negligible economic importance.

Psocodea are a remarkably old group, with phylogenomic estimates suggesting their first radiations started in the Carboniferous (330 mya), if not earlier. Modern families start appearing in the Cretaceous.

One of the most peculiar branches of the Psocodea are undoubtedly the true lice, which are among the most prevalent ectoparasites in birds and many mammals (but not bats). It seems odd that there are no lice on bats, one of the most diverse groups of mammals on earth today, and there is no clear explanation available. A reasonable hypothesis suggests that although animals can, of course, 'pick up' lice from other species, the aerial lifestyle of bats does not bring them into intimate contact with other animals from which they could acquire lice secondarily. Furthermore, the habitats of true lice ancestors may not have favoured an association with bats: true lice evolved from barklice-like ancestors who had probably become adapted to eating the detritus in nests, lairs, and burrows. Indeed, true lice are most closely related to the barklice family Liposcelididae (suborder Troctomorpha), whose members are still frequently associated with vertebrate nests, and some have even been found on the fur or plumage of the animals that constructed them. A close association of true lice ancestors with animal nests, coupled with the loss of wings, could easily have led to a more intimate and permanent relationship with the animal itself. Early mammals and birds arose from the end of the Cretaceous and, by the end of first part of the Tertiary Period (the

Palaeocene and Eocene 65–53 mya), the forerunners of modern mammals and most of the main groups of birds had evolved. The earliest recognizable parasitic louse comes from an Eocene fossil (44 mya) with chewed up feathers still in its foregut. Molecular divergence estimates push the origins of true lice back to the Cretaceous (up to 140 mya) when mammals and birds started radiating, showing a clear co-evolution between host and parasite. Older estimates going back to the Permian (> 200 mya) have been suggested but need to be further re-evaluated.

Suborders of the Phthiraptera

Suborder	Feeding method	Food	Hosts
Amblycera	Biting	Skin, secretions, feathers, (blood*)	Birds and mammals
Ischnocera	Biting	Skin, secretions, feathers, (blood*)	Birds and mammals
Anoplura	Sucking	Blood	Mammals, including humans
Rhynchophthirina	Biting	Skin, debris, etc.	Warthogs and elephants

* Some species have mandibles modified to pierce the skin of their hosts.

The order has been split into four suborders within the Phthiraptera (see Box), although their validity remains to be confirmed by molecular data. We mention these suborders and the term Phthiraptera primarily for historical and practical reasons, as most of the published literature prior to the 2000s employs this classification system.

Amblyceran and ischnoceran lice use birds and mammals as hosts, while the only hosts of Rhyncophthirina and anoplurans are placental mammals. All lice are **wingless** and have tough, flexible, **dorso-ventrally flattened bodies** (unlike fleas which are laterally flattened). Additionally, the **eyes are very small or totally absent**, there are **no ocelli**, and the **antennae, which are short and stout, have between three and five segments.** In some species, the antennae are used for holding onto the host. **The three pairs of legs are short and robust** with the tarsi and claws usually highly modified for grasping hair or feathers. The surface of the body is well endowed with sensory hairs of various kinds.

Lice eggs are typically stuck to the hair or feather with a waterproof, fast-acting glue. The nymphs pass through three nymphal stages, taking anything from two weeks to a few months to reach adulthood. Many lice have symbiotic relationships with bacteria, which live in special mycetocytes associated with the digestive system. These bacteria allow the lice to digest feather protein (keratin) and blood. Unlike fleas (Siphonaptera), which are holometabolous and have larvae that are not attached to the host, the entire life cycle of lice is tied to the host animals, and they will die if separated from their host for too long.

The mouthparts of species in the suborders Amblycera and Ischnocera are very similar in design to those of non-parasitic Psocodea in that they have short biting mandibles, and the head is braced against the substrate by a stiff, rod-like component of the maxillae. The head is generally broader than the thorax, and the antennae are less than half the length of the head. These lice are mostly host specific on a variety of bird and mammal species. It is estimated that all bird species and a quarter of all mammal species are used as hosts by ischnoceran and amblyceran lice. Some species feed on the blood of the host, but the great majority eat hair, feather, and skin fragments, the products of sebaceous glands, and other dermatological detritus.

Species of the family Menoponidae are typical amblyceran lice. They are confined to birds and most feed on feathers or live in the feather shafts, but some suck blood. The head is large, broadly triangular, and expanded behind the eyes, and the mandibles bite horizontally. The short antennae are slightly clubbed and can be concealed in grooves on the underside of the head. The abdomen is broadly oval, and the legs are short and stout, each with two specially adapted claws for holding on to their host's feathers. Some species are quite serious pests of poultry. The Large Hen or Chicken Body Louse (*Menacanthus stramineus*) and the Shaft Louse (*Menopon gallinae*) are two well-known pest species. Infestations of these poultry lice lead to feather loss and reduced host health. Species in the genus *Piagetiella* actually live inside the throat pouches of cormorants and pelicans where they suck blood. Oddly, the females leave the pouch to lay their eggs on neck feathers. Studies of the lice on nestling swifts have shown that higher than normal parasite loads do not seem to affect their growth rate or success, confirming the prediction that a vertically transmitted parasite (transmitted between generations) would not be successful if it adversely affected the reproductive success of its hosts. Another study on cliff swallows showed that fumigation of birds to rid them of a suite of ectoparasites, including blood-sucking cimicid bugs, fleas, and chewing lice, increased adult survivorship by 12% over non-treated birds.

Mammal-chewing lice of the family Trichodectidae are typical ischnocerans. The head is not triangular, but rather squarish in appearance. The antennae are prominent and have three segments, and the mandibles bite vertically. The legs are short and stout, and each bears a single tarsal claw. Trichodectids are parasitic on Carnivora as well as rodents, hyraxes, lorises, and some monkeys. They live in the fur of their hosts and feed on skin fragments, hair, secretions, and blood. If populations of these small insects become large, the hair or fur of the host can become very damaged. Some species in this family are pests of domesticated animals. The Cattle Biting Louse (*Bovicola bovis*), the Horse Biting Louse (*Bovicola equi*), and the Dog Louse (*Trichodectes canis*) are typical examples and all can cause severe irritation to their hosts, as well as loss of hair due to prolonged scratching. The Dog Louse is known to transmit tapeworms from dog to dog, but no other species in this family is a known disease vector.

The mouthparts of species in the suborder Anoplura (sucking lice) are modified to form a piercing structure composed of three stylets, which can be retracted into a pouch in the head when not in use. In these lice, the head is generally narrower than the thorax, and the antennae are about the same

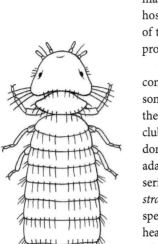

∧ A typical biting louse of the suborder Amblycera.

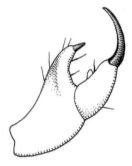

Modifications to the leg of the Head Louse to enable it to grip human hair.

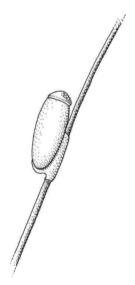

The eggs or 'nits' of the Head Louse are glued firmly to the shaft of human hair.

The Pubic Louse has a characteristic broad appearance.

length as the head. Anoplurans are all attached to mammals (including humans) where they suck blood, and some can be serious vectors of several human and animal diseases. Body lice belong to the family Pediculidae and have pear-shaped, flattened bodies. Like all lice, they are well adapted to their hosts and have short, strong, inwardly curved legs, each armed with a large claw for grasping and climbing through hair.

The family contains only one genus *Pediculus*, which comprises two species. *Pediculus humanus* has two subspecies that live on humans (*Pediculus humanus humanus* (the Body Louse) and *Pediculus humanus capitis* (the Head Louse)), which also occurs on monkeys from the New World. The other species, *Pediculus schaeffi*, is found on gibbons and apes. Human head lice and body lice are sometimes referred to by trinomens (as mentioned), or they are treated as sibling species. They are very similar in appearance but have very different habits. The Head Louse is confined to head hair and rarely acts as a vector of disease. The female attaches its eggs (nits) to hair shafts with a strong glue, and a complete cycle from egg to adult takes about 3–4 weeks.

The Head Louse is transmitted from host to host by the exchange of head wear, combs, brushes, direct contact, and fallen hair carrying unhatched eggs. Bites can be very irritating and can become infected or cause allergic reactions. Outbreaks of human head lice are an increasingly common occurrence in schools, especially among younger children, who play and work together in very close contact. The use of insecticidal shampoos has inevitably led to resistance developing in the lice, but regular washing and combing with a fine nit or 'dust' comb is very effective.

The Body Louse attaches its eggs to clothing, especially along the seams, and lives mainly in the fibres of the clothing. It leaves the clothing to feed on blood and then returns to its hiding place. Each meal may take a few minutes and they feed at frequent intervals. The whole life cycle from egg to adult may take anywhere between eight days or as long as five weeks. The Body Louse is transmitted by changing or sharing infested clothes or bedding, can survive longer off the host than the Head Louse, and is capable of producing many more offspring. This species has had an immense impact on the course of human history, and its effects in determining the final result of many conflicts has been documented since the fifteenth century. Until the invention of DDT during the Second World War, many more people died of louse-borne diseases, such as epidemic typhus and relapsing fever, than were killed by their opponents. The best-known example is the defeat of Napoleon's army, not by the Russians but by the loss of hundreds of thousands of men to typhus, spread by the infestations of Human Body Louse. Only a few thousand of more than half a million men that started the campaign, survived. Lice carry diseases caused by bacteria-like organisms called rickettsiae (singular rickettsia). Epidemic typhus is caused by *Rickettsia prowazeki*. These organisms develop inside gut cells of the louse and are passed out in the excrement. They can then be scratched into louse bites, or can be inhaled as a component of dust, or swallowed.

The Pubic Louse (*Phthirus pubis*) belongs to the anopluran family, the Pthiridae, and is superbly adapted for clinging to the thick pubic hair of humans.

The tarsal claws of the middle and hind legs are large and fold back to grip the hair shaft firmly. Pubic lice do not move around a great deal and are certainly incapable of the prodigious feats of athleticism alluded to in the work of toilet-wall poets. The lice can only move between humans who are engaged in sexual activity. The imaginative sexual behaviours of humans have led pubic lice to occasionally infest other parts of the body, such as beards and eyelashes. The family also contains another species, *Phthirus gorillae*, which, as the name implies, is confined to gorillas.

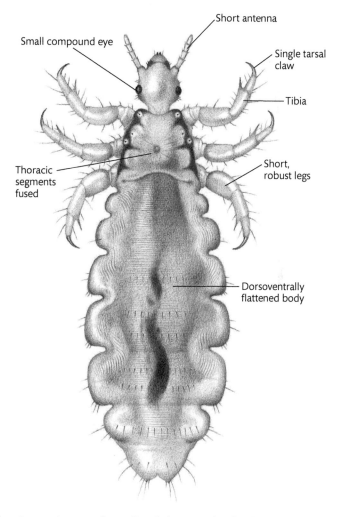

Short antenna

Small compound eye

Single tarsal claw

Tibia

Thoracic segments fused

Short, robust legs

Dorsoventrally flattened body

❯ **Pediculus humanus capitis—the Human Head louse.**

Other interesting anopluran lice belong to the family Echinophthiriidae, whose member species are blood-sucking ectoparasites on sea mammals. For instance, *Lepidophthirus macrorhini* is found on the hind flippers of elephant seals, and the hosts of *Arctocephalus ogmorhini* are Weddell seals. Louse reproduction takes place when the hosts are breeding on land, and egg laying is stimulated by temperature rise. Mammals like dolphins, porpoises, and whales that are permanently at sea do not have parasitic lice.

The fourth and smallest phthirapteran suborder is the Rhyncophtherina, which contains just two species in the family Haematomyzidae. One of these lice, whose mouthparts, carried at the end of an extended proboscis, is specific to elephants, the other to the African warthog.

The degree of host specificity in lice is high, and many species are restricted to particular areas of the body, especially where the host finds it difficult to groom. In this way, more than one louse species can inhabit the host simultaneously. Fahrenholtz's Rule states that ectoparasites, such as lice, have co-evolved with their hosts so closely that when the host speciates the lice do as well. This has been termed phyletic tracking. If Fahrenholtz's Rule always held true, then you would expect there to be exactly the same number of parasites as there are hosts, and that there would be a one-to-one match between parasites and hosts. In other words, the phylogenies of the two groups would be pretty much the same. However, when comparing actual phylogenies, many discrepancies appear. There are various reasons to explain this. A species of louse may have speciated on a single host, or a host species may have speciated independently of its louse. Lice may have switched to an additional, unrelated host species, either because it was geographically close or in some way readily accessible because the hosts interacted, perhaps socially. These phenomena have been called resource tracking. Further complications arise when a host becomes extinct, and its louse survives on a secondary host.

The lousiness of an animal might have an effect, not only on its own survival, but also on its inclusive fitness. Hamilton and Zuk (1982) put forward the idea that certain showy, secondary sexual characteristics might enable the selection of mates that were less infested with parasites. In birds, high loads of parasitic lice might lead to the birds having damaged or less bright plumage, and they might spend more time in preening activities. A study of pigeons did indeed show that females picked mates that had no lice in preference to ones that had artificially high numbers of lice. Although the clean males did not seem to groom themselves any less than the lousy males, they did spend significantly more time displaying to females.

Louse-infested animals can remove lice by biting them through preening or grooming. Birds may get some relief from their parasites by dust bathing, but 'anting' is certainly more effective. Here, birds spread their wings out and flap them over the nest of a formicine ant species. The workers swarm over the bird ejecting formic acid, which is an effective louse deterrent. However, as fast as hosts evolve parasite avoidance behaviours, it is sure that the lice co-evolve strategies to minimize their effects.

Key reading

Barker, S. C. (1994). Phylogeny and classification, origins and evolution of host associations of lice. *International Journal of Parasitology* **24**: 1285–91.

Broadhead, E. and Richards, A. M. (1982). The Psocoptera of East Africa—a taxonomic and ecological survey. *Biological Journal of the Linnean Society* **17**: 137–216.

Brown, C. R., Brown, M. B., and Rannala, B. (1995). Ectoparasites reduce long-term survival of their avian host. *Proceedings of the Royal Society of London B* **262**: 313–19.

Clayton, D. H. (1990). Mate choice in experimentally parasitized rock doves: lousy males lose. *American Zoologist* **30**: 251–62.

Greenwood, S. R. (1988). Habitat stability and wing length in two species of arboreal Psocoptera. *Oikos* **52**: 235–8.

Grimaldi, D. and Engel, M. S. (2006). Fossil Lipsocelididae and the lice ages (Insecta: Psocodea). *Proceedings of the Royal Society B: Biological Sciences* **273**: 625–33.

Hamilton, W. D. and Zuk, M. (1982). Heritable true fitness and bright birds: a role for parasites. *Science* **218**: 384–7.

Hellenthal, R. A. and Price, R. D. (1991). Biosystematics of the chewing lice of pocket gophers. *Annual Review of Entomology* **36**: 185–203.

Johnson, K. P., Dietrich, C. H., Friedrich, F., et al. (2018). Phylogenomics and the evolution of hemipteroid insects. *Proceedings of the National Academy of Sciences of the United States of America* **115**: 12775–80.

Lyal, C. H. C. (1986). Coevolutionary relationships of lice and their hosts: a test of Fahrenholtz's Rule. In: *Coevolution and systematics*. Stone, A. R. and Hawksworth, D. L. (eds), pp. 77–91. Systematics Association, Oxford.

Lyal, C. H. C. (1987). Co-evolution of trichodectid lice (Insecta: Phthiraptera) and their mammalian hosts. *Journal of Natural History* **21**: 1–28.

Loye, J. E. and Zuk, M. (eds) (1991). *Bird-parasite interactions*. Ornithology series **2**. Oxford University Press, Oxford.

Murray, M. D. (1976). Insect parasites of marine birds and mammals. In: *Marine insects*. Cheng, L. (ed), pp. 79–96. North Holland Publishing Co., Amsterdam.

Smith, V. S., Ford, D., Johnson, K. P., Johnson, P. C. D., Yoshizawa, K., and Light, J. E. (2011). Multiple lineages of lice pass through the K–Pg boundary. *Biology Letters* **7**: 782–5.

Thornton, I. W. B. (1985). The geographical and ecological distribution of arboreal Psocoptera. *Annual Review of Entomology* **30**: 175–96.

Yoshizawa, K., Ferreira, R. L., Yao, I., Lienhard, C., and Kamimura, Y. (2018). Independent origins of female penis and its coevolution with male vagina in cave insects (Psocodea: Prionoglarididae). *Biology Letters* **14**: 20180533.

Yoshizawa, K., Ferreira, R. L., Kamimura, Y., and Lienhard, C. (2014). Female penis, male vagina, and their correlated evolution in a cave insect. *Current Biology* **24**: 1006–10.

Yoshizawa, K. and Lienhard, C. (2010). In search of the sister group of the true lice: a systematic review of booklice and their relatives, with an updated checklist of Liposcelididae (Insecta: Psocodea). *Arthropod Systematics and Phylogeny* **68**: 181–95.

Yoshizawa, K. and Johnson, K. P. (2010). How stable is the 'Polyphyly of Lice' hypothesis (Insecta: Psocodea)?: a comparison of phylogenetic signal in multiple genes. *Molecular Phylogenetics and Evolution* **55**: 939–51.

Wappler, T., Smith, W. S., and Dalgleish, R. C. (2004). Scratching an ancient itch: an Eocene bird louse fossil. *Proceedings of the Royal Society B: Biological Sciences* **271**: 255–8.

Wearing-Wilde, J. (1996). Mate choice and competition in the barklouse *Lepinotus patruelis* (Psocoptera: Trogiidae): the effect of diet quality and sex ratio. *Journal of Insect Behaviour* **9**: 599–612.

Hemiptera
(true bugs)

Common name	Bugs, aphids, hoppers, etc.	Metamorphosis	Incomplete (egg, nymph, adult)
Derivation	Gk. *hemi*-half; *pteron*-a wing		
		Distribution	Worldwide
		Number of families	About 170
Size	Body length 1–120 mm. Mostly under 50 mm	Known world species	> 107,000 (9.45%)

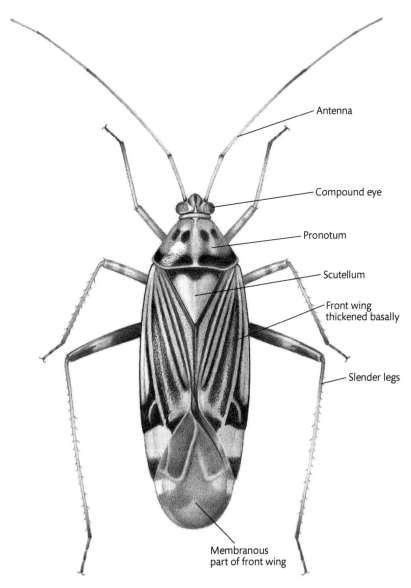

Rhabdomiris striatellus – **Hemiptera**

Essential Entomology. Second Edition. George C. McGavin and Leonidas-Romanos Davranoglou, Oxford University Press.
© George C. McGavin and Leonidas-Romanos Davranoglou (2022). DOI: 10.1093/oso/9780192843111.001.0001

Key features

- mouthparts forming a piercing/sucking beak or rostrum for liquid feeding
- stink glands and sound producing organs sometimes present
- many are significant crop pests, and some transmit human and animal diseases

Spittle bug nymphs hide in a mass of their liquid waste colloquially known as 'cuckoo spit'.

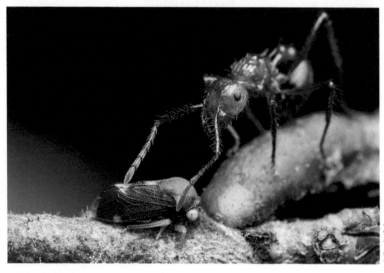

Ants frequently tend hemipteran insects for their honeydew. Here, an ant is tapping a treehopper (Membracidae), requesting a drop of sugary liquid. Sri Lanka.

© Zestin Soh

© Zestin Soh

© Zestin Soh

The nymphs of many assassin bugs (Reduviidae) camouflage themselves by attaching debris and discarded prey on their body. In this way, they are both invisible to their unsuspecting prey, as well as larger predators. Singapore.

Bugs, which comprise about 9.5% of all insect species, are the largest and most successful of the exopterygote orders. They range from minute, wingless scale insects, hardly resembling insects at all to giant water bugs with raptorial front legs capable of catching fish and frogs. Virtually every type of terrestrial and freshwater habitat has a particular and characteristic bug fauna, and ocean striders of the gerrid genus *Halobates* can be found on the sea, hundreds of miles from land.

The order is split into four suborders: the Auchenorrhyncha (cicadas, froghoppers, lantern bugs, leafhoppers, planthoppers, and treehoppers), the Coleorrhyncha (moss bugs), the Heteroptera (the true land and water bugs), and the Sternorrhyncha (jumping plant lice, whiteflies, conifer woolly aphids, phylloxerans, aphids, scale insects, and mealy bugs). Older classifications

© Anne Riley

Pond skaters are predators who locate their prey by detecting the ripples they produce on the water surface when drowning. Here, a group of *Gerris lacustris* are feeding on a dead *Bombus pascuorum* bumble bee. UK.

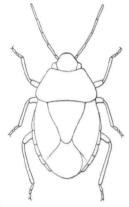

> Female aphid bearing live young. *Megoura viciae*, UK.

A typical heteropteran bug (wings held flat over body).

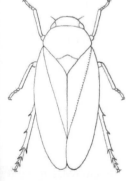

A typical auchenorrhynchan bug (wings held roof-like over body).

lumped the Auchenorrhyncha, Coleorrhyncha, and Sternorrhyncha into a single suborder known as the Homoptera, which has now been shown by molecular and morphological studies to be paraphyletic and is therefore obsolete. **Two pairs of wings** are usually present and in heteropterans, the front part of the front wings is toughened, leaving a membranous region at the rear.

In Auchenorrhyncha, the front wings and hind wings may be membranous, or the front wings may be entirely toughened.

The earliest bugs were probably plant feeders and lived on land, similar to many species today. All coleorrhynchans, auchenorrhynchans, and sternorrhynchans are herbivorous, feeding on the sap or cell contents of vascular plants. While the Heteroptera contains many predacious species, mixed feeders, and a few specialist blood feeders, the majority of species are herbivorous.

Bugs are sometimes called the 'perfect suckers' because the single unique character linking them all is the possession of **piercing, sucking mouthparts in the form of a long, typically four-segmented beak (known as the labium or rostrum in entomological terminology).** In heteropteran bugs, the labium arises from the front part of the head and can be hinged forward to point down or forwards, in front of the head.

Suborders of Hemiptera

Suborder	Number of families	Feeding habits
Coleorrhyncha	1 (Peloridiidae)	Herbivorous on mosses and liverworts
Heteroptera	About 88	Herbivorous, predacious, parasitic
Auchenorrhyncha	33	Herbivorous
Sternorrhyncha	About 49	Herbivorous

(i)

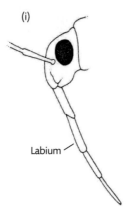

Labium

(iii)

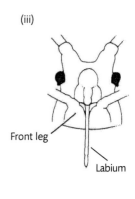

Front leg

Labium

(ii)

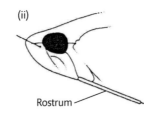

Rostrum

⟩ **Position of labium in the three main bug suborders.**

(i) Heteropteran bug: labium arises from front of head.
(ii) Auchenorrhynchan bug: labium arises from rear of head.
(iii) Sternorrhynchan bug: labium arises nearly between front legs.

This allows much greater flexibility and a larger choice of food. In the Auchenorrhyncha and Sternorrhyncha, the labium, which arises from the posterior part of the head, or seemingly from between the front legs, is permanently directed backwards.

With the exception of non-feeding male scale insects and the sexual forms of a few aphids, whose mouthparts are vestigial or lacking, the bug labium is similar throughout the order. The outer covering of the labium, (with 1–4 segments), is grooved for most of its length and surrounds the slender, toughened feeding stylets. The stylet bundle is made up of a pair of mandibular and a pair of maxillary stylets. The mandibular stylets enclose the maxillary stylets and can be closely connected by means of longitudinal ridges and grooves on their surfaces fitting together like the seal of a zip-lock plastic bag. The two pairs of stylets can slide freely on each other but are difficult to pull apart. The mandibular stylets have saw-like serrations, teeth, and, sometimes, barbs to penetrate plant and animal tissues. Predacious bugs penetrate the cuticle of their prey through a weak spot and use their long stylets and saliva to macerate the internal tissues before they are sucked out. The inner surfaces of the maxillary stylets are folded into longitudinal ridges and grooves, which firmly unite the two and provide two very fine, parallel canals running along their entire length. The ventral canal is the salivary canal, which carries digestive enzymes from the salivary glands in the anterior part of the thorax, the other is the food canal. Bug saliva is a complex mixture of a number of different enzymes, toxins, lubricants, and other substances. Herbivorous bugs need enzymes, such as pectinases, to break down plant cell walls, while the saliva of carnivorous heteropteran bugs may contain powerful venoms causing the instantaneous paralysis, death, and internal liquefication of prey.

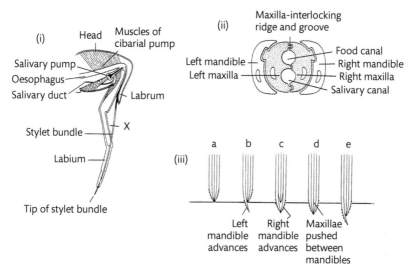

(i) **Diagrammatic section through a bug's head.**
(ii) **Cross-section through stylet bundle at X.**
(iii) **Stages in the insertion of a bug's stylet bundle into food source. The process may be repeated many times until the required location is reached.**

In plant sap-feeding bugs, such as aphids, the site of the phloem vessels in the plant may be some distance from the surface, and the stylet bundle has to wander between the tough-walled cells of the plant's epidermis before reaching a feeding site. The stylets are protected by the formation of a proteinaceous sheath formed by the hardening of special salivary gland secretions, which are produced throughout the course of penetration. Inside the head, powerful muscles operating the sucking or cibarial pump draw the liquid or pre-dissolved food up the stylet bundle and pass it into the pharynx. Bugs feeding on the sap in phloem vessels do not require very strong cibarial pumps as their diet is under slight positive pressure.

As plant material is not generally very nutritious, having at most 4% of its dry weight as nitrogen, there are two options open to herbivorous bugs. They can consume large volumes or grow slowly. The sap of phloem vessels, fed on by the great majority of herbivorous bugs, usually contains much less than 0.5% weight to volume of soluble nitrogen, but up to 9–10% weight to volume of sugars. Xylem sap, fed on by cicadas, froghoppers, and some leafhopper species, is even poorer in nutrients than phloem sap, and some bugs that specialize on this diet may take anything from several months to many years to complete their nymphal development.

A modification of bugs' guts seen in many sap suckers, such as the Cicadidae, Cercopidae, some Cicadellidae, and virtually all the coccoid families, is the filter chamber, formed by the intimate association of epithelial tissues from the rear of the midgut or the front of the hindgut with those of the anterior foregut. This allows large amounts of water and low molecular weight molecules to bypass the midgut entirely and pass straight into the rectum, while the rest of the food carries on through the midgut to be processed more efficiently.

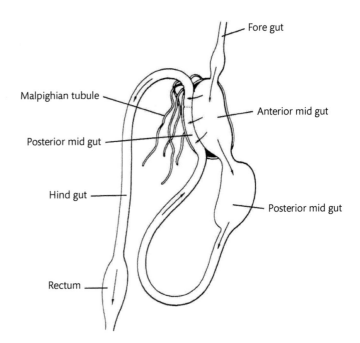

Schematic representation of the digestive tract and filter chamber of an auchenorrhynchan bug. Most of the water in its diet passes directly from the interior midgut into the posterior midgut and hindgut via the filter chamber.

Some heteropteran bugs rely on symbiotic microorganisms to provide them with essential nutrients, and these may be contained within special bodies in the haemocoel or inside the lumen of the midgut. In blood-feeding species, bacteria are responsible for providing the bugs with B vitamins. However, auchenorrhynchan, coleorrhynchan, and sternorrhynchan bugs that feed generally on a diet low in nutrients, are especially well known for their bacterial (Proteobacteria) symbionts, which are contained inside special cells called mycetocytes. The mycetocytes, which may be aggregated into bodies known as mycetomes, are free in the haemolymph of the body cavity or associated with the fat body. The symbionts are important in the recycling of nitrogen, providing their hosts with vitamins, certain essential amino acids such as tryptophan, which are absent from the diet, and certain lipids, such as the sterols, which insects are unable to manufacture themselves and are vital to the production of ecdysone (the insect moulting hormone). The symbionts of a few bug groups have been examined. Aphid symbionts belong to the *Buchnera aphidicola* complex and these differ from those seen in whiteflies or scale insects. Others, such as whiteflies and coleorrhynchans, share the same type of endosymbionts, indicating that their association is very ancient. The success and radiation of sap-feeding auchenorrhynchan and sternorrhynchan bugs was made possible by the acquisition of bacterial symbionts. A unique case of insect–plant coevolution can be found in the whitefly *Bemisia tabaci*, a major agricultural pest. This species acquired a detoxification gene from its host plant through horizontal gene transfer (non-sexual movement of DNA between two different species) millions of years ago. This 'stolen' plant gene allows *B. tabaci* to neutralize the chemical compounds produced by plants to ward off pests, and to feed on a variety of species that other insects cannot.

Predatory bugs can either be active hunters, like many assassin bugs that stalk and pounce on their prey, or ambushers, that wait for their prey. Timid predators, like anthocorids and water bugs such as hydrometrids, do not use their legs to hold prey, preferring to stab very small or slow-moving insects, larvae, eggs, and pupae with their rostrum. Predatory stink bugs (Pentatomidae) have been shown to locate their caterpillar prey by responding to specific chemical odours as well as the vibrations produced when the caterpillars are chewing on leaves. Some intriguing examples of prey specialization can be seen in several bugs associated with webs. The reduviid subfamily Emesinae contains a number of elongate, thread-legged species that feed on insects caught in spider webs or on the spiders themselves. *Stenolemus* species have specially modified antennae, which are used to send and receive vibrational signals through the web. When the bug moves on to a web, it causes the web to sway in a manner to simulate wind motion, therefore being acoustically invisible to its spider prey. The bug also actively manipulates the web by cutting silk threads in its path, therefore covering its tracks. Other interesting prey-capturing techniques are found in assassin bugs that may have special hairs on various parts of their bodies, which are designed to secrete sticky substances. In the nymphs of many genera, the dorsal body surface hairs are used to hold debris or the sucked-out bodies of their prey as a camouflage. Some adult assassin bugs produce sticky secretions on their front legs and use them to catch small insects. Others, particularly those belonging to the subfamily Apiomerinae, go one step further and collect certain plant resins on brush-like organs on their front tibia, and use this to trap stingless bees and beetles. Even more interesting is the use of tools by some assassin species in the subfamily Salyavatinae that prey on termites. Nymphs of the neotropical species *Salyavata variegata* disguise themselves heavily by sticking debris to glandular hairs on their backs and go fishing for their prey using the sucked-out bodies of previous victims as bait. The nymphs dangle the bait in front of holes in the termite nest, and the workers that come out to investigate and often grasp the lure are seized and eaten by the assassin bug.

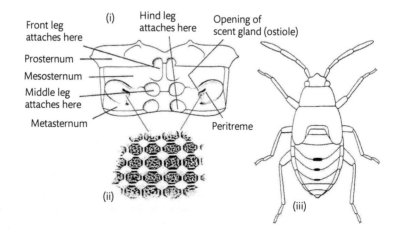

(i) View of the underside of an adult heteropteran bug's thorax, showing location of metathoracic scent gland openings.
(ii) Magnification of the surface sculpturing of the 'evaporative' or scent retentive area around the gland opening.
(iii) Heteropteran nymph with three dorsal abdominal gland openings (Lygaeidae).

A characteristic defence system of heteropterans is the production of mixtures of noxious or repellent organic compounds from scent glands in the metathorax of adults and the abdomen of nymphs. The secretions may also contain components which function as aphrodisiacs and sexual attractants, or promote aggregation, dispersal, and alarm. The secretions produced by adult bugs are generally more complex and varied than those produced by nymphs, and fall into several major chemical categories, such as the aliphatic aldehydes (alkanals), aliphatic alcohols (alkanols), and dicarbonyl compounds. The compounds can be active as vapours, soluble liquids, sprayed into the eyes or mouth of an attacker, or wiped by the bug on to the surface of the attacker, where they may penetrate beneath the cuticle or skin. Some scents panic ants by imitating their own alarm pheromones. Of course, evolution can turn the tables on the best laid plans, and certain parasitic flies (Tachinidae) and a number of hymenopteran parasitoids use the defensive scents and sexual odours of nymphal and adult bugs as cues to locate and thus parasitize them or their eggs.

Many seed-feeding lygaeid bugs, such as the red-and-black-coloured Milkweed bug (*Oncopeltus fasciatus*), are able to store toxic cardenolides obtained from their host plants in special compartments of their body, which they release rapidly when attacked.

Many bugs have evolved mutualistic relationships with ants. Among the heteropterans, ground-living nymphs and sometimes adults of a few families, mimic ants and the similarity enables them to prey on their host's larvae and pupae. Plant-feeding heteropterans, such as plataspids (Plataspidae) and leaf-footed bugs (Coreidae), are known to provide food for ants in return for protection from predators and parasites. However, it is among the auchenorrhynchan and sternorrhynchan families that we find some of the most complicated ant relationships. Many of these interactions, known as trophobioses, are based on the simple principle of reward (usually honeydew) for services rendered (usually protection). When sap-sucking bugs feed, they extract the dilute nitrogen-containing nutrients they require, and their watery wastes are high in sugars. This therefore represents a very valuable food resource to ants, and even to humans. The manna that sustained the Israelites on their mythical flight from Egypt is very likely to have been the honeydew of scale insects feeding gregariously on Tamarisk bushes. The ants, in return for honeydew, act as bodyguards, and when protecting their particular colony of aphids or scale insects, they will attack any predator or parasitic wasp that comes near. Some ants even treat their colonies of bugs like domesticated animals, building carton nests over them or taking them into their own nests to protect them from frosts. Some mealy bug species (Pseudococcidae) will climb voluntarily onto the backs of their bodyguards or are carried to safety in the ants' jaws.

For a long time, the only insects thought to show sociality in varying degrees were social wasps and bees (Hymenoptera) and termites (Isoptera). The discovery, some 20 years ago, of a special non-reproducing soldier caste in species belonging to the aphid subfamily Eriosomatinae was of great interest, for it showed that eusociality also occurred in the Hemiptera. For example, first instar nymphs of the Sugar Cane Woolly aphid (*Ceratovacuna lanigera*) serve as colony defenders by using sharply pointed frontal horns to attack predators like hover fly larvae. In addition, the soldiers secrete droplets from their abdominal cornicles, which are spread on the attacker's body. The secretion contains an alarm pheromone, which causes all other aphids in the colony to back away and recruits many more of the sterile soldier nymphs. Several gall-inhabiting aphid species have also been shown to have a first instar soldier caste. In these species, the soldiers may have distinctly thickened or lengthened legs, which are used to kick and pierce the bodies of predators, such as anthocorid bugs, ladybirds, and hover fly larvae, as well as herbivorous competitors such as moth caterpillars. The fighting techniques vary from species to species: some use their hind legs to tear cuticle, others have sharp horns, and many use their short, stout rostrum to stab. Although soldiers may spend much longer in the first instar and are able to feed, in the vast majority of species that have them, they do not ever live past the first instar and are thus effectively sterile. In the aphid *Pemphigus spirothecae*, spirally shaped galls are formed on the leaf petioles of the host plant, Black Poplar (*Populus nigra*). Unusually, the soldiers that survive their first instar develop into wingless adults and give rise parthenogenetically to more soldier first instar nymphs, as well as normal first instar nymphs that will ultimately become the winged generation with the job of producing the sexual forms.

Most of the soldier aphids that attack enemies die in the process. Other aphid species can protect themselves, and many species are able to produce secretions from a pair of abdominal cornicles. Droplets are exuded when the aphids are attacked, and these can spread onto the predator's body and mouthparts, where it hardens into a thick glue. Some aphids are polymorphic, the red and green forms of the pea aphid (*Acyrthosiphon pisum*) being an example. Recently, it has been shown that red morphs are less vulnerable to parasitic wasps, but more likely to be eaten by predators, and green morphs were safer from predators than from parasitoids.

Many sternorrhynchan bugs protect themselves by secreting waxes or resins. The well-known Woolly Apple Aphid (*Eriosoma lanigerum*) is easily recognizable by the tangled masses of white wax that it secretes. Scale insects in several families yield commercial quantities of waxes and resins in a very pure state. The wax is produced from special wax glands located in the epidermis. The products of these glands are carried to the outside through ducts to variously shaped pores. Scale insect wax covering may take the form of short, hollow or solid curls with a circular cross-section. Others may produce shapeless masses, and some produce very delicate solid threads less than 0.1 μm in diameter, which can be formed into hollow cylinders themselves. Macaroni-like or even more complicated extrusions with a diameter of up to 7 μm are also produced. In the Lac insect (*Kerria lacca*) nearly 80% of the secretions are resins, whereas in some species of ground pearl (Margarodidae), the secretions are essentially pure wax.

Most bugs reproduce bisexually, but a few species of scale insect, such as the Cottony Cushion Scale (*Icerya purchasi*), are hermaphrodites. Due to their profound impact on commercially important plants, and their phenomenal

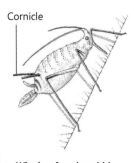

Cornicle

∧ Wingless female aphid giving birth to a first instar nymph.

powers of reproduction and dispersal, aphids have been the subject of considerable study over the years. Life cycles and reproduction are usually relatively simple, but around 10% of aphid species have complex life cycles involving sexual and asexual forms and alternations of generations between two unrelated host plants (one woody and one herbaceous) at different times of the year. In sexual forms, wings and mouthparts may be entirely lacking. In a simple, single host plant case, overwintering eggs hatch in the spring and give rise to winged or wingless, parthenogenetically reproducing females. These females, known as fundatrices, produce nymphs that develop into parthenogenetic adult females called virginoparae. Feeding continues and a variable number of generations exhibiting asexual reproduction are produced until the beginning of the autumn, when the nymphs will develop into wingless reproductive females and males. If the host plant has developed a very high aphid population, many will have already flown off to seek out other, less-crowded host plants of the correct species. Sexual reproduction and mating take place and the resultant eggs are laid on the host plant.

Mate finding in bugs can simply involve males locating females at oviposition or feeding sites, or it may require sound or vibrational signalling or pheromones. Often difficult to see due to their cryptic colouration and behaviour, cicadas (Cicadidae) are well known for their ability to produce very loud acoustic songs. The songs, which are produced by males, originate from a pair of organs called tymbals located on either side of the abdomen. The tymbal organs are resilin-containing cuticular membranes stiffened by a number of sclerotized ribs. As the tymbal muscle, attached to the inside of each tymbal, contracts, the tymbal is deformed inwards and each of the ribs buckle producing a loud click. The resonating tymbal organs drive an abdominal air sac resonator, and the resultant sound is radiated to the outside via the tympanic membranes. The tympanic membranes, or 'hearing organs', which are present in males and females, are on the underside of the abdomen, protected by a cuticular flap called the operculum. Another fascinating aspect of the biology of some cicadas that has intrigued entomologists for more than 300 years is the synchronous mass emergence of periodical cicadas in the genus *Magicicada*. Like all cicadas, the females lay eggs on the twigs of trees and hatching nymphs drop to the ground where they enter the soil. The nymphs spend the next 13 or 17 years feeding on the very poor diet of xylem sap. The three species of periodical cicada in North America are different from each other in terms of morphology and behaviour, and all have both 13- or 17-year life cycle forms, the longer life cycle forms occurring more in northerly regions. Four-year accelerations can occur, changing a 17-year life cycle form to a 13-year life cycle form. When the adults emerge *en masse*, often within a matter of a few days of each other, the densities reached can be as high as 3.5 million insects per hectare. The mechanisms by which the cycle is timed, and emergence is triggered, are unknown but might involve annual cycles of food quality, temperature, or a molecular clock. The reasons why these cicadas take so long to develop remain unknown. One hypothesis suggests that the mass emergence of billions of cicadas may overwhelm predators, which leads to less overall mortality for adult cicadas, and also prevents their enemies from becoming accustomed to hunting cicadas every year. Another hypothesis states that such long life cycles evolved to prevent the hybridization with other species of cicadas.

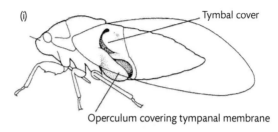

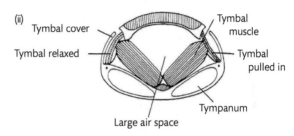

▶ (i) Location of the tymbal (sound-producing) and tympanal (hearing) organs on one side of a cicada. (ii) Schematic cross section through cicada at base of abdomen.

Most auchenorrhynchans, coleorrhynchans, and heteropterans produce substrate-borne vibrational songs that cannot be heard by the human ear, using various abdominal organs, some of which being smaller, 'silent' versions of the tymbals of cicadas. The vibrational energy of these small bugs is transmitted as mechanical vibrations through their legs to the foliage, while the acoustic signals of cicadas are primarily transmitted through the air. Water bugs, such as gerrids, catch prey by detecting the ripples they make and also attract mates by sending out pulsed surface ripples with their legs.

When copulation occurs, males introduce sperm by means of an aedeagus or penis via the female's vagina. Copulation usually begins with the male bug on top of the female and holding her back with his front legs; however, the final positions taken by mating bugs can be quite different. Once the genitalia are locked in place, sperm transfer can begin. To avoid other males displacing sperm with their own, males may remain in copula for a long time, or may guard their mate until just before or even after she lays the eggs. In many bugs, the female (or less commonly the male) parent may invest quite a great deal of energy in protecting their eggs and nymphs from attack. The occurrence of female brood guarding is documented in several families, but it is characteristic of treehoppers (Membracidae) and certain stink bugs (Acanthosomatidae). Females guard eggs or young nymphs with vigorous wing fanning, buzzing, rapid movements, physical contact, and butting. In some species of shield bug (Acanthosomatidae), female guarding, while effective in deterring predators, does not seem to reduce the levels of attacks from parasitoids. If a female bug protects her offspring, she can lay fewer eggs, but the survivorship of those eggs is much greater.

While it is usually the job of the female to protect the brood, males of some assassin bugs take over. In *Rhynocoris* (Reduviidae), when the female has finished laying, the male takes his position astride the egg mass. Over the course

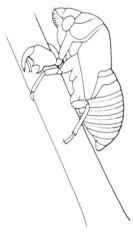

◢ A final instar cicada nymph crawls up a stem where it will moult into the adult form.

of the next few days, the female will mate several times with the guarding male and add to the original egg mass by laying additional egg batches. During the time that the male is guarding, he will chase off parasitic wasps and other enemies, but may not always be completely successful in defending his progeny for, when guarding a large egg mass, some of the eggs on the outside of the mass may be left exposed and become parasitized. When the eggs have all hatched, the male loses interest and moves away. The most famous cases of sexual role-reversal in bugs are to be found in certain giant water bugs (Belostomatidae). The females of many species, such as those belonging to the genera *Belostoma* and *Abedus* take a major part in courting and competing for available males, onto whose folded front wings they glue their eggs. The cost to the male of this brooding service may be quite high for, while carrying a heavy egg mass, he is less mobile, cannot fly, and is less able to catch prey. It is therefore important that the male makes sure that the eggs he is aerating and protecting from going mouldy are his. There would be no point in putting himself at risk for another male's genes. Due to the widespread phenomenon of sperm precedence, the sperm from the last mating is generally going to fertilize a much larger proportion of the eggs laid by the female. Males will only allow the females to glue eggs on their backs once they have mated with them and further enhance their assurance of paternity by letting the female glue only a few eggs at a time before copulating again. When all the eggs are laid the female leaves the male to look after their brood. There is recent evidence of this kind of egg brooding behaviour to protect against predators and parasites in one species of terrestrial bug, *Phyllomorpha laciniata* (Coreidae). Both males and females carry eggs on their backs, but males carry twice as many as females.

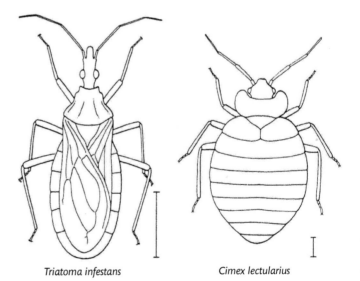

❯ Two notorious human blood feeders. Approximate body lengths are shown beside each bug.

Triatoma infestans   *Cimex lectularius*

Bed bugs (Cimicidae) have had a long history of association with people, and probably became intimately dependent on human blood when early humans settled down to live in caves with fires for warmth during cold spells. Bed bugs are a very old group, being dated to about 100 mya, meaning that they prob-ably were parasitic on both dinosaurs (including birds) and early mammals. The most common bed bug feeding on humans, *Cimex lectularius*, initially parasitized bats and switched to hosts of the genus *Homo* sometime between 99,000–867,000 years ago. Adult bed bugs are flattened, reddish brown, wing-less, nocturnally active insects with a body length of 4–6 mm. During the day they hide in bedding, mattresses, or cracks in floors and walls. There are five nymphal stages, and the entire life cycle can take anything from 2–10 months. Hosts are located at close range by body temperature, odours, and carbon dioxide. The adults can live for several months, and sometimes longer than a year without a blood meal. When they feed, adult bed bugs can consume more than five times their own weight in blood in a matter of 15 minutes, after which they hide away.

Bugs are not generally thought of as disease carriers, but there are excep-tions. Chagas' disease, prevalent in Central and South America, is caused by the microscopic protozoan organism, *Trypanosoma cruzi*, that is carried by certain kissing bugs (Reduviidae: Triatominae). The disease organism and the bugs that carry it have been present in Central and South America for a long time and, before the arrival of people, infection was confined to a variety of other mammalian species, birds, and a marsupial. At the beginning of the twenty-first century, Brazilian doctor Carlos Chagas first recognized and de-scribed the disease in humans and investigated the life cycle and other aspects of the protozoan and its insect vectors. Certain estimates indicate that up to 8 million humans may be affected by this disease, and that many more are at risk. Species of *Triatoma*, *Panstrongylus*, and *Rhodnius* have become inti-mately associated with humans, their houses, and their domestic animals, and the three most important in human disease transmission are *Triatoma infes-tans*, *Triatoma dimidiata*, and *Rhodnius prolixus*. The disease is much more common in rural areas where houses may harbour up to 1,000 bugs. During the day the bugs hide in thatch or in cracks and crevices in walls and emerge to feed at night, biting sleeping people, especially around the face and mouth, hence the common name of kissing bugs. An adult triatomine bug can ingest anything up to ten times its own weight in a single blood meal. The risk of infection from a single bite is low, but as people may receive tens of bites a night and thousands in a year, the risk of infection by *Trypanosoma cruzi* are considerable. When they feed, the bugs leave excrement behind on the skin and transmission of the protozoan occurs when people scratch their bites. As many as 28,000 people may contract Chagas' disease annually, most of these being children (congenital transmission), who may die in the early critical phase. Those that survive are infected for life, and the parasite, circulating in the bloodstream, infects, multiplies, and ultimately destroys muscle and nerve cells all around the body, particularly in the heart and digestive system. The medical problems associated with the parasite are many and varied, ranging from general weakness to serious problems of the alimentary tract, cardiac ar-rhythmia, heart failure, and death. It has been suggested that Charles Darwin contracted Chagas' disease while working in the New World.

While some heteropteran bugs, such as the Southern Green Stink Bug (Pentatomidae: *Nezara viridula*), can be widespread and polyphagous plants pests, it is auchenorrhynchan and sternorrhynchan bugs that are responsible for the biggest crop losses worldwide. Their feeding causes physical damage, metabolic disorders, and directly infects or makes the plants prone to a plethora of viral, rickettsial, mycoplasmal, and fungal diseases. Pest whitefly species (Aleyrodidae) attack citrus fruit trees and crops grown under glass. Scale insects and mealy bugs (superfamily Coccoidea) attack a wide range of economically important plant species. The viral diseases that are carried and transmitted to plants by sucking bugs probably do much more damage than the removal of nutrients. Aphid pests, for instance, are capable of transmitting a number of different viruses to their host plants simultaneously. The incredibly polyphagous species *Myzus persicae* have been shown to act as a vector for more than 110 different plant viruses. Aphids, commonly called blackfly or greenfly, are incredibly abundant and in one acre (0.4 hectare) there may be 2,000 million individuals on green parts of plants, and a further 260 million underground on roots. Aphids evolved 280–300 mya, but really became a significant group with the rise of the flowering plants in the Cretaceous (60–100 mya). Aphids are small and do not fly very fast (1–3 km per hour), and, as wind speeds are typically more than 3 km per hour, they have little control over where they go. They can, however, travel hundreds of miles. During the day, as the ground heats up, convection currents are set up which carry huge numbers of aphids up into the air. They have been recorded at 600 m (2,000 feet) or more. In the evening, as the ground cools, the air above the ground gets colder than the air higher above it. Convection ceases and the aphids come down wherever they happen to be. Billions may perish at sea or never find a host plant, but those pest species that find themselves settling on a suitable crop can build up to damaging numbers very quickly. The number of aphids flying each day is determined by factors such as how many are becoming adult, how overcrowded they are, the quality of the food plant, and climatic conditions like temperature and light intensity.

More than 21,000 world species of leafhopper (Cicadellidae) have been named and many are serious crop pests. Leafhopper saliva is toxic to the plants and phloem-feeding species are the next most notorious vectors of plant viral diseases to aphids. Many major crop plants, such as cotton, potatoes, beans, rice, and alfalfa, are at risk and resistant plant strains are being developed. Some hemipterans are consumed by humans, and in fact the stink glands of many heteropterans are prized food items. The ancient Greeks had a taste for egg-filled female cicadas, while giant water bugs (Belostomatidae) are regularly eaten in Southeast Asia, and their extract is often used as seasoning for other dishes. Some stink bugs (e.g., *Edessa mexicana*) are a delicacy in certain areas of Mexico and are said to taste like cinnamon.

Hemiptera are a remarkably old group, with origins in the Carboniferous (about 360 mya), perhaps even earlier. Most modern families started diversifying in the Jurassic–Cretaceous. Extinct Hemiptera, like their modern relatives, were among the most abundant insect groups in virtually every habitat, and their fossils are remarkably well represented.

Key reading

Akhoundi, M., Sereno, D., Durand, R., Mirzaei, A., Bruel, C., Delaunay, P., Marty, P., and Izri, A. (2020). Bed bugs (Hemiptera, Cimicidae): overview of classification, evolution and dispersion. *International Journal of Environmental Research and Public Health* **17**(12): 4576.

Andersen, N. M. (1982). *The semiaquatic bugs (Hemiptera, Gerromorpha)*. Entomonograph series **3**. Scandinavian Science Press, Denmark.

Bennet-Clark, H. C. (1997). Tymbal mechanics and the control of song frequency in the cicada Cyclochila australasiae. *Journal of Experimental Biology* **200**: 1681–94.

Blackman, R. L. and Eastop, V. F. (2000). *Aphids on the world's crops: an identification and information guide* (2nd edn). John Wiley, Chichester.

Bramer, C., Friedrich, F., and Dobler, S. (2016). Defence by plant toxins in milkweed bugs (Heteroptera: Lygaeinae) through the evolution of a sophisticated storage compartment. *Systematic Entomology* **42**(1): 15–30.

Buckley, R. C. (1987). Interactions involving plants, Homoptera and ants. *Annual Review of Ecology and Systematics* **18**: 111–35.

Cobben, R. H. (1968). *Evolutionary trends in Heteroptera. Part 1: Eggs, architecture of the shell, gross embryology and eclosion*. Centre for Agricultural Publishing and Documentation, Wageningen.

Cobben, R. H. (1978). *Evolutionary trends in Heteroptera. Part 2: Mouthpart structures and feeding strategies*. V. Veenman and Zonen bv, Wageningen.

Davranoglou, L.-R., Cicirello, A., Taylor, G. K., and Mortimer, B. (2019). Planthopper bugs use a fast, cyclic elastic recoil mechanism for effective vibrational communication at small body size. *PLoS Biology* **17**(3): e3000155.

Dixon, A. F. G. (1997). *Aphid ecology: an optimization approach*. Chapman and Hall, London.

Foster, W. A. (1990). Experimental evidence for effective and altruistic colony defence against natural predators by soldiers of the gall forming aphid *Pemphigus spirothecae* (Aphididae: Pemphigidae). *Behavioral Ecology and Sociobiology* **27**: 421–30.

Itô, Y. (1989). The evolutionary biology of sterile soldiers in aphids. *Trends in Ecology and Evolution* **4**: 69–73.

Kaitala, A. (1997). Oviposition on the backs of conspecifics: an unusual reproductive tactic in a coreid bug. *Oikos* **77**: 381–9.

McGavin, G. C. (1993). *Bugs of the world*. Blandford Press, London.

Pfannenstiel, R. S., Hunt, R. E., and Yeargan, K. V. (1995). Orientation of a hemipteran predator to vibrations produced by feeding caterpillars. *Journal of Insect Behavior* **8**: 1–9

Rotheray, G. E. (1989). *Aphid predators*. Naturalists handbook series **11**. Richmond Publishing, Slough.

Santos-Garcia, D., Latorre, A., Moya, A., Gibbs, G., Hartung, V., Dettner, K., Kuechler, S. M., and Silva, F. J. (2014). Small but powerful, the primary endosymbiont of moss bugs, *Candidatus Evansia muelleri*, holds a reduced genome with large biosynthetic capabilities. *Genome Biology and Evolution* **6**(7): 1875–93.

Schuh, T. and Weirauch, C. (2019). *True bugs of the world (Hemiptera: Heteroptera): classification and natural history* (2nd edn). Siri Scientific Press, Castleton.

Soley, F. G., Jackson, R. R., and Taylor, P. W. (2011). Biology of *Stenolemus giraffa* (Hemiptera: Reduviidae), a web-invading, araneophagic assassin bug from Australia. *New Zealand Journal of Zoology* **38**: 297–316.

Stonedahl, G. M. and Dolling, W. R. (1991). Heteroptera identification: a reference guide with special emphasis on economic groups. *Journal of Natural History* **25**: 1027–66.

Xia, J., Guo, Z., Yang, Z., Han, H., Wang, S., Xu, H., Yang, X., Yang, F., Wu, Q., Xie, W., Zhou, X., Dermauw, W., Turlings, T. C. J., and Zhang, Y. (2021). Whitefly hijacks a plant detoxification gene that neutralizes plant toxins. *Cell* **184**(7): 1693–1705; e17.

Walker, A. A., Weirauch, C., Fry, B. G., and King, G. F. (2016). Venoms of heteropteran insects: a treasure trove of diverse pharmacological toolkits. *Toxins* **8**(2): 43.

Weirauch, C. (2006). Anatomy of disguise: camouflaging structures in the nymphs of some Reduviidae (Hemiptera: Heteroptera). *American Museum Novitates* **3542**: 1–18.

Wilcox, R. S. (1979). Sex discrimination in *Gerris remigis*: role of a surface wave signal. *Science* **206**: 1325–7.

World Health Organization (2021). Chagas disease (also known as American trypanosomiasis). Retrieved from: https://www.who.int/en/news-room/fact-sheets/detail/chagas-
disease-(american-trypanosomiasis)

Young, D. and Bennet-Clark, H. C. (1995). The role of the tymbal in cicada sound
production. *Journal of Experimental Biology* **198**: 1001–19.

Thysanoptera
(thrips)

Common name	Thrips	Metamorphosis	Incomplete but with resting pupa-like stages
Derivation	Gk. *thysanos*–fringe, tassel; *pteron*–a wing	Distribution	Worldwide
Size	Body length 0.5–12 mm. Mostly under 3 mm	Number of families	9
		Known world species	6,200 (0.55%)

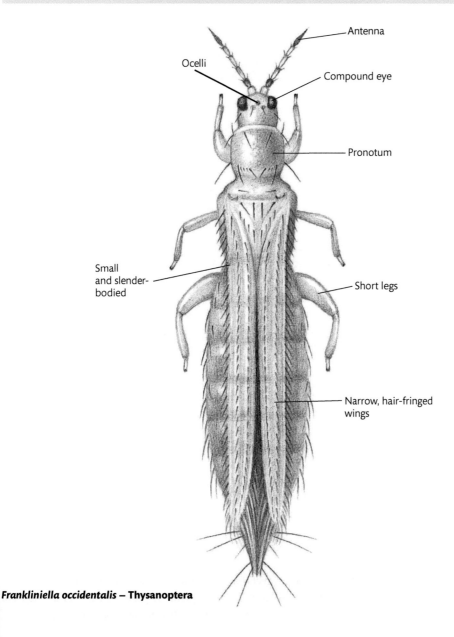

Antenna

Ocelli

Compound eye

Pronotum

Small and slender-bodied

Short legs

Narrow, hair-fringed wings

Frankliniella occidentalis – **Thysanoptera**

Essential Entomology. Second Edition. George C. McGavin and Leonidas-Romanos Davranoglou, Oxford University Press.
© George C. McGavin and Leonidas-Romanos Davranoglou (2022). DOI: 10.1093/oso/9780192843111.001.0001

Key features

- small to very small, slender
- narrow wings with dense fringes of hairs
- mouth with only a single mandible (left one)
- arolia modified into protrusible vesicles
- many species are serious crop pests

❯ Thrips are very small insects with oar-like hairy wings, and asymmetrical mouthparts. *Megalurothrips sjostedti* can be a serious pest of cowpeas. Saudi Arabia.

︿ A typical adult thysanopteran.

Thrips are generally small or very **small, slender-bodied,** readily recognizable insects. **The two pairs of wings are very narrow** and, in addition to bearing **long fringes of hairs on their margins (known as 'cilia'), they have little or no obvious venation.**

Thrip wings can be of normal size, reduced, vestigial, or totally lacking. The head carries a pair of short antennae, each with between four and nine segments, and conspicuous compound eyes, and three ocelli in winged individuals.

The piercing and sucking mouthparts of thrips are unusual in that they are asymmetrical. The right mandible is extremely reduced and non-functional, while the left mandible is sharp and stylet-like, and it is used to penetrate plant tissue or the bodies of minute insects. The remaining mouthparts form long hemiptera-like stylets and are used to suck up the liquid foodstuff or, in some cases, fungal spores. The three pairs of legs are characterized by the presence of specialized devices that are used for attachment to plant and other surfaces: the tarsi of all species have a large adhesive, bladder-like protrusible vesicle contained in a hollow between the claws. When the tarsus contacts a surface, the bladder is pumped up with haemolymph. The front pair of legs of most thrips are usually greatly enlarged, making thrips look like the Popeye of the insect world.

Although these insects have a hemimetabolous mode of development and are most closely related to the Hemiptera, they are unusual in that there are one or more pupa-like resting stages between the two, true nymphal stages

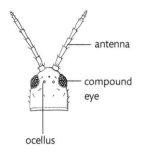

antenna

compound
eye

ocellus

⌃ **Detail of the head of a thrip.**

❯ **arrhenotoky**
parthenogenesis where
haploid males are produced
from unfertilized eggs

and the adult. In some cases, there are three pre-adult stages, of which the first may still be capable of feeding. The next two pre-adult stages become more pupa-like with a degree of tissue reorganization, and a cocoon may even be spun.

Like the Hymenoptera, thrips are haplodiploid. That is, female thrips, which result from fertilized eggs, are diploid, however, males are haploid. This means that unfertilized virgin females can give birth to males, in a process known as **arrhenotoky**. However, there are also some thelytokous species, where females are produced from unfertilized eggs.

Although Thysanoptera is a fairly small order, their life cycle can be quite complex. A few species show sub-social behaviour and parental care. In the Australian species, *Oncothrips tepperi* and *Oncothrips habrus* (Phlaeothripidae), sub-fertile soldier morphs defend their colony and siblings from attack by other insects, as well as take-overs by other thrips species.

Even in small insects like thrips, sex and violence can be inextricably linked. In some species of *Hoplothrips* (colonial fungus-feeding thrips), males fight, often to the death, using enlarged front legs to grapple and stab each other in order to control access to prime egg-laying sites and therefore access to females. Victorious males, which tend to be larger than vanquished males, get to mate with the females just before they lay eggs, and are able to assess the reproductive potential of their mates by 'measuring' the females' abdomen with their front legs. Fights between males usually result from challenging males sneaking in to mate with females. Fighting is clearly worth the risk because defending males may be defeated through accumulated injuries from previous fights, and the rewards of becoming a defender are high in terms of inclusive fitness. Development from egg to adult stage may be very rapid, taking less than a fortnight, and there may be more than one generation during a single year. Herbivorous and pollen-feeding species can be very destructive and are responsible for the transmission of many plant diseases, while predacious species may be of benefit in controlling pests. Some phytophagous species are of use in the control of weeds. Around 300 species in more than 50 genera induce galls on their host plants. In these species, galls are formed by the action of feeding by the thrips on growing tissue, but the causal mechanism is not known.

There are two suborders: the Terebrantia, which includes large families such as the Aeolothripidae and the Thripidae, and the Tubulifera, which contains a single family, the Phlaeopthripidae (tube-tailed thrips). In Terebrantia, the wings are held parallel at rest, and females lay eggs in plant tissue with serrated, blade-like ovipositors. In Tubulifera, the wings, when present, overlap each other when folded, and females lay their eggs in crevices or on plants.

Aeolothripidae are commonly called banded thrips due to the fact that the wings may have cross bands or mottled patches. Some unusual species are nearly wingless and may look very ant-like. They are generally yellowish-brown to dark brown in colour, and in the females, the ovipositor is curved upwards. Aeolothripids can be found on cruciferous and leguminous plants, grasses, and conifers. Although also known collectively as predacious thrips, most species are pollen feeders. The best-known and commonest genus is

Aeolothrips. The species *Aeolothrips fasciatus* has a very wide distribution, is yellow to dark brown with banded wings, and is commonly found in the heads of clovers. The nymphs are yellow or orange. The fully developed nymphs form a silken cocoon underground before emerging as adults. Predacious species are of use in controlling natural populations of other thrips, aphids, and other small insects and mites.

The ovipositor in common thrips (Thripidae) bends downwards, not upwards, and these insects can be found on the leaves and flowers of a vast range of plants. The vast majority are plant sap suckers, but some feed on fungal spores, or are predacious. Reproduction can occur parthenogenetically, and in some pest species (e.g., Common Greenhouse Thrip, *Heliothrips haemorrhoidalis*), males have been collected rarely. Females lay their smooth eggs inside plant tissues. This family contains several genera (*Limothrips*, *Thrips*, and *Taeniothrips*) with many pest species, some of which can do serious damage by their feeding and the transmission of viruses to a wide range of important crops. Crops affected include peas, beans, citrus trees, cereals, coffee, tea, tobacco, onions, cultivated flowers, and greenhouse produce. The Grain Thrip (*Limothrips cerealium*) is a widespread pest that breeds in the ears of cereal crops and emerges in vast numbers when the crop is ripe. *Frankliniella* contains several pest species such as *Frankliniella occidentalis* (the Western Flower Thrip). This species, which attacks vegetables and ornamental crops grown under glass, is thought to be the most significant pest thysanopteran in Europe. *Thrips tabaci* is a widespread and polyphagous pest that attacks crops, such as cotton, tomatoes, and, as the name implies, tobacco. Apart from being the vector for a number of plant viruses, this species feeds mainly in the young flowers and causes the young fruit to abort.

Tube-tailed thrips, which can be recognized by their tubular, pointed abdominal end, are rather larger and stouter-bodied than aeolothripids or thripids, and most are dark brown or black. This family contains some of the world's largest thrip species; a few tropical genera such as *Phasmothrips*, have species measuring up to 12 mm in length. Tube-tailed thrips can be found on a wide range of herbaceous plants, shrubs, and trees, in flowers, on twigs, under bark, in soil, and in leaf litter. The majority of species feed on fungal threads or spores, but there are predacious species that feed on mites and small, soft insects like whiteflies. *Liothrips vaneeckei* damages lily bulbs, while other species attack olives and grasses.

As expected, the parasitoids that use thrips as hosts are small species. Larval thrips are parasitized mainly by wasp species in the Eulophidae, and thrip eggs by wasp species in the Trichogrammatidae and Mymaridae. Some of these parasitoids may have a significant role in controlling greenhouse pest thrip species.

Thrips are a very old group, with putative fossils from the Permian (about 290 mya). The earliest definitive thrip fossils come from the Late Triassic, about 200 mya. Some remarkable amber fossils from the Cretaceous of Spain (100 mya) have trapped thrips covered in pollen grains of a cycad or gingko tree. The thrips deliberately covered themselves with pollen, which they carried off along the plant to be used as nourishment for their larvae. As the thrips moved along the flowers, they pollinated them, suggesting that these insects played a vital role in these ancient ecosystems—in fact, this fossil is the earliest record of pollination in existence. In addition, the fact that these thrips were deliberately collecting pollen for their young shows that they were likely sub-social, like many thrips species today.

Key reading

Ananthakrishnan, T. N. (1993). Bionomics of thrips. *Annual Review of Entomology* **38**: 71–92.

Crespi, B. J. (1988). Risks and benefits of lethal male fighting in the colonial, polygynous thrips *Hoplothrips karnyi* (Insecta: Thysanoptera). *Behavioral Ecology and Sociobiology* **22**: 293–301.

Crespi, B. J. (1992). Eusociality in Australian gall thrips. *Nature* **359**: 724–6.

Crespi, B. J., Carmean, D. A., and Chapman, T. W. (1997). Ecology and evolution of galling thrips and their allies. *Annual Review of Entomology* **42**: 51–71.

Grimaldi, D., Shmakov, A., and Fraser, N. (2004). Mesozoic thrips and early evolution of the order Thysanoptera (Insecta). *Journal of Paleontology* **78**: 941–52.

Van der Kooi, C. J. and Schwander, T. (2014). Evolution of asexuality via different mechanisms in grass thrips (Thysanoptera: *Aptinothrips*). *Evolution* **68**: 1883–93.

Lewis, T. (ed) (1997). *Thrips as crop pests*. CAB International, Wallingford.

Loomans, A. J. M. and Van Lenteren, J. C. (1995). Biological control of thrips pests: a review of thrips parasitoids. *Wageningen Agricultural University Papers* **95**: 89–201.

Mound, L. A. and Kibby, G. (1998). *Thysanoptera: an identification guide* (2nd edn). CAB International, Wallingford.

Mound, L. A., Heming, B. S., and Palmer, J. M. (1980). Phylogenetic relationships between the families of recent Thysanoptera (Insecta). *Zoological Journal of the Linnean Society* **69**: 111–41.

Palmer, J. M., Mound, L. A., and Du Heaume, G. J. (1989). *CIE Guides to insects of importance to man: 2. Thysanoptera*. CAB International, Wallingford.

Peñalver, E., Labandeira, C. C., Barrón, E., Delclòs, X., Nel, P., Nel, A., Tafforeau, P., and Soriano, C. (2012). Thrips pollination of Mesozoic gymnosperms. *Proceedings of the National Academy of Sciences of the United States of America* **109**: 8623–8.

Tommasini, M. G. and Maini, S. (1995) *Frankliniella occidentalis* and other thrips harmful to vegetable and ornamental crops in Europe. *Wageningen Agricultural University Papers* **95**: 1–42.

Division **Endopterygota** (= Holometabola)

This monophyletic group contains the most successful insect orders. The young stages are called larvae and look very different from the adults they will become. Their wings develop internally (endopterygote) and metamorphosis is complete (holometabolous). The total transformation from larva to adult takes place during a pupal stage.

Neuropterida

This superorder comprises the Raphidioptera (snakeflies), the Megaloptera (alderflies, dobsonflies, and fishflies), and the Neuroptera (lacewings, antlions, and allies). The Neuropterida are generally considered as one of the earliest lineages of Holometabola, and likely represent the sister-group to the Coleopterida, the superorder that includes Coleoptera (beetles) and the Strepsiptera (twisted-wing insects). With more than 10,000 described species, Neuropterida represents the fifth-largest superorder of Holometabola.

Essential Entomology. Second Edition. George C. McGavin and Leonidas-Romanos Davranoglou, Oxford University Press.
© George C. McGavin and Leonidas-Romanos Davranoglou (2022). DOI: 10.1093/oso/9780192843111.001.0001

Raphidioptera
(snakeflies)

Common name	Snakeflies	Metamorphosis	Complete (egg, larva, pupa, adult)
Derivation	Gk. *Rhaphis*–a needle; *rhaphe*–a seam; *pteron*–a wing	Distribution	Northern hemisphere
		Number of families	2
Size	6–28 mm (body length)	Known world species	248 (0.022%)

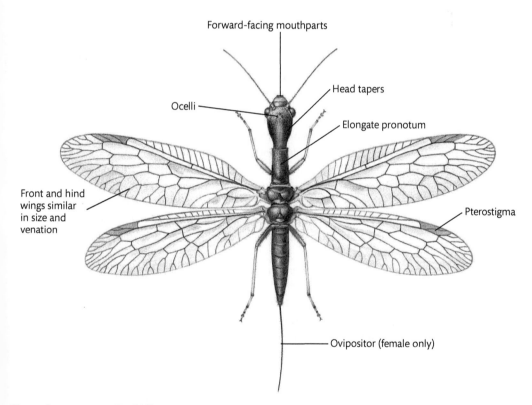

Forward-facing mouthparts

Head tapers

Ocelli

Elongate pronotum

Front and hind wings similar in size and venation

Pterostigma

Ovipositor (female only)

Phaeostigma notata – **Raphidioptera**

Essential Entomology. Second Edition. George C. McGavin and Leonidas-Romanos Davranoglou, Oxford University Press.
© George C. McGavin and Leonidas-Romanos Davranoglou (2022). DOI: 10.1093/oso/9780192843111.001.0001

Key features

- prothorax very elongate, appearing as a long 'neck'
- head elongate and slightly flattened
- ovipositor about as long or longer than body
- wings transparent

© Paul Brock

> Snakeflies such as this female *Atlantoraphidia maculicollis* take their name from their elongate head and prothorax. UK.

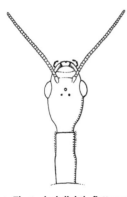

⌃ The typical slightly flattened head with biting, forwards-facing mouthparts, supported by an elongate prothorax of the Raphidiidae.

These moderately shiny, dark-greyish, or dark reddish-brown insects are the sister-group to all other Neuropterida. With fewer than 300 species, snakeflies represent the smallest holometabolan order. The name snakefly comes from the snake-like way in which the adults catch their prey. The **head, which is slightly flattened with biting, forward-facing mouthparts**, is supported by an **elongate prothorax** and can be raised and extended forward to seize food. The **head is broadest across the eyes and tapers behind** to meet the prothorax. The antennae are half as long (Raphidiidae) or as long (Inocellidae) as the body.

Ocelli are present in the Raphidiidae but are absent in the Inocellidae. The **two pairs of wings are similarly sized**, transparent, and have veins that are forked close to the wing margins. Both pairs of wings have a small dark or pale mark called the pterostigma on the front edge towards the wing tip. The females, which are a little larger than males, have a long, slender, and conspicuous ovipositor, which is about as long as the body, or distinctly longer.

Snakeflies are unusual among most insects in that they reach their greatest diversity in cool, temperate regions of the Northern Hemisphere, particularly in the Mediterranean, Central Asia, and North America. The reasons underlying this peculiar distribution are unknown, although the observation that snakefly larvae require cool temperatures to pupate might represent an adaptation to a temperate climate.

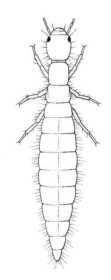

Snakefly larvae are terrestrial and live under bark or in rotting wood (*Raphidia* sp.).

Raphidioptera are usually found in wooded areas especially among rank vegetation. Adults and larvae are active during the day, and they are voracious predators of aphids and other small soft-bodied arthropods such as spiders and springtails. However, adult Raphidiidae may occasionally consume nectar and pollen, whereas the Inocellidae may not feed at all.

Courtship in these insects involves prolonged vibrational signal duets between males and females. These 'songs' are produced by sustained abdominal percussion and wing fluttering. Inocellid males may also fasten their head on the female's abdomen during mating, using unique abdominal organs. Several hundred eggs are laid, in groups of up to 100, in slits and cracks in tree bark or in rotting timber. The long, slender larvae have well-developed heads and legs, are very agile, and hunt for food under loose bark, in rotting wood, or in leaf litter. The short, curved, and pointed mandibles are similar to those of the adults.

Pupation takes place in a simple chamber in rotten wood or leaf litter, and, as there is no cocoon, the pupa can move around. The commonest genera are *Raphidia*, *Agulla*, and *Xanthostigma*.

Although definitive snakefly fossils are scarce, molecular estimates suggest that the earliest ancestors of extant species appeared during the early Cretaceous (about 136 mya). One would expect snakeflies to have appeared much earlier, given that they are the sister-group to all other Neuropterida, which are thought to have evolved more than 239 mya. However, diverse and ancient lineages of tropical, warm-adapted snakeflies may have lived in the Mesozoic but did not survive the climatic catastrophes of the Cretaceous–Tertiary extinction event caused by the meteor impact that also killed the non-avian dinosaurs. As a result, only their poorly diverse, cool-adapted relatives survived, resulting in the temperate snakefly fauna we observe today.

The economic importance of Raphidioptera has not been sufficiently researched. However, given that snakeflies are effective predators of aphids and other arthropod pests, they may act as natural agents of pest control in certain parts of the world.

Key reading

Aspöck, H. (1998). Distribution and biogeography of the order Raphidioptera: updated facts and a new hypothesis. *Acta Zoologica Fennica* **209**: 33–44.

Aspöck, U. and Aspöck, H. (2007). Verbliebene Vielfalt vergangener Blüte. Zur Evolution, Phylogenie und Biodiversität der Neuropterida (Insecta: Endopterygota). *Denisia* **20**: 451–516.

Aspöck, U., Aspöck, H., and Rausch, H. (1994). Die Kopulation der Raphidiopteren: eine zusammenfassende Übersicht des gegenwärtigen Wissensstandes (Insecta: Neuropteroidea). *Mitteilungen der Deutschen Gesellschaft für Allgemeine und Angewandte Entomologie* **9**: 393–402.

Engel M., Pérez de la Fuente, R., Peñalver, E., and Delclòs, X. (2012). Snakefly diversity in Early Cretaceous amber from Spain (Neuropterida, Raphidioptera). *ZooKeys* **204**: 13–40.

Henry, C. S. (2006). Acoustic communication in neuropterid insects. In: *Insect sounds and communication*. Drosopoulos, S. and Claridge, M. F. (eds), pp. 153–66. Taylor and Francis, Boca Raton.

Oswald, J. D. (2015). Neuropterida species of the world: a catalogue of the species-group names of the extant and fossil Neuroptera, Megaloptera, Raphidioptera and Glosselytrodea (Insecta: Neuropterida) of the World. Lacewing Digital Library v. 4.0. Retrieved from: http://lacewing.tamu.edu/SpeciesCatalog/Main.

Vasilikopoulos, A., Misof, B., Meusemann, K., Lieberz, D., Flouri, T., Beutel, R., Niehuis, O., Wappler, T., Rust, J., Peters, R., Donath, A., Podsiadlowski, L., Mayer, C., Bartel, D., Böhm, A., Liu, S., Kapli, P., Greve, C., Jepson, J., and Aspöck, U. (2020). An integrative phylogenomic approach to elucidate the evolutionary history and divergence times of Neuropterida (Insecta: Holometabola). *BMC Evolutionary Biology* **20**: 64.

Megaloptera
(alderflies, dobsonflies, and fishflies)

Common name	Alderflies, dobsonflies, and fishflies	**Metamorphosis**	Complete (egg, larva, pupa, adult)
Derivation	Gk. *megalo*-large, great; *pteron*-a wing	**Distribution**	Widespread but mainly temperate regions
Size	Body length 10–150 mm. Wingspan 18–170 mm	**Number of families**	2
		Known world species	373 (0.033%)

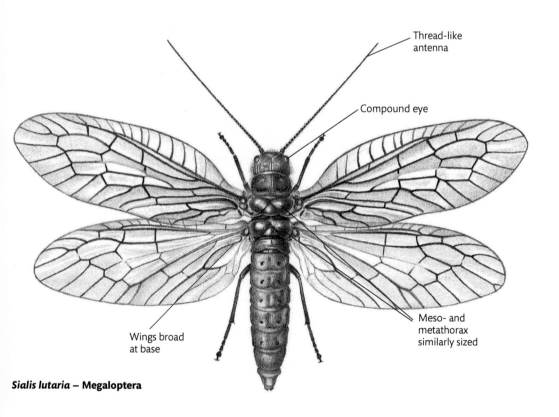

Thread-like antenna

Compound eye

Meso- and metathorax similarly sized

Wings broad at base

Sialis lutaria – **Megaloptera**

Essential Entomology. Second Edition. George C. McGavin and Leonidas-Romanos Davranoglou, Oxford University Press.
© George C. McGavin and Leonidas-Romanos Davranoglou (2022). DOI: 10.1093/oso/9780192843111.001.0001

Key features

- aquatic larvae with distinct lateral abdominal gills
- wings of roughly equal size
- medium-sized to very large insects
- adults always near water

❯ Dobsonflies, such as this *Protohermes motuoensis* represent some of the largest Neuroptera. Although they are voracious aquatic predators as larvae, adults probably do not feed at all. Northern India.

© George McGavin

There are two families, the alderflies (Sialidae) and the dobsonflies and fish-flies (Corydalidae). The head carries a pair of **prominent compound eyes** and unmodified, **thread-like antennae**. Ocelli are present in corydalids but absent in sialids. Despite their **conspicuous jaws**, adult megalopterans do not feed. In some male Corydalidae, the jaws may be hugely enlarged to several times the length of the head and are used in male-to-male combat or for grasping the female during mating. These are known as the dobsonflies. Despite the alarming proportion of their jaws, dobsonflies are harmless, although it is just conceivable that their larvae, known as hellgrammites or toe-biters, might occasionally earn their vernacular name. Male Corydalidae of the subfamily Chauliodinae are known as the fishflies and lack the distinctly enlarged jaws.

Megalopterans do not fly readily, and when they do, they flutter rather weakly and alight readily. There are **two pairs of wings with a branching pattern of venation**. The **hind wings are broad at their bases**, due to the presence of large anal lobes, and are usually larger than the front wings. At rest, the wings may be held roof-like over the body or may appear to be almost wrapped around the body. Megalopteran adults are found near cool, clean streams. Dobsonflies prefer running water, while alderflies can be found in ponds and canals as well as streams.

Mating in Megaloptera takes place on marginal vegetation or on the ground. In Sialidae, both sexes locate each other and communicate by rhythmically tapping their abdomen and wings on the substrate, which produces species-specific low-frequency vibrations that they sense through their legs. In Corydalidae, only males produce these vibrational signals, as part of a pre-mating ritual. The males of certain male dobsonfly (*Protohermes*) and fishfly species (*Parachauliodes*, *Neochauliodes*) are known to attach a jelly-like external spermatophore to the genitalia of the female. The spermatophore contains bundles of sperm in seminal fluid, which are transferred to the female over the course of several hours. Males remain in copula throughout this time, for, if they separate, the female may eat the spermatophore before all the sperm have been transferred successfully. Females produce large numbers of eggs, which they stick to reeds and other plants near or hanging over the water. The egg masses may frequently resemble bird droppings, which provides an effective camouflage against many predators.

The larvae of all megalopterans are aquatic, highly predacious, with simple or branched abdominal gills, and can be very similar in general appearance to larvae of some water beetle families. When the young larvae hatch, they drop or crawl into the water using their well-developed legs. In alderflies, the larval abdomen has a single terminal filament and seven pairs of lateral filaments.

The abdomen of a dobsonfly larva has a terminal pair of small hook-like **prolegs**.

The larvae hide or burrow, and ambush a range of arthropod prey, particularly small crustaceans, and will also take worms, amphibians, and even small fish. Larval development goes through a maximum of 11 instars and can take anything from 12 months in alderflies, but sometimes more than 48 months in dobsonflies. The time spent as a pupa is relatively short and can be as little as two weeks. Unlike species in the closely related order Neuroptera, the final larval stage does not spin a cocoon in which to pupate. Instead, pupation takes place on land within a simple chamber made in moist soil, sand, mossy vegetation, or under rotting wood. The pupae have functional jaws, can move around freely, and may even protect themselves if attacked.

The fossil record of Megaloptera is rather poor, with specimens that can be confidently assigned to Megaloptera being known from the Early and Middle Jurassic (about 200–170 mya). Molecular divergence estimates suggest that the common ancestor of extant Megaloptera appeared in the Triassic about 239 mya. Most morphological and molecular studies suggest that Megaloptera represent the sister-group to the Neuroptera. Megaloptera are of no direct economic importance, although they may serve as bioindicators of freshwater ecosystems.

⌃ At rest, the wings are held together roof-like over the body.

❯ proleg
a fleshy abdominal limb of insect larvae

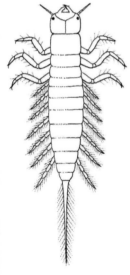

⌃ The aquatic larva of an alderfly, showing the single terminal filament and seven pairs of lateral filaments, *Sialis* sp.

Key reading

Hayashi F. (1996). Insemination through an externally attached spermatophore: bundled sperm and post-copulatory mate guarding by male fishflies (Megaloptera: Corydalidae). *Journal of Insect Physiology* **42**: 859–66.

Henry, C. S. (2006). Acoustic communication in neuropterid insects. In: *Insect sounds and communication*. Drosopoulos, S. and Claridge, M. F. (eds), pp. 153–66. Taylor and Francis, Boca Raton.

Liu, X., Wang, Y., Shih, C., Ren, D., and Yang, D. (2012). Early evolution and historical biogeography of fishflies (Megaloptera: Chauliodinae): implications from a phylogeny combining fossil and extant taxa. *PLoS ONE* **7**: e40345.

Mangan, B. P. (1994). Pupation ecology of the dobsonfly *Corydalus cornutus* (Corydalidae: Megaloptera) along a large river. *Journal of Freshwater Ecology* **9**: 57–62.

Merritt, R. W. and Cummins, K. W. (eds) (1996). *An introduction to the aquatic insects of North America* (3rd edn). Kendall/Hunt, Dubuque.

New, T. R. and Theischinger, G. (1993). *Megaloptera (alderflies, dobsonflies)*. Handbuch der Zoologie Band IV, Arthropoda, Insecta, Tome **33**. de Gruyter, Berlin.

Oswald, J. D. (2015). Neuropterida species of the world: a catalogue of the species-group names of the extant and fossil Neuroptera, Megaloptera, Raphidioptera and Glosselytrodea (Insecta: Neuropterida) of the World. Lacewing Digital Library v. 4.0. Retrieved from: http://lacewing.tamu.edu/SpeciesCatalog/Main.

Rupprecht, R. (1975). Die Kommunikation von *Sialis* (Megaloptera) durch Vibrationssignale. *Journal of Insect Physiology* **21**: 3057–313, 3157–320.

Vasilikopoulos, A., Misof, B., Meusemann, K., Lieberz, D., Flouri, T., Beutel, R., Niehuis, O., Wappler, T., Rust, J., Peters, R., Donath, A., Podsiadlowski, L., Mayer, C., Bartel, D., Böhm, A., Liu, S., Kapli, P., Greve, C., Jepson, J., and Aspöck, U. (2020). An integrative phylogenomic approach to elucidate the evolutionary history and divergence times of Neuropterida (Insecta: Holometabola). *BMC Evolutionary Biology* **20**: 64.

Whiting, M. F. (1994). Cladistic analysis of the alderflies of America North of Mexico (Megaloptera: Sialidae). *Systematic Entomology* **19**: 77–91.

Neuroptera
(lacewings, antlions, and mantidflies)

Common name	Lacewings, antlions, and mantidflies	Metamorphosis	Complete (egg, larva, pupa, adult)
Derivation	Gk. *neuron*–sinew, nerve; *pteron*–a wing	Distribution	Worldwide
		Number of families	18
Size	Body length 2–90 mm. Wingspan 1.8–150 mm	Known world species	> 5,820 (0.51%)

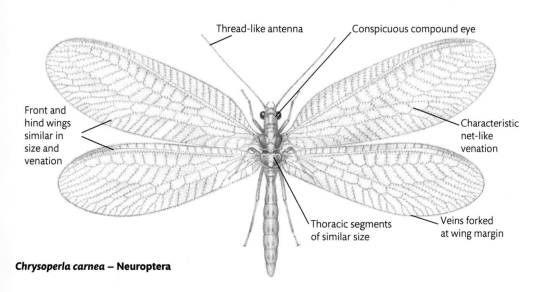

Thread-like antenna

Conspicuous compound eye

Front and hind wings similar in size and venation

Characteristic net-like venation

Thoracic segments of similar size

Veins forked at wing margin

Chrysoperla carnea – **Neuroptera**

Essential Entomology. Second Edition. George C. McGavin and Leonidas-Romanos Davranoglou, Oxford University Press.
© George C. McGavin and Leonidas-Romanos Davranoglou (2022). DOI: 10.1093/oso/9780192843111.001.0001

Key features

- elongate bodied
- prominent eyes
- wing venation complex and net-like
- larval mandibles and maxillae form sucking mouthparts

© Rupert Soskin

❭ Ascalaphids are among the
most beautiful neuropterans.
Libelloides coccajus, France.

Members of this order have a worldwide distribution, although many families
are restricted to particular regions. The vast majority of adult neuropterans
are predatory (with a few nectar-feeding exceptions) and are mainly active in
the evening or after dark. The larvae of all species are highly predacious and
can be found in a wide range of habitat types, even freshwater, where larvae
of the Sisyridae predate on sponges.

The species of some families exhibit remarkable convergent similarities
to other insect groups. Adult Mantispidae have enlarged, raptorial front legs
similar in design and function to those of praying mantids. Owlflies (Ascalaphidae) are efficient aerial predators which look similar to dragonflies in
general appearance, and also catch prey while flying.

In general, adult neuropterans have **biting, chewing mouthparts** and a pair
of **conspicuous, laterally placed compound eyes**, which, in some species, appear to shine. Ocelli may or may not be present. The antennae can be short,
about as long as the head and thorax combined, but are generally much longer
and thread-like. In some owlflies and antlions, the end of the antennae may be
swollen to form a club. The adults of some families have prothoracic glands

© George McGavin

> Lacewing larvae consume vast amounts of aphids and other soft-bodied insect during their lifetime. UK.

capable of producing substances that repel some predators. There are **usually two pairs of similarly sized wings** held roof-like over the body at rest. The **venation in most families is net-like** with the main veins forking at the wing margins in many families. Exceptions include the Coniopterygidae, which comprise very small species with a white, powdery covering, where the hind wings are often reduced, and the Nemopteridae, whose hind wings are developed into long, tail-like structures. In these species, it has been shown that the long hind wings may deter smaller predatory insects such as robber flies (Asilidae) by making the nemopterid appear bigger than its actual size.

Although courtship and mating have been described in relatively few species, it seems that all families produce species-specific vibrational songs and pheromones. Male and female green lacewings produce low-frequency vibrations by vertical abdominal vibration. These vibrations, which are produced in short bursts every second with a frequency typically less than 80 Hz, are transmitted through the legs to the foliage. The legs of lacewings have different types of **chordotonal organs**, of which the subgenual organ in the tibiae, for the reception of substrate-borne vibration, is the most important. The males usually sing first, and the females respond at species-specific timeframes. These courtship duets are a prelude to mating and have been shown to play a major role as a reproductive isolation mechanism in the genus *Chrysoperla* and may be the primary force driving the rampant sympatric speciation observed in this genus.

> **chordotonal organs** subcuticular mechanoreceptors that pick up vibrational signals—made up of one or more units called scolopidia

Larval morphology is varied and very much linked with lifestyle. Free-living hunters are slender with longish legs, and many have specialized adhesive disc-like structures on the last two abdominal segments for clinging onto foliage. The bodies of pit-building or ambushing species tend to be fat and squat with short legs, very large jaws, extended necks, and other modifications. The strong, sickle-shaped mouthparts of the larvae are highly modified

⌃ A lacewing larva (*Chrysopa* spp.) sucks the body fluids from an aphid using sharp, curved mandibles.

as sucking needles, a character that clearly separates this order from the Mega-loptera. The sharply pointed and curved mandibles have a groove that runs along their internal surface. Pressed tightly against the groove, an elongated maxilla forms a hollow tube, through which the larvae can inject enzymes and suck the juices of their prey.

The alimentary tract of larval neuropterans is peculiar in that the hindgut is not united with the midgut. During larval life, only fluids are expelled and any solid material that accumulates remains in the midgut, only to be passed as a **meconium** when the adult first emerges from the pupa with its gut joined up. This characteristic is also found in most members of the Hymenoptera.

> **meconium**
the first excrement passed by
a newly emerged adult

The larvae of Mantispidae are parasitic on spider egg sacs or the larvae of hymenopterans, such as social wasps or bees. In *Mantispa uhleri*, the mobile first instar larvae find an adult spider and climb aboard to await the pro-duction of an egg sac, inside which they will feed. To keep going until this happens, the young mantispid sucks the spider's haemolymph. Once inside the egg sac or host cell, the larva moults and becomes more maggot-like with short legs. The larvae of Ascalaphidae are 'sit-and-wait' predators that lie in debris or rest on plants to seize anything suitable that comes within range of their jaws. A unique predatory strategy may be found in the larvae of the beaded lacewing (Berothidae) species *Lomamyia latipennis*. This species lives in the colonies of the termite *Reticulitermes hesperus*. When a hungry larva encounters a termite, it raises the tip of its abdomen, and waves it in front of the termite's head, releasing an unknown chemical. The effect of this chemi-cal is to paralyse the termite, which allows the larva of *Lomamyia* to feed on it undisturbed. This phenomenon has fascinated entomology enthusiasts, who refer to this predatory strategy as the 'death fart'. However, observation of this strategy has never been repeated since the original study, and the chemical involved is unknown.

In the majority of Neuroptera, larvae are mature after three instars and pupate inside a fragile, spherical, silk cocoon spun from special organs (modified malpighian tubules) at the end of their abdomens. Sometimes the cocoon will incorporate small sand grains or soil particles. The pupae are similar to those of the Megaloptera in that they have free appendages and functional mouthparts, which are used to open the cocoon before the adult emerges from its pupal exuvia.

Many neuropterans are directly or indirectly beneficial in controlling pop-ulations of mites and small, soft-bodied animals, such as aphids, mealybugs, and scale insects. The two biggest neuropteran families are the common or green lacewings (Chrysopidae) and antlions (Myrmeleontidae). Common or green lacewings are generally green in body colour, the wings often with del-icate tints of pink, green, and blue. The wings of chrysopids are distinctive in that they have two seemingly zigzag veins and lack any dark patterning. The eyes are bright golden, brassy, or reddish. Both adults and their larvae are vo-racious predators of aphids, thrips, psyllids, scale insects, coccids, and mites, although, in some species, the adults will eat pollen, nectar, and honeydew. The adults are mainly nocturnal in habits, and species of the largest subfam-ily, the Chrysopinae, have bat-detecting, ultrasonic sound receptors located in the large veins of the wings. Chrysopid eggs are laid on vegetation and, like the eggs of most species, are supported by distinctive long, delicate stalks.

The larvae, which are often pale yellow, white, or greenish, with brown or black markings, can disguise themselves by attaching the sucked-out bodies of their victims on to special, hooked dorsal hairs. The disguise helps them to move about, camouflaged in the colony and may protect them from birds. Some larvae that are predacious on mealy bug colonies gather the mealy bugs' waxy coverings and use them to build up a protective cover on their own backs. Although soft-bodied and weak, aphids can fight back. Many species of aphid produce a sticky substance from their abdominal cornicles, which is designed to harden inside and block the hollow jaws of lacewing larvae. Green lacewings pupate in pea-shaped, silk cocoons attached to the under-sides of leaves. Adults will often enter houses to hibernate and are attracted to lights. *Chrysopa*, the largest genus in the entire family, is very common and widespread. *Chrysoperla carnea* and other species have been reared in large numbers for use as a biological control agents in various crops.

Brown lacewings (Hemerobiidae) are less common than green lacewings, and can be found in deciduous woodlands, gardens, and hedgerows. The eggs are laid on plants but are not stalked. The larvae do not have any body warts or tubercles, and their body hairs are simple and not hooked.

Adult antlions (Myrmeleontidae) superficially resemble large damselflies but are softer and have short knobbed or club-ended antennae, which are about twice as long as the head. The wings are long and narrow and may be clear or have distinctive brown or black patterns. The abdomen is elon-gate and, in males, has terminal sexual clasping organs that look a bit like the forceps of earwigs. Open woodland, scrub grassland, dunes, and warm, dry sandy areas are typical habitats for antlions. Although a few eat pollen, most adult antlions are carnivorous, and seize insects from vegetation or on the wing. The females lay single eggs in soil or sand. Some larvae construct insect-trapping pits in loose sand, while others live as ambush prey on the trunks of trees in soil or under stones and debris. All antlion larvae, sometimes called doodlebugs, are carnivorous. They are broadly oval, with very large jaws armed with sharp spinous teeth for preying on a wide range of insects and spiders. In the best-known species, such as those of the genus *Myrmeleon*, the larvae live in sand at the bottom of conical pits with only their sharp jaws showing. Where the antlions construct their pit is dependent on a number of abiotic factors, such as shelter and substrate type. If population densities are high, some species will space each other by means of sand flicking and other forms of display. If too much sand is flicked into a pit from a neighbouring pit, the antlion will usually relocate. Ants or any other insect stumbling into the pit will tumble to the bottom or be showered with sand grains flicked up by the antlion. Once at the bottom of the pit, the prey is seized and, when sucked dry, is flicked out before the larva renovates its pit. The size of the sand grains is important. Ants are much more likely to escape from the pit if the grains are too big. It is not surprising that, in such a specific interac-tion, ants have learned to avoid antlions, or areas where the larvae are likely to be lurking. Some experiments on ground-foraging ant species show that

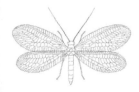
A typical lacewing showing the complex wing venation.

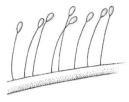
Chrysopid eggs supported above the leaf surface by long delicate stalks.

experienced ants will not be tempted into high-risk areas even by attractive baits and will continue to avoid areas from which antlions have been removed. The larvae of some Australian antlions, *Callistoleon manselli* and *Callistoleon illustris*, have taken the art of pit building a stage further by adding furrows or trenches radiating from the central pit. The extra construction work is worth the effort as the furrows increase the prey capture rate. However, the building behaviour is plastic, and because the furrows are added after the central pit is dug, the larvae will not bother to make them if they are supplied with plenty of prey.

The evolutionary history of Neuroptera goes very far back in time, with phylogenomic estimates suggesting an origin in the Permian (about 280 mya), which is in agreement with the available fossil evidence. Many of the modern families and their behaviours appeared early on. For example, lacewing larvae from Early Cretaceous Lebanese amber (about 130 mya), were found to be camouflaged with bits of debris attached on their body through specialized hairs. The debris particles formed a protective dome-like structure that shielded the larvae from attacks (presumably from their hosts), a behaviour that is found in many lacewing larvae today.

Key reading

Canard, M., Semeria, Y and New, T. R. (eds) (1984). *Biology of the Chrysopidae.* Entomologica series **27**. Junk Publishers, The Hague.

Brooks, S. J. and Bernard, P. C. (1990). The green lacewings of the world: a generic review. *Bulletin of the British Museum (Natural History) Entomology* **63**: 137–210.

Devetak, D. and Pabst, M. A. (1994). Structure of the subgenual organ in the green lacewing, *Chrysoperla carnea. Tissue and Cell* **26**: 249–57.

Engel, M. S., Winterton, S. L., and Breitkreuz L. C. V. (2018). Phylogeny and evolution of Neuropterida: where have wings of lace taken us? *Annual Review of Entomology* **63**: 531–51

Gotelli, N. J. (1996). Ant community structure: effects of predatory antlions. *Ecology* **77**: 630–8.

Henry, C. S. (1977). The behaviour and life histories of two North American ascalaphids. *Annals of the Entomological Society of America* **70**: 117–95.

Henry, C. S. (1979). Acoustical communication during courtship and mating in the green lacewing *Chrysopa carnea* (Neuroptera: Chrysopidae). *Annals of the Entomological Society of America* **72**: 68–79.

Henry, C. S. (1980). Acoustical communication in *Chrysopa rufilabris* (Neuroptera: Chrysopidae), a green lacewing with two distinct calls. *Proceeding of the Entomological Society of Washington* **82**: 1–8.

Henry, C. S. (1985). Sibling species, call differences, and speciation in green lacewings (Neuroptera: Chrysopidae: *Chrysoperla*). *Evolution* **39**: 965–84.

Henry, C. S. (1994). Singing and cryptic speciation in insects. *Trends in Ecology and Evolution* **9**: 388–92.

Henry, C. S. (2006). Acoustic communication in neuropterid insects. In: *Insect sounds and communication.* Drosopoulos, S. and Claridge, M.F. (eds), pp. 153–66. Taylor and Francis, Boca Raton.

Johnson, J. B. and Hagen, K. S. (1981). A neuropterous larva uses an allomone to attack termites. *Nature* **289**: 506–7.

New, T. R. (1975). The biology of the Chrysopidae and Hemerobiidae (Neuroptera), with reference to their usage as biocontrol agents: a review. *Transactions of the Royal Entomological Society of London* **127**: 115–40.

Oswald, J. D. (2015). Neuropterida species of the world: a catalogue of the species-group names of the extant and fossil Neuroptera, Megaloptera, Raphidioptera and Glosselytrodea (Insecta: Neuropterida) of the World. Lacewing Digital Library v. 4.0. Retrieved from: http://lacewing.tamu.edu/SpeciesCatalog/Main.

Pérez-de la Fuente, R., Peñalver, E., Azar, D., and Engel, M.S. (2018). A soil-carrying lacewing larva in Early Cretaceous Lebanese amber. *Scientific Reports* **8**: 16663.

Redborg, K. E. (1982). Interference by the mantispid *Mantispa uhleri* with the development of the spider *Lycosa rabida*. *Ecological Entomology* **7**: 187–96.

Vasilikopoulos, A., Misof, B., Meusemann, K. Lieberz, D., Flouri, T., Beutel, R., Niehuis, O., Wappler, T., Rust, J., Peters, R., Donath, A., Podsiadlowski, L., Mayer, C., Bartel, D., Böhm, A., Liu, S., Kapli, P., Greve, C., Jepson, J., and Aspöck, U. (2020) An integrative phylogenomic approach to elucidate the evolutionary history and divergence times of Neuropterida (Insecta: Holometabola). *BMC Evolutionary Biology* **20**: 64.

Matsura, T. and Kitching, R. L. (1993). The structure of the trap and trap building behaviour in *Callistoleon manselli* New (Neuroptera: Myrmeleontidae) *Australian Journal of Zoology* **41**: 77–84.

Wells, M. M. and Henry, C. S. (1992). The role of courtship songs in reproductive isolation among populations of green lacewings of the genus *Chrysoperla* (Neuroptera: Chrysopidae). *Evolution* **46**: 31–42.

Coleopterida

This grouping, which includes beetles (Coleoptera) and twisted-wing insects (Strepsiptera), is the most diverse animal lineage, with more than 400,000 species.

Essential Entomology. Second Edition. George C. McGavin and Leonidas-Romanos Davranoglou, Oxford University Press.
© George C. McGavin and Leonidas-Romanos Davranoglou (2022). DOI: 10.1093/oso/9780192843111.001.0001

Coleoptera
(beetles)

Common name	Beetles	Metamorphosis	Complete (egg, larva, pupa, adult)
Derivation	Gk. *Koleos*-sheath; *pteron*-a wing	Distribution	Worldwide
Size	Body length 0.1–180 mm. Mostly under 25 mm	Number of families	170
		Known world species	About 400,000 (35.34%)

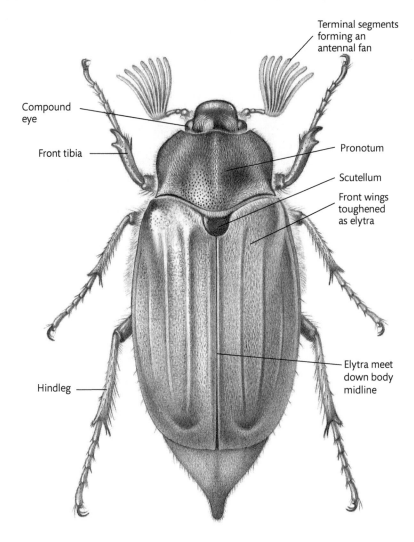

Melolontha melolontha – Coleoptera

Essential Entomology. Second Edition. George C. McGavin and Leonidas-Romanos Davranoglou, Oxford University Press.
© George C. McGavin and Leonidas-Romanos Davranoglou (2022). DOI: 10.1093/oso/9780192843111.001.0001

Key features

- very abundant and ubiquitous in all terrestrial and freshwater ecosystems
- front wings modified as rigid elytra covering hind wings
- prothorax usually large, and distinct from the head and rest of the body
- many species are crop pests
- the largest order

❯ The stag beetle's (*Lucanus cervus*) impressive mandibles are used in male-to-male combat. UK.

❯ Click beetles (Elateridae) are able to escape from predators by using a jumping mechanism on the underside of their thorax. Singapore.

© Zestin Soh

> Giraffe weevils (*Trachelophorus giraffa*) from Madagascar need no explanation as to their common name. The long 'neck' of the male is an extension of the head, used to construct nests for the females and in male-to-male combat.

Chitin, the major component of arthropod cuticle, was first described as proteinaceous in nature from a study of the wing cases of the cockchafer *Melolontha* published by Odier in 1823. The protective front wing cases have contributed greatly to the success of beetles, and their tough nature has a useful spin-off for archaeologists and palaeoecologists. The wing cases, or even just fragments of wing cases, and other body parts in sediments and deposits are often recognizable to the level of species. As the host plants and habits of these beetles are unlikely to have changed much in the last 1.5 million years, building up a picture of exactly which species were present can provide invaluable details of what plants were growing at the time and what the habitat was like. Identifying beetle communities in archaeological excavations serves as an invaluable tool for understanding past climate, habitat modification by humans, their diet and subsistence methods, as well as methods of food storage.

© Anne Riley

> Many male beetles such as this cockchafer, have large comb-like antennae, which are covered with olfactory receptors that allow them to locate females at a distance. UK.

© Rupert Soskin

❯ Beetles play an essential role in the recycling of nutrients in virtually every terrestrial ecosystem. Cockchafer larvae decompose plant debris, allowing for living plants to extract the nutrients from decaying matter. UK.

© Rupert Soskin

❯ The corn weevil (*Sitophilus zeamais*) is a notorious pest of cereal products. France.

There is now some doubt as to whether the British polymath J. B. S. Haldane really said 'an inordinate fondness for beetles', in reply to being asked what could be inferred of the Creator by studying the works of creation, but it is a good story. There can be no doubt that beetles really are the most successful multicellular animals. They comprise at least 40% of all known insect species and about 25% of all known animals. There are more species of beetles than there are species of plant, and, like all other insect groups, many species await discovery and description. It is estimated that the total number of beetle species (including those yet undescribed) ranges between 1.5 to 2.1 million.

> Cerambycid beetles like *Rhagium sycophanta* are the primary decomposers of live and decaying wood. France.

© Rupert Soskin

> Ladybirds are voracious predators of aphids and other agricultural pests, and for this reason they can be exceptionally beneficial for gardens and eco-friendly crops. *Coccinella septempuncta*, UK.

© Anne Riley

The species comprising the order are classified into four very unequal suborders (in terms of species number): the Archostemata, Myxophaga, Adephaga, and Polyphaga (see Box).

Coleopteran suborders

Suborder	Major families	Habits
Archostemata	3 Primitive families: Cupedidae, Micromalthidae (1 sp.), Omatidae	Mostly found in decaying wood, fungivorous
Myxophaga	4 Families: Cyathoceridae, Hydroscaphidae, Microsporidae, Torridincolidae	Aquatic or associated with wet habitats, algal feeders
Adephaga	11 Families including Carabidae, Dytiscidae, Gyrinidae	Terrestrial and aquatic, mostly predacious
Polyphaga	> 149 Families including Anobiidae, Bruchidae, Buprestidae, Cerambycidae, Chrysomelidae, Coccinellidae, Curculionidae, Dermestidae, Elateridae, Scarabaeidae, Scolytidae, Staphylinidae, Tenebrionidae	Variable: mostly herbivorous, some saprophagous, wood-boring, predacious, fungivorous, etc.

Beetles can be found in every conceivable terrestrial and freshwater habitat, from the equator to the polar regions. One of the features that has enabled them to become so successful, and the most dominant insect taxon, is **the front pair of wings, which are toughened and act as wing cases or elytra** to protect the much **larger, membranous hind wings**, which are folded underneath at rest. In all species, the elytra meet down the midline of the body, and, in some species, they are permanently fused together. In most species, the elytra cover most of the thorax and abdomen, but, in some (e.g., the Staphylinidae), they are shortened (as in earwigs) to allow greater abdominal flexibility. In these species, the hind wings are folded double, and the dorsal surface of the abdominal segments are specially toughened to compensate. The strength of the body and the mechanical protection afforded by the sclerotized elytra has enabled the beetles to safely burrow, dig, and squeeze into places that other insects cannot reach. Wherever you look, under bark and stones, in the ground, inside plants, in leaf litter, in caves, in dung, and in carcasses, beetles have been able to colonize successfully. Additionally, the elytra provide protection from predators, and, as they also cover the abdominal spiracles, they cut down on water loss. The space beneath the elytra allows aquatic species to maintain an air store while submerged. Beetles have **chewing-biting mouthparts** and usually have **antennae with fewer than eleven segments**. They are generally compact and tough-bodied, and some species can be so armoured that entomologists find great difficulty in pushing even the sharpest stainless-steel pin through their body. The joints where body and limb segments meet are often specially strengthened, recessed, or retractile, so that not even the tiniest chink is available for predatory jaws or an ovipositor to gain access. The body shape has evolved to suit a multitude of functions and environments. Long-legged species for desert life; smooth, streamlined species for aquatic life; robust, tank-like species for earth moving; seed-shaped mimics to deceive predators; and so on.

Among the beetle multitudes there are many scavengers, predators, and a few specialized parasites, but the vast majority of beetle species are herbivorous and here lies the second major reason for their great success. The first beetles that arose in the Permian were detritivores or fungivores, but it did not take them long (50 million years or so) to move on to conifers and cycads, which were spreading in the Triassic and Jurassic. The early herbivorous beetles chewed particular plant parts, such as the pollen-bearing structures, and the numbers of these tissue-feeding taxa rose steadily. The rise to dominance of the flowering plants (Angiospermae) in the Cretaceous provided beetles with a multiplicity of opportunities to radiate. Already well adapted to eating plant tissues, more and more herbivorous beetles appeared. As quickly as the evolution equipped the plants with chemicals and other defences against herbivores, the herbivores evolved counter-measures. Enzymes appeared to deal with plant poisons, or the beetles simply shifted their attention to a different, less well-defended plant part. Each evolutionary thrust and parry led to the appearance of more and more species. The 100-million-year-long co-evolutionary contest between beetles and flowering plants has produced very species-rich superfamilies, such as the leaf beetles (Chrysomeloidea) and the weevils (Curculionoidea), and today their interactions represent the most familiar biological association on the planet.

Most beetles reproduce sexually. Mate attraction and courtship involves sound production, pheromones, and light displays. Fireflies and glow worms belong to the family Lampyridae. The most distinctive feature of these beetles is their ability, shared with some click beetles (Elateridae) and other related families (e.g., Phengodidae), to produce light from special luminous organs. The adults of some species and the larvae of all species are predacious, feeding mostly on land snails. The luminous organs on the underside of the abdomen are covered by transparent cuticle, and the beetles regulate the flash interval and duration by controlling the oxygen supply to the organs, where a chemical reaction involving a substance called luciferin and the enzyme luciferase, produces a cold, greenish light.

Mating in beetles always takes place with the male holding on to the female's back. Sometimes the male will remain in copula for many hours or days, even though sperm transfer has been completed. This post-insemination mate guarding is just one form of paternity assurance seen in beetles. One odd example concerns the incredibly long front legs of the male Harlequin Beetle, *Acrocinus longimanus* (Cerambycidae). Before naturalists saw living specimens, they imagined that the beetles used their legs to brachiate through vegetation like miniature gibbons, but their real purpose is just as unusual. As in most insects, sperm is used on a last-in, first-out basis, and it is important that the female does not mate with another male before she lays her eggs. The last male to mate before egg laying will fertilize the lion's share of the eggs. To protect his mate from rivals, and therefore his investment, the male Harlequin Beetle uses his long front legs as a moveable corral until the eggs are laid.

The females of most beetles lay eggs. Some species are ovoviviparous, and the eggs hatch inside the reproductive tract of the female and she gives birth to larvae. Beetle larvae are very variable in appearance, but all have toughened head capsules with forward-facing, biting mouthparts, and short antennae having a maximum of four segments. There are four distinctive larval types: **campodeiform** larvae, like those of ground beetles or water beetles, are active predators with well-developed legs and large jaws; **eruciform** larvae, like those of the leaf beetles, are herbivorous with short or very short legs; soft, white, C-shaped, **scarabaeiform** grubs are typical of scarabaeid beetles and related groups and have three pairs of thoracic legs and are found in rotting plant tissue or animal dung; and legless **apodous** grubs are typical of weevils, click-beetles, and other beetles that bore or gall plants. There are never any leg-like structures on the abdominal segments, and, when larval development is complete, pupation occurs and the transformation to the adult stage begins. Beetle pupae have immobile mouthparts and free appendages (they are not firmly attached to the body surface). Many species pupate inside a chamber underground or inside the larval food plant, and, sometimes, a cocoon is made.

Beetle larvae can be herbivorous, carnivorous, or saprophagous, and many species develop in wood, dung, carrion, and fungi. There are a number of interesting parasitic families. Fewer than 60 species belong to the Rhipiceridae, whose larvae are endoparasites of cicada nymphs. The young active beetle larvae grab hold of a cicada nymph before it goes underground. The 2,000 or so world species of oil or blister beetles (Meloidae) are parasitic, or rather predacious, on eggs of grasshoppers, or the larvae of some Hymenoptera. The larvae of bee parasites wait on flowers for the correct species of host to visit, and then cling to the body hairs to enable transportation back to the nest. Fewer than 500 species belong to the Rhipiphoridae, the larvae of which are endoparasites of cockroaches, wasps, and bees.

The vast majority of beetles are terrestrial and less than 3% are aquatic. Beetles in some families (Gyrinidae, Haliplidae, Hygrobiidae, Noteridae, and Dytiscidae) spend most of their life history in aquatic environments, while others may be aquatic only as larvae or as adults. Pupation is terrestrial. Species in the Hydraenidae are only aquatic as adults, and some beetles belonging to families such as the Curculionidae and Chrysomelidae are only aquatic as larvae.

There are more than 4,000 species of diving beetles (Dytiscidae), a widespread family of water beetles where both larvae and adults are carnivorous. The adults have a streamlined body shape and long, slightly flattened, hair-fringed hind legs for efficient swimming. The males of many species have special, swollen front tarsi, which are used to grip the female during mating.

These beetles store air under their wing cases and when they need to obtain more air, they rise to the water's surface tail first (the tip of their abdomen).

Diving beetles have fully developed wings and can fly very well.

The larvae and adults of whirligig beetles (Gyrinidae) eat mosquito larvae. There are about 800 species of these small beetles, which get their common name from the characteristic circular, jerky way in which they move over the surface of ponds and slow-moving streams. At first glance, they look as if they only have front legs as the middle and hind legs are very short, stout, and

⌃ Detail of the front leg of the *Dytiscus marginalis* male showing the enlarged tarsal segments.

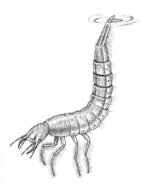

Larva of *Dytiscus marginalis* taking air at the water surface.

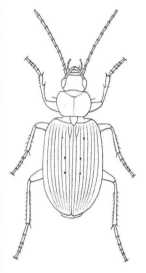

A typical ground beetle (Carabidae).

The larvae of tiger beetles (Cicindellinae) live in vertical tunnels in soil or sand.

paddle-like. The front pair of legs is used for catching food. The eyes of these beetles are interesting in that they are divided horizontally into two zones: one for above-water vision and one for below-water vision.

In the same suborder (Adephaga), the Carabidae, or ground beetles, are a large family of well over 40,000 species worldwide.

Although most of these beetles are dark coloured, some species can be metallic or brightly coloured. As major predators in many natural and managed habitats, ground beetles are important elements in food chains. They regulate pests and disseminate insect viruses and other entomopathogens, and they also serve as useful indicator species in conservation studies. Many carabids are specialist hunters and can, for example, follow the mucus trails left by snails and slugs. The larvae of tiger beetles (Cicindellinae, which are often considered a family in their own right), live in vertical tunnels in soil or sand. From these tunnels they ambush passing prey and drag them underground. The larvae have a flattened top of the head and pronotum, which form the lid of the burrow.

More than 500 species of bombardier beetles (Brachininae) are known worldwide. The most common genus is *Brachinus*. These beetles have a highly unusual and impressive defence mechanism. Accidentally standing barefoot on two bombardier beetles simultaneously on a shower room floor in the tropics was, for one of us (G McG), an unforgettable lesson. Like both barrels of a miniature shotgun being discharged, the beetles escaped but left the author with two patches of discoloured skin, which persisted for weeks. This is due to the fact that bombardier beetles are able to direct a stream of boiling-hot quinones from the rear of their abdomens to ward off predators. In some species, the stream consists of up to 500 pulses per second. The hot quinones result from a chemical reaction between hydroquinones and hydrogen peroxide in the abdomen. These compounds are secreted by cells and fill a reservoir from where they are passed through a non-return valve to a cuticle-lined reaction chamber. Here the mixture is acted on by oxidative enzymes, which causes a very rapid exothermic reaction that releases oxygen, and, in doing so, boils part of the contents. The rapid rise in pressure due to the production of gases forces the liquid out of the chamber with a relatively loud popping sound—like a mini-explosion. The effect on predators, both small and large, is immediate. Creationists have stated gleefully that these beetles provide them with an irrefutable example of 'special creation'. However, their argument is easily dismissed in much the same way as their creationist view of the evolution of the vertebrate eye. They argue that it is impossible to see how such a complex system, which requires more than two essential components, could have evolved by chance alone. They point out that the reaction chamber or either of the two chemicals are useless in isolation, but their inability to see how the system could have evolved by natural selection stems as much from a lack of imagination as a lack of knowledge.

This is how the system may have evolved by natural selection. In fact, quinones are not at all uncommon in arthropods and, as they are non-palatable, they commonly have a defensive function. From these, hydroquinones evolved as a superior line of defence; indeed, many arthropods accumulate defensive compounds for later use. Predation is an immensely strong selective pressure, and any small adaptation that improves survivability will be passed on to future generations. Many insects, including beetles, have a simple defensive system of secretory glands that fill a reservoir, the end of which is closed by a valve so that the stored contents do not leak to the outside before they are needed. Hydrogen peroxide might easily become mixed with hydroquinones in any secreted defensive liquid because it is a normal metabolic by-product and would increase the liquids' predator-deterring quality. A mixture of hydroquinones and hydrogen peroxide reacts very slowly to produce quinines, which would further improve the defensive function, but the reaction would be faster in the presence of enzymes. Catalases and peroxidases, the enzymes involved, are commonplace and, were they able to act on the secretions as they were expelled from the reservoir, they could increase the output temperature of the liquid and produce even more quinones. This is yet another small improvement to the system which could easily have evolved by natural selection, and which would give those beetles possessing it a significant adaptive advantage. The tube through which the secretions leave the reservoir increases in size and gets more enzyme-secreting capacity. The temperature of the defensive secretion goes up. The tube becomes a thick-walled reaction chamber and so on. Each little stage along the way is advantageous, and thousands upon thousands of generations of beetles, each one put to the ultimate test of leaving offspring before they get eaten, will produce better and still better defences.

Another remarkable lineage of ground beetle, with about 600 species, is the Paussinae—commonly known as the ant-nest beetles. Most of these beetles look nothing like other carabids, as they are generally very robust, with flattened antennae that may take various shapes—they can even be flat and rounded, like elephant ears, or with horns, like deer antlers. As their name suggests, they are obligate or facultative parasites of ant nests, where they feed on ant larvae and even ant workers. Specialized glands on their body surface of ant-nest beetles exude chemical compounds that imitate the ant pheromones that allow the latter to communicate with each other. In this way, ant-nest beetles are chemically invisible to their ant hosts and are not attacked. Using stridulatory organs, ant-nest beetles are also able to speak three different 'languages' by imitating the sounds produced by different ant castes (workers, soldiers, queen). In this way, the ant-nest beetles are treated as royalty by their hosts, being constantly taken care of, and gaining access to any chamber of the ant nest.

Ninety per cent of all beetle families are placed in the suborder Polyphaga, and some of these are extremely species rich. More than 63,000 species belong to the family Staphylinidae. Rove beetles, as they are called, have a characteristic, elongate, parallel-sided shape and very short wing cases. Most species are small predators or scavengers, and live in a vast range of habitats, from seaweeds to the fur of mammals. Being the dominant beetle lineages in most leaf litter habitats, rove beetles come in frequent contact with ants. This has led numerous lineages, such as the Aleocharinae and many Pselaphinae, to

^ With its jaws held wide open and abdomen raised, a staphylinid beetle displays a typical threat posture.

evolve into obligate parasites of ant nests. Staphylinidae contain the largest number of **myrmecophilous** species than any other insect group.

> **myrmecophilous**
> Greek for ant-loving. Any species that lives with and depends on ants for food, protection, and care

Species of *Nicrophorus*, which belong to a small, closely related family of fewer than 300 species, the burying beetles (Silphidae) are well known for their habit of locating a carcass of an animal and burying it by digging out the soil beneath. Other insects are highly attracted to carcasses of dead animals, and the burying behaviour has evolved to prevent competition from fly maggots. It has been shown that the cooperation of the two parent beetles leads to greater breeding success. As in many species, single mothers who have to compete for resources with conspecifics do worse than those with partners. Usually, a monogamous pair will work together, but in cases where the carcasses are big enough it might pay a male to try to attract a second mate. In this way, he will father more offspring, but more brood means that the reproductive success of the first female declines. Males in this position do try to attract other females by emitting sexual odours, but their partners do all they physically can to stop them from signalling. Sometimes, where there is a large corpse and likelihood of intense competition from flies and other carrion insects, it is advantageous for silphids to breed communally. There appears to be an agreed truce between females who would normally compete, and, in these examples, cooperative behaviour extends to females caring for each other's offspring. Take-overs are common at the height of the breeding season. Losing pairs will be ejected from the carcass, and if any eggs have been laid, they will be killed before the new female lays her own eggs. The guts of burying beetles host symbiotic microbes, such as certain yeasts. When these microbes are deposited on carrion, they prevent the growth of harmful microbes that would otherwise lead to further putrefaction of the carcass and the build-up of toxic substances. Using this approach, parent burying beetles are able to maintain the carcass in a fresh state that provides a high-quality nutritional resource for their larvae.

More than 30,000 species of scarab beetles, chafers, and related species make up the Scarabaeidae, which includes some of the largest insects of all. The family is divided into several important subfamilies: Scarabaeinae (dung beetles or scarabs), Aphodiinae (small dung beetles), Cetoniinae (fruit and flower chafers), Dynastinae (Hercules and rhinoceros beetles), Rutelinae (leaf chafers), and Melolonthinae (leaf chafers). Adults and larvae feed on foodstuffs ranging from dung and carrion to living or dead plant material, and many herbivorous species are pests. Rhinoceros beetles (*Oryctes rhinoceros*), for instance, can be a serious pest in coconut palm plantations. Insecticides are too expensive and difficult to apply, but the problem has been solved by the discovery of a suitable baculovirus biocontrol agent.

Despite enormous variation in shape, colour, and size between species, a single character, the distinctive antennal club, unites these species. The club is made up of three to seven flat, expanded, moveable plates, which can be closed or opened out. Scarabaeid larvae are typical white grubs with well-developed jaws, and many live in the soil and feed on roots.

Dung-burying beetles are extremely important recyclers in many parts of the world and can clear vast quantities of dung in a relatively short time, returning valuable nutrients to the soil; a job they have been doing for a long time. Cretaceous trace fossils found have been interpreted as the remains of herbivorous dinosaur dung with evidence of dung beetle workings. Species in a dung beetle assemblage vary greatly in size and the manner in which they utilize the resource. Several size classes of species roll balls of dung away using their hind legs, bury it, and lay an egg inside the ball. The dung is used as a food source by the larvae and the adults alike. Species in the genera *Scarabeus* and *Kheper* are well-known dung rollers and can move balls as big as 50 mm across. Others tunnel into and under the dung, some breed inside the dung pad, and a few (called kleptocoprids) lay their eggs in the dung buried by other species. Competition for dung can be intense and, in some species, the females cling to the anal hairs of mammals and are thus in a good position to deposit their eggs in fresh dung. In Australia, settler introduction of non-native animals like cattle, horses, and sheep caused immense problems. The Australian dung beetle community could not handle the vast quantities of dung produced, which lay on the ground, degrading pastureland and providing a breeding ground for enormous numbers of flies. The introduction of 16 or so species of European and African dung beetles alleviated the problem.

The females of some species (*Copris*) stay with their developing brood until they are ready to emerge as adults themselves. In these beetles, the parental care has an important outcome for the survivorship of the young. In particular, balls deserted by females prior to pupation are likely to become infected by species of fungi, which cover the outside and invade the interior.

Treatment of cattle with avermectins to control parasitic infections is having a serious negative effect on dung beetles and other insects of dung communities in many parts of the world. These pesticides are excreted in the faeces and remain active for a long time. The result is the greatly delayed disappearance of the dung beetle and unknown knock-on effects in associated communities.

The scarab beetle, one of the most important religious symbols of ancient Egypt, has been depicted in countless hieroglyphs, paintings, amulets, and pieces of jewellery. Khepri (The Being), depicted as a man with a scarab as his head, was the god of the morning sun.

The first scarab beetle to be worshipped by the ancient Egyptians was probably the large and brightly coloured species *Kheper aegyptiorum*. Initially, the symbolism interpreted the movement and daily rebirth of the sun in terms of the beetle rolling its ball. It has now been suggested that priests dug up the dung balls, and probably observed the process of metamorphosis 5,000 years before it was described by the French entomologist Jean-Henri Fabre. They developed their ideas along the lines that if the sun and the scarab could 'die' (enter the ground), be transformed, and then be reborn, the same might be possible for humans. It is conceivable that complex procedures of mummification represent an imitation of the dung beetles' pupal stage, designed to protect the body underground during its transformation and ultimate rebirth.

The 15,000 or so species of jewel beetles (Buprestidae) are typically inhabitants of woodland and forest, particularly in tropical regions. The majority are very attractive metallic green, blue, and red, with contrasting markings. The adults are tough, slightly flattened, and taper towards the rear of the body. The larvae have a distinctive shape with a large, expanded prothorax and tapering abdomen.

They are commonly called flat-head borers, and most species chew oval cross-sectioned tunnels in the wood of dead or dying trees. Some species lay their eggs exclusively in freshly burned wood, and the adults are attracted from long range to forest fires. Species of the genus *Melanophila* have been shown to have paired infra-red detectors very close to the coxae of their middle legs. These incredible heat-sensing pit organs are each made up of 50–100 sensillae located at the bottom of a pit in the cuticle. The sensitivity of the organs, which can respond to fires generating temperatures between 425–1150°C, and their ventral location enables flying beetles to detect suitable mating and egg-laying sites from a distance of several kilometres. A few buprestid species bore in the pith or roots of their host plants, and some are gall formers. The common name of the family refers to the fact that, in many parts of the world, the highly coloured and beautifully patterned wing cases are used as decoration in embroidery and as jewellery.

The Cerambycidae is a large family with about 37,000 species worldwide. Longhorn beetles, as they are often known, are typically elongate with long or very long antennae.

The larvae of almost all cerambycids feed by boring into live or dead wood, and several species can be forestry and timber pests. The larvae have symbiotic bacteria within special cells in the haemolymph that provide them with essential nutrients lacking in their diet, and in many cases larval development takes several years.

The leaf beetles (Chrysomelidae) are a large family consisting of more than 37,000 species. Leaf beetles tend to have a large, fairly smooth, rounded appearance, and many are very brightly coloured. All species are herbivorous, the adults chewing leaves and flowers and the larvae feeding externally or boring into stems, roots, and leaves. A great many species, particularly the flea beetles (subfamily: Alticincae), are damaging to crops. The notorious Colorado Potato Beetle, *Leptinotarsa decemlineata*, which feeds on potatoes, tomatoes, and aubergines, and spreads plant diseases, is a very serious pest. *Criocerus asparagi*, the Asparagus Beetle, and *Oulema melanopa*, the Cereal Leaf Beetle, are examples of two serious pest species that were introduced accidentally to North America. A few chrysomelids have their uses and have been successful in biological control programmes against weeds. St John's Wort or Klamath Weed (*Hypericum perforatum*), brought to North America by early European settlers, became a serious weed but was eventually controlled by a European chrysomelid beetle. Related to the leaf beetles are the Bruchidae, sometimes called pea and bean weevils. The adults of all species lay their eggs on seeds, and the larvae burrow inside, feed, and pupate just below the surface. Many species such as *Bruchus pisorum*, *Callosobruchus chinensis*, *Callosobruchus maculatus*, and *Acanthoscelides obtectus* are serious pests of stored peas, beans, and leguminous field crops.

Ladybirds or ladybug beetles (Coccinellidae) are favourite insects with people of all ages. They have inspired poetry, become characters in folk

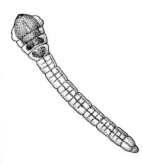

⌃ The expanded and flattened thorax of jewel beetle larvae is a very characteristic feature, and results in the tunnels they bore having an oval cross-section.

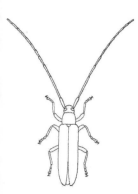

⌃ A typical longhorn beetle (Cerambycidae).

⌃ A typical ladybird beetle (Coccinellidae).

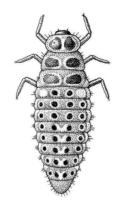

A typical ladybird larva.

A typical ladybird pupa.

The wood-boring larvae of the Deathwatch Beetle can cause immense structural damage to timber.

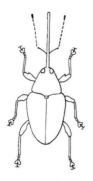

A typical weevil (Curculionidae).

stories, and are symbols of health and happiness. Most of these small, brightly coloured beetles are predacious on soft-bodied insects, such as aphids, mealybugs, and scale insects, but there are some herbivorous species (*Epilachna*), which can sometimes be crop pests. The predacious species can consume vast quantities of prey during the course of their life cycle. Ladybird larvae are very active with well-developed legs and seek out colonies of aphids on their host plants using visual and olfactory cues.

When fully grown, the larvae pupate on foliage. Adult ladybirds show a phenomenon known as reflex bleeding, whereby toxic fluids ooze out of joints, especially the knees, and serve to deter would-be predators. Ladybirds are very beneficial in controlling the natural populations of many pest insects, and several species have been used as biological control agents. *Rodolia cardinalis*, for instance, has been used in many parts of the world to control the serious citrus pest, Cottony Cushion Scale (Margarodidae: *Icerya purchasi*). However, one species, the Harlequin Ladybird (*Harmonia axyridis*), has become a major invasive insect worldwide, replacing native ladybird species in the areas it invades, and occasionally harming agriculture. This species is particularly harmful to viticulture, as large aggregations of ladybirds alter the taste of wine.

The 2,200 or so small, light-brown to black beetle species that comprise the Anobiidae feed on a wide range of animal and plant materials, including wood, fungi, and dried organic foods like biscuits, bread, and foodstuffs. Some anobiid species develop on stored, dry plant products, and can be pests in stores, herbaria, and houses. Well-known cosmopolitan pest species include the Drugstore Beetle (*Stegobium paniceum*), the larvae of which feed on a wide range of organic materials, *Lasioderma serricorne*, the cosmopolitan Cigarette Beetle, which attacks stored tobacco, spices, and drugs, and *Anobium punctatum*, the Furniture or Cabinet Beetle, whose larvae (woodworm) can destroy structural wood and wooden objects of all kinds. The adult beetles normally emerge from pupa just below the surface of the wood during the summer. Sometimes, larvae boring through wooden shelving will continue through books (bookworms). The gut of all these larvae contains symbiotic yeasts which are responsible for breaking down the otherwise indigestible cellulose to sugars. The Deathwatch Beetle (*Xestobium rufovillosum*) gets its name from the sexual signals that both sexes of this species generate in late spring. They prefer hardwood timbers previously attacked by fungus, and the adults, bracing themselves with their legs, tap their heads against the substrate to attract mates. Males are able to orientate themselves to females by tapping and waiting for a reply before moving in the direction of the vibrations. The origin of the common name derives from the fact that the faint sounds can be heard in a quiet room or church where the dead are laid out.

Of all the phytophagous beetles that have had a significant impact on humans since agriculture first developed, the weevils (Curculionidae), with more than 86,000 species worldwide, are the most economically important.

These beetles are characterized by the possession of a snout-like extension of the head called the rostrum, which carries the jaws and the characteristically elbowed antennae.

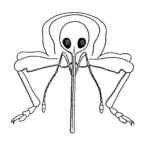

The front view of a weevil showing the elongate rostrum and the elbowed antennae.

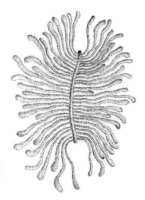

The patterns made by developing bark beetle larvae as they tunnel away from the egg chamber.

Sometimes, the rostrum can be as long as the rest of the body. Nearly all species are herbivorous, and the larvae, depending on the species, can be found eating every plant part from roots to seeds. Very many species are serious pests, and cause great economic losses, some of which deserve special mention. The Granary Weevil, *Sitophilus granarius*, has been recorded from the time of the Pharaohs. A related species, the Rice Weevil, *Sitophilus oryzae*, is also a serious pest of grains. *Anthonomus grandis*, the Cotton Boll Weevil, is a scourge in cotton plantations. As for many crops, there is an increasing demand for non-chemical pest control, and the use of parasitoids and other natural enemies is meeting with some success. Massive releases of the pteromalid wasp *Catolaccus grandis*, for example, have killed over 90% of Boll Weevils in certain areas.

Bark beetles (Scolytidae), which are closely related to weevils, are the most important pests of forests. Adult beetles bore into the cambium of trees just under the bark, where they fed and lay eggs. The developing larvae tunnel outwards from the egg chamber leaving characteristic tracks as they go.

The adults of some species (*Xyleborus* spp.) carry fungal spores, and inoculate the tunnels where the fungus grows, providing food. Bark beetles of the genus *Scolytus* were the vectors of the Dutch elm disease fungus (*Ceratostomella ulmi*) that killed elm trees in many parts of Britain and North America. Species of the genera *Dendroctonus* and *Ips* are widespread pests of coniferous plantations across the northern hemisphere.

The darkling beetles (Tenebrionidae) are a large family of around 20,000 species whose larvae are associated with soil, rotting wood, fungi, and litter. Some species, such as those of the genus *Tenebrio*, can be pests in grain, and species of *Tribolium* are pests of flour, cereals, dried fruit, and meal.

In the Namib desert, some tenebrionid species climb to the tops of dunes in the morning and dip their heads into the moist prevailing wind. Tiny droplets of water condense on their elytra, which are specially sculptured with hydrophobic grooves and hydrophilic bumps. These droplets coalesce to produce larger droplets that run forward. A bed of hairs at the junction of the prothorax and the head channels the water from the dorsal surface to the ventral surface and thus towards the mouth. This, one among many adaptations for survival, is why beetles are the most abundant and successful of all insects.

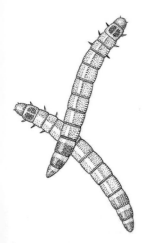

Some species, such as mealworms (of the genus *Tenebrio*), can be pests in grain.

Key reading

Bouchard, P., Bousquet, Y., Davies, A. E., Alonso-Zarazaga, M. A., Lawrence, J. F., Lyal, C. H., Newton, A. F., Reid, C. A., Schmitt, M., Slipiński, S. A., and Smith, A. B. (2011). Family-group names in Coleoptera (Insecta). *ZooKeys* **88**: 1–972.

Caltagirone, L. E. and Doutt, R. L. (1989). The history of the Vedalia beetle importation to California and its impact on the development of biological control. *Annual Review of Entomology* **34**: 1–16.

Crowson, R. A. (1981). *The biology of the Coleoptera*. Academic Press, London.

Dean, J., Aneshansley, D. J., Edgerton, H. E., and Eisner, T. (1990). Defensive spray of the bombardier beetle: a biological pulse jet. *Science* **248**: 1219–21.

Desender, K., Dufrêne, M., Loreau, M., Luff, M. L., and Maelfait, J.-P. (eds) (1994). *Carabid beetles: ecology and evolution*. Entomologica series **51**. Kluwer Academic Publishers, Dordrecht.

Di Giulio, A., Maurizi, E., Barbero, F., Sala, M., Fattorini, S., Balletto, E., and Bonelli, S. (2015). The pied piper: a parasitic beetle's melodies modulate ant behaviours. *PloS One* **10**: e0130541.

Elias, S. A. (1994). *Quaternary insects and their environment*. Smithsonian Press, Washington, DC.

Erwin, T. L., Ball, G. E., Whitehead, D. L., and Halpern, A. L. (eds) (1979). *Carabid beetles: their evolution, natural history and classification*. Junk Publishers, The Hague.

Evans, W. G. (1966). Perception of infrared radiation from forest fires by *Melanophila acuminata* De Geer (Buprestidae, Coleoptera). *Ecology* **47**: 1061–5.

Evans, M. E. G. (1973). The jump of the click beetle (Coleoptera: Elateridae)—energetics and mechanics. *Journal of Zoology London* **169**: 181–94.

Farrell, B. D. (1998). 'Inordinate fondness' explained: why are there so many beetles? *Science* **281**: 555–9.

Goulson, D., Birch, M. C., and Wyatt, T. (1994). Mate location in the Deathwatch beetle, *Xestobium rufovillosum* De Geer (Anobiidae): orientation to substrate vibrations. *Animal Behaviour* **47**: 899–907.

Hanski, I. and Cambefort, Y. (eds) (1991). *Dung beetle ecology*. Princeton University Press, Princeton, NJ.

Hare, J. D. (1990). Ecology and management of the Colorado Potato beetle. *Annual Review of Entomology* **35**: 81–100.

Hodek, I. and Honek, A. (eds) (1996). *Biology of Coccinellidae*. Entomologica series **54**. Kluwer Academic Publishers, Dordrecht.

Jolivet, P. H., Petitpierre, E., and Hsiao, T. H. (1988). *Biology of the Chrysomelidae*. Entomologica series **42**. Kluwer Academic Publishers, Dordrecht.

Jolivet, P. H., Cox, M. L., and Petitpierre, E. (1994). *Novel aspects of the biology of Chrysomelidae*. Entomologica series **50**. Kluwer Academic Publishers, Dordrecht.

Lawrence, J. F. and Newton, Jr., A. F. (1982). Evolution and classification of beetles. *Annual Review of Entomology* **13**: 261–90.

Lloyd, J. E. (1983). Bioluminescence and communication in insects. *Annual Review of Entomology* **28**: 131–60.

Lövei, G. L. and Sunderland, K. D. (1996). Ecology and behaviour of ground beetles (Coleoptera: Carabidae). *Annual Review of Entomology* **41**: 231–56.

Majerus, M. and Kearns, P. (1989). *Ladybirds*. Naturalists' handbooks **10**. Richmond Publishing, Slough.

Matthew, C. (2018). On the significance of insect remains and traces in archaeological interpretation. *Global Journal of Archaeology & Anthropology* **2**: 555–93.

Mizell III, R. F. (2007). Impact of *Harmonia axyridis* (Coleoptera: Coccinellidae) on native arthropod predators in pecan and crape myrtle. *Florida Entomologist* **90**: 524–36.

Morris, M. G. (1991) *Weevils*. Naturalists' handbooks **16**. Richmond Publishing, Slough.

Parker, J. (2016) Myrmecophily in beetles (Coleoptera): evolutionary patterns and biological mechanisms. *Myrmecological News* **22**: 65–108.

Pearson, D. L. (1988) Biology of tiger beetles. *Annual Review of Entomology* **33**: 123–47.

Pickering, G., Lin, J., Riesen, R., Reynolds, A., Brindle, I., and Soleas, G. Influence of *Harmonia axyridis* on the sensory properties of white and red wine. *American Journal of Enology and Viticulture* **55**: 153–9.

Robinson, M. (2001). Insects of palaeoenvironmental indicators. In: *Handbook of Archaeological Sciences*. Brothwell, D. R. and Pollard, A. M. (eds), pp. 119–31. Wiley, Chichester.

Shukla, S. P., Plata, C., Reichelt, M., Steiger, S., Heckel, D. G., Kaltenpoth, M., Vilcinskas, A., and Vogel, H. (2018). Microbiome-assisted carrion preservation aids larval development in a burying beetle. *Proceedings of the National Academy of Sciences* **115**: 11274–9

Stork, N. E., McBroom, J., Gely, C., and Hamilton, A. J. (2015). New approaches narrow global species estimates for beetles, insects, and terrestrial arthropods. *Proceedings of the National Academy of Sciences of the United States of America* **112**: 7519–23.

Strong, L. (1992). Avermectins: a review of their impact on insects of cattle dung. *Bulletin of Entomological Research* **82**: 265–74.

Vondran, T., Apel, K.-H., and Schitmz, H. (1995). The infrared receptor of *Melanophila acuminata* De Geer (Coleoptera: Buprestidae): ultrastructural study of a unique insect thermoreceptor and its possible decent from a hair mechanoreceptor. *Tissue and Cell* **27**: 645–58.

Strepsiptera
(strepsipterans)

Common name	Strepsipterans, twisted-wing insects	Metamorphosis	Complete (egg, larva, pupa, adult)
Derivation	Gk. *Streptos*–twisted; *pteron*–a wing	Distribution	Worldwide
		Number of families	10
Size	Body length 0.4–35 mm (usually under 6 mm)	Known world species	About 630 (0.056%)

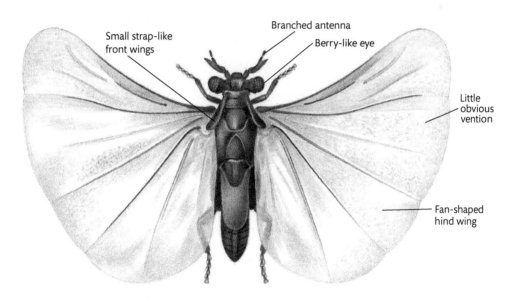

Small strap-like front wings

Branched antenna

Berry-like eye

Little obvious vention

Fan-shaped hind wing

Stylops melittae (male) – **Strepsiptera**

Essential Entomology. Second Edition. George C. McGavin and Leonidas-Romanos Davranoglou, Oxford University Press.
© George C. McGavin and Leonidas-Romanos Davranoglou (2022). DOI: 10.1093/oso/9780192843111.001.0001

Key features

- endoparasites of insects
- rarely seen
- sexes extremely sexually dimorphic

> A male *Stylops ater* mating with a female that is parasitising an *Andrena vaga* bee (female is the yellow spot at the end of the bee's abdomen). Netherlands.

© John Smit

The first members of this strange order that were described belonged to the family Stylopidae, and, for this reason, the common name 'stylops' is often used to refer to any strepsipteran species, and the insects they parasitize are described as having been 'stylopized'. However, the common name twisted-wing insects is becoming more widespread in recent years.

Strepsipterans are highly specialized endoparasites of other insects in at least seven orders and 34 insect families belonging to the orders Thysanura, Blattodea, Mantodea, Orthoptera, Hemiptera, Diptera, and Hymenoptera. There has even been a report of a microcaddisfly being used as a host. Despite this broad host range, the insects most commonly stylopized are bugs, wasps, and bees. Males and females may parasitize different insect hosts.

The order is divided into two suborders—the Mengenillidia (one small family, the Mengenillidae) and the Stylopidia (eight families)—as well as a family of uncertain subordinal placement, the Bahiaxenidae. In the Mengenillidae, the fully grown male and female larvae leave the hosts, which are species of silverfish (Lepismatidae), and pupation takes places outside. In this family, both adult males and females are free-living, a unique condition

© John Smit

⟩ A male *Stylops ater* at rest.
Netherlands.

⟩ neoteny
the retention of immature
characteristics in the adult
stage

among the Strepsiptera. The males are winged, but the females are **neotenous**, grub-like, and wingless, although they possess legs and antennae.

The females of all other Strepsiptera are totally endoparasitic, as they never leave the confines of their host's body, and they are surrounded by the cuticle of their own pupal stage. The **females do not move and have no eyes, antennae, mouthparts, legs, wings, nor external genitalia.** They do not go through a pupal stage, and remain as legless larviform adults, their body almost entirely filled with eggs. The only bit of them that is visible and protrudes through the body wall of the host is the relatively featureless, chitinized cephalothorax. **Adult males**, on the other hand, are **dark coloured, free-living, winged**, do not feed, and are short lived (typically for just a few hours). Their heads have **prominent, raspberry-like eyes** and **strange, branched antennae**, and the **thorax bears a pair of small strap-like front wings** and much larger, **fan-shaped hind wings with little obvious venation**. The legs of some males lack claws. Interestingly, the front wings of male strepsipterans, which are reduced to flattened stalk-like or strap-like structures, have been shown to function in a manner analogous to dipteran halteres. As the front wings oscillate rapidly in flight, any rotational movement of the insect will induce inertial (Coriolis) forces in the wings, perpendicular to the plane of oscillation. These forces are thought to be sensed by two fields of campaniform sensillae at the bases of the front wings, and the nerves from these run to the thoracic ganglia. Mechanical stimulation of the front wings produces compensatory movements of the head and abdomen.

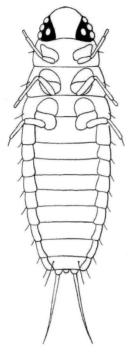

∧ Mobile first instar larva (triungulin) of a stylopid.

Normal male

Stylopized male

Normal female

Stylopized female

∧ Reversal of secondary sexual characteristics in a bee (*Andrena chrysoceles*) parasitized by a strepsipteran.

Males are able to detect and respond to sexual pheromones emitted by virgin females and, after locating them, the male clings to the body of the host and mates with her. The precise way mating is achieved in Strepsiptera is contentious, but the few known examples are fascinating. In Mengenill-idae, the male uses his sharp penis to pierce the female in any body part other than the head, and ejaculates directly in her blood stream (haemo-coel), where fertilization takes place. At least in some Stylopidia, mating also takes place via traumatic insemination, but the male pierces a specific invagi-nation of the female cephalothorax to inject his semen. This invagination is specifically adapted for this purpose, and likely serves to protect the fe-male from the damage caused by being pierced by the male's penis. Other Stylopidia may mate somewhat more conventionally by pushing their penis through an opening in the pupal skin that surrounds the female, although this requires further confirmation. The production of the male-attracting pheromone ceases when copulation takes place, and in some species the males are able to detect whether a female has mated previously with more than one partner. One species avoids the hassle of being marauded by males by following a parthenogenic approach, where only females are born. The num-ber of eggs a female strepsipteran produces varies from about a thousand to many hundreds of thousands, depending on her size, which has been deter-mined by the size of the host insect inside which she has developed. In one species, a single egg may give rise to more than one embryo by subdividing (polyembryony). The first instar larvae develop within the body of the female and emerge via the brood canal opening around the cephalothorax. Females may give birth to many thousands of tiny, six-legged, first instar larvae called primary larvae (also known by certain authors as triungulins or planidia), whose main job is to find immature stages of the appropriate host insect species.

If their hosts are exopterygote insects, such as grasshoppers or bugs, the primary larvae are shed as the host moves about. In the case of species that parasitize social Hymenoptera, such as wasps and bees, the primary larvae emerge from the brood canal of their mother when the host visits a flower. Here they will sit, ready to grab hold of a suitable adult who will unwittingly transport them back to the nest, where they will find the right sort of eggs or larvae inside which to develop. The primary larvae produce enzymes that soften the cuticle of the host and, once inside, they moult immediately to a legless endoparasite. After five or more grub-like instars, the fully grown lar-vae pupate, with one end of the pupa sticking out from the body wall of the host. If the pupa contains a male, the cephalothorax end of the pupal cuticle is pushed off and the male flies away, whereas females remain *in situ*. Hosts of males usually die when the strepsipteran emerges but hosts of females live on until she has mated and given birth. In all cases, hosts suffer what is known as parasitic castration and are unable to reproduce. However, stylopized in-sects live longer than usual, and have larger fat stores as well, which allow their strepsipteran parasites to complete their life cycle and have abundant resources throughout their development. Furthermore, stylopized social in-sects like wasps display altered gene expression patterns in their brain, which suggests that Strepsiptera are able to manipulate the behaviour of their host for their own benefit.

The type of host and the position in which the female strepsipteran pupates is characteristic of particular families. Members of the Elenchidae are often found on grasshoppers; Halticophagidae on treehoppers, leafhoppers, spittle bugs, and mole crickets; and Stylopidae on bees (Adrenidae and Halictidae) and wasps (Sphecidae and Vespidae).

> *Stylops* sp. (to same scale)

Female in abdomen of bee

Male

In most known life cycles, both males and females of any strepsipteran species parasitize the same host species. However, there are some exceptions, where males and female utilize very different hosts. The family Myrmecolacidae contains a little over 120 species, mostly known from the collection of male specimens. The females of only a handful of species have been found, and, in the few cases where males and females have been linked, they have been found in completely different hosts; the females attacking mantids, grasshoppers, or crickets, and the males using ants as their hosts. The differences in size, both of the sexes and their respective hosts, is remarkable. Male myrmecolacids are less than 1.5 mm in length, whereas the females are much larger, some species, ranging in length from 20–35 mm, being the largest strepsipterans in the world. The females are endoparasitic in several species of phytophagous cricket, belonging to the genera *Segestidea* and *Segetes* that can cause damage to oil and coconut palms in Papua New Guinea. The introduction of a strepsipteran, *Stichotrema dallatorreanum*, to islands and areas where it is absent may be an effective means of biologically controlling pest crickets in plantations. One of the effects of stylopization in female crickets concerns the host's eggs. More than half of the eggs from parasitized females are darker and misshaped compared to normal eggs. In addition, electron microscopic examination of the eggs reveals that their shells lack the fine structures necessary for normal gaseous exchange. In this way, the populations of these pest crickets are greatly reduced.

One species that has a significant role in pest control is *Elenchus japonicus* (Elenchidae). This species is parasitic in two important pests of rice in Asia, the White-backed Planthopper, *Sogatella furcifera*, and the notorious Brown Planthopper, *Nilaparvata lugens*. The planthoppers are obviously killed when males emerge, but the effects of stylopization on the host are greater because the genitalia of the bugs are affected, or sometimes even lacking. The planthoppers, therefore, cannot reproduce, but are still able to disperse to new areas, further spreading the parasite. The species *Stichotrema dallatorreanum* is also used as a pest control agent in Papua New Guinea against bush crickets, as they are parthenogenic and can be readily mass reared.

Due to the highly modified morphology of both sexes of Strepsiptera, the relationships of this order had been highly controversial ever since its discovery. Due to morphological characteristics these insects share with some families of beetles, such as the parasitic Rhipiphoridae, the Strepsiptera have sometimes been regarded as a superfamily of the Coleoptera. Early molecular sequence data instead indicated a close relationship with the Diptera. However, detailed morphological and phylogenomic studies have conclusively demonstrated that Strepsiptera are the sister group to Coleoptera, and any similarities with the Rhipiphoridae arose independently, due to their similar parasitic lifestyle.

The fossil record of Strepsiptera is relatively scarce, with the oldest fossils of either adult males or primary larvae dating to the Cretaceous. This shows that this group has been following a remarkably constant parasitic way of life for at least 100 million years. Molecular estimates suggest that Strepsiptera arose at least 120 mya, although the ultimate origin of the group may go back to the Permian, when their lineage split from the common ancestor of beetles.

Key reading

Beani, L., Dallai, R., Cappa, F., Manfredini, F., Zaccaroni, M., Lorenzi, M. C., and Mercati, D. (2021). A Strepsipteran parasite extends the lifespan of workers in a social wasp. *Scientific Reports* **11**: 7235.

Geffre, A. C., Liu, R., Manfredini, F., Beani, L., Kathirithamby, J., Grozinger C. M., and Toth, A. L. (2017). Transcriptomics of an extended phenotype: parasite manipulation of wasp social behaviour shifts expression of caste-related genes. *Proceedings of the Royal Society B* **284**: 20170029.

Kathirithamby, J. and Hamilton, W. D. (1992). More covert sex: the elusive females of the Myrmecolacidae. *Trends in Ecology and Evolution* 7: 349–51.

Kathirithamby, J. (2009). Host-parasitoid associations in Strepsiptera. *Annual Review of Entomology* **54**: 227–49.

Kathirithamby, J. (2018). Biodiversity of Strepsiptera. In: *Insect biodiversity: science and society*, volume II. Foottit, R. G. and Adler, P. H. (eds), pp. 673–703. John Wiley & Sons Ltd., Hoboken.

Kinzelbach, R. (1990). The systematic position of the Strepsiptera (Insecta) *American Entomologist* **36**: 292–303.

Kinzelbach, R. and Pohl, H. (1994). The fossil Strepsiptera (Insecta: Strepsiptera). *Annals of the Entomological Society of America* **87**: 59–70.

Misof, B., Liu, S., Meusemann, K., Peters, R. S., Donath, A., Mayer, C., Frandsen, P. B., et al. (2014). Phylogenomics resolves the timing and pattern of insect evolution. *Science* **346**: 7639–7767.

Niehuis, O., Hartig, G., Grath, S., Pohl, H., Lehmann, J., Tafer, H., Donath, A., Krauss, V., Eisenhardt, C., Hertel, J., Petersen, M., Mayer, C., Meusemann, K., Peters, R. S., Stadler, P. F., Beutel, R. G., Bornberg-Bauer, E., McKenna D. D., and Misof, B. (2013). Genomic and morphological evidence converge to resolve the enigma of Strepsiptera. *Current Biology* **23**: 1388.

Peinert, M., Wipfler, B., Jetschke, G., Kleinteich, T., Gorb, S. N., Beutel, R. G., and Polh, H. (2016). Traumatic insemination and female counter-adaptation in Strepsiptera (Insecta). *Scientific Reports* 6: 25052.

Pix, W., Nalbach, G., and Zeil, J. (1993). Strepsipteran forewings are haltere-like organs of equilibrium. *Naturwissenschaften* **80**: 371–4.

Pohl, H., Batelka, J., Prokop, J., Müller, P., Yavorskaya, M. I., and Beutel, R. G. (2018). A needle in a haystack: Mesozoic origin of parasitism in Strepsiptera revealed by first definite Cretaceous primary larva (Insecta). *PeerJ* **6**: e5943.

Solulu, T. M., Simpson, S. J., and Kathirithamby J. (1998). Effects of strepsipteran parasitism on a tettigoniid pest of oil palm in Papua New Guinea. *Physiological Entomology* **38**: 388–98.

Antliophora

This superorder comprises scorpionflies (Mecoptera), fleas (Siphonaptera), and true flies (Diptera). While scorpionflies are modestly diverse, flies are among the richest insect orders, with at least 158,000 species.

Essential Entomology. Second Edition. George C. McGavin and Leonidas-Romanos Davranoglou, Oxford University Press.
© George C. McGavin and Leonidas-Romanos Davranoglou (2022). DOI: 10.1093/oso/9780192843111.001.0001

Mecoptera
(scorpionflies)

Common name	Scorpionflies	Metamorphosis	Complete (egg, larva, pupa, adult)
Derivation	Gk. *Mekos*-long; *pteron*-a wing		
		Distribution	Worldwide
Size	Body length 3-28 mm	Number of families	9
		Known world species	About 750 (0.066%)

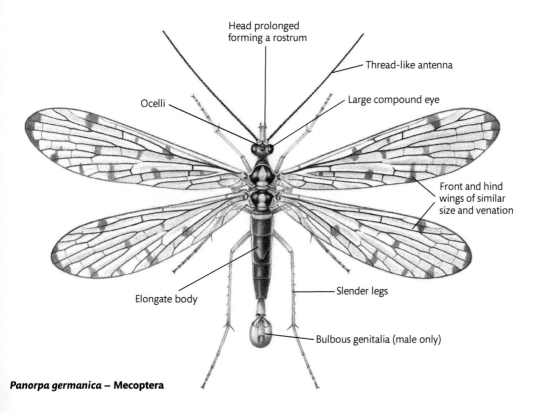

Head prolonged forming a rostrum

Thread-like antenna

Ocelli

Large compound eye

Front and hind wings of similar size and venation

Elongate body

Slender legs

Bulbous genitalia (male only)

Panorpa germanica – **Mecoptera**

Essential Entomology. Second Edition. George C. McGavin and Leonidas-Romanos Davranoglou, Oxford University Press.
© George C. McGavin and Leonidas-Romanos Davranoglou (2022). DOI: 10.1093/oso/9780192843111.001.0001

Key features

- distinctive elongated face
- abdominal tip of males often resembling a scorpion sting
- mostly in damp wooded areas

> A male scorpionfly browsing on pollen. The scorpion-like abdominal tip of many mecopterans gives them their common name. *Panorpa communis,* Hungary.

© Anne Riley

The common name of the order is derived from the occurrence in the males of some species of swollen genitalia, which are positioned at the end of a slender abdomen and held curved upwards in a scorpion-like fashion.

Many features of these insects, such as the two pairs of equally sized and richly veined membranous wings, render them superficially similar to neuropterans, but, in fact, they are most closely related to the Diptera and Siphonaptera. Some mecopteran families (Choristidae, Nannochorsitidae, Notiothaumidae, and Meropeidae) comprising only dozen or so primitive species are found only in the southern hemisphere. The Bittacidae (hang-ingflies), a larger family, is also confined to southern continents, while the Panorpidae (common scorpionflies) and the Boreidae (snow scorpionflies or snow fleas) occur only in the northern hemisphere. With the exception of boreids, scorpionflies are generally found in damp, wooded, or shady areas

Female

Male

^ Details of the abdominal
tips of male and female
scorpionflies showing
external genitalia.

with an abundance of vegetation. The body is generally elongate and cylindrical, and the **large, narrow wings** may be clear or marked with spots or bands of dark colour. Some species are short-winged or completely wingless. The head has **large compound eyes, three ocelli, and is characteristically extended downwards in the form of a beak or rostrum**, at the end of which are **biting mouthparts. The antennae, which are thread-like and have as many as 60 segments,** are typically longer than half the length of the body. The legs are long and slender in most species, but, in the hangingflies (Bittacidae), they are very long indeed and used for prey capture. Bittacids, which resemble crane flies (Tipulidae), hang from vegetation, holding on with their front legs, or fly, trailing their slender, hind legs in the air. The fifth tarsal segments of the hind legs are enlarged and raptorial to seize small insects. The majority of other scorpionflies, typified by the Panorpidae, feed on dead or dying insects, and will also consume carrion, nectar, sap flows, and fruit juice. There may be intense interspecific competition for food, and it is the larger species which get the biggest share. An unusual feeding strategy, seen in panorpids, is the robbing of prey from orb webs. These scorpionflies are able to walk over webs to feed on insects that have been trapped, or even feed on silk-wrapped insects that the spider has stored. Despite being careful, they do sometimes get trapped, but are able to free themselves using salivary secretions. Snow scorpionflies (Boreidae) feed on mosses, and members of the family Panorpodidae are herbivorous. Unusually, the larvae of the southern hemisphere family Nannochoristidae are aquatic and live around the margins of lakes and in swampy areas, where they prey on larval non-biting midges (Chironomidae).

Prior to mating, which often takes place after dark, Mecoptera frequently engage in complex courtship rituals that involve slow wing movements that display species-specific wing pigmentation patterns, as well as rhythmic abdominal vibrations, which generate low-frequency vibrational songs. Mating often involves pheromones and the offer of nuptial gifts. In some species, sexual pheromones only attract conspecific females, while in others, both males and females are attracted. Mating behaviour has been studied in depth in a few species, mostly in the genus *Panorpa*. Males use a variety of courtship techniques, of which the commonest is the offering of gifts, which may take the form of a dead insect, or a mass of saliva produced by the males which is stuck to the surface of a leaf. The male then emits his sexual pheromone to attract a mate to his offering. In scorpionflies, size is everything, and females will quickly reject males whose gifts are too small. In the case of salivary masses, the size of the gift is a good indication of how well the male has fed prior to mating. As would be expected, females prefer males who provide large nutritious gifts and, as a result, they are able to lay more eggs. Having accepted the gift, the female will feed on it during the process of copulation, which may last more than an hour. Competition for females, and the gifts to offer them, can be fierce. Males who do not find good enough gifts may take females by force, and after grabbing a potential mate using their genital claspers, they employ the notal organ (a clamp-like structure located on the dorsal surface of the abdomen) to restrain one of the female's front wings during mating. The no-tal organ is also used during normal mating. Males will often try to wrestle rival males from females while they are copulating. Females may mate with several males during their adult life. In hangingflies (*Harpobittacus* and *Bitta-cus*), the picture is further complicated by males stealing each other's nuptial gifts, either aggressively or by stealth. In the latter case, a male will imitate the behaviour of a responsive female, and before the duped male discovers his mistake, they seize the nuptial gift for their own use. In some species, part of the male genitalia takes the shape of a long, coiled 'penisfillum', and it is likely this structure might be used to remove sperm from previous matings.

Eggs are laid in small groups in the soil. The larvae of most species re-semble small caterpillars or beetle grubs. The larvae usually have compound eyes (not a collection of simple eyes), which vary in the number of om-matidia from fewer than 10 to more than 30, but some species are eyeless. There are three pairs of thoracic legs and a number of abdominal prolegs in those larvae resembling caterpillars (Panorpidae, Bittacidae, and Cho-ristidae). The soil surface-dwelling larvae of some hangingflies are able to camouflage themselves by eating soil and then covering themselves with the excrement produced. After three or four moults, pupation takes place in an underground cell, or among damp soil or vegetation. The pupae can move their jaws, and their legs and antennae are not stuck down to the body surface.

Snow scorpionflies (Boreidae) are dark brown or black in colour and can be seen in autumn and winter on the surface of snow, among mosses, or under stones, particularly in cold or mountainous regions. These insects are rarely encountered, but, when present, are noticeable on account of their colour and rapid walking. Males use their reduced, hook-like wings to hold and carry the females during mating. The eggs are laid in moss, and the larvae, which have well-developed thoracic legs, feed on moss and other plant matter. Su-perbly adapted for life in cold conditions, these insects contain antifreeze compounds and can tolerate temperatures down to −6°C, and, when walking on the surface of snow, the body of species such as *Boreus brumalis* absorbs radiation from the sun. When threatened, snow scorpionflies use a catapult-like mechanism on their mid- and hind legs to make long-distance jumps that allow them to escape from potential predators.

Scorpionflies are a particularly old order, with fossils dating to the Early Permian (about 290 mya). It appears that extinct mecopterans were up to three times more diverse than their living relatives, although this could be artefactual, as many of these fossils might represent early Antliophora and not true Mecoptera. The relationships of scorpionflies are debated, especially with regards to fleas (Siphonaptera), which may represent their sister group, or a lineage within Mecoptera (discussed in more detail in Siphonaptera (fleas)). What is certain, is that flies (Diptera) are the sister group to Mecoptera and Siphonaptera.

The economic importance of Mecoptera is negligible, although they might have a useful ecological role as recyclers in many ecosystems.

Key reading

Burrows, M. (2011). Jumping mechanisms and performance of snow fleas (Mecoptera, Boreidae). *Journal of Experimental Biology* **214**: 2362–74.

Byers, G. W. and Thornhill R. (1983). Biology of the Mecoptera. *Annual Review of Entomology* **28**: 203–28.

Courtin, G. M., Shorthouse, J. D., and West, R. J. (1984). Energy relations of the snow scorpionfly *Boreus brumalis* (Mecoptera) on the surface of snow. *Oikos* **43**: 241–5.

Hartbauer, M., Gepp, J., Hinteregger, K., and Koblmüller, S. (2015). Diversity of wing patterns and abdomen-generated substrate sounds in 3 European scorpionfly species. *Insect Science* **22**: 521–31.

Penny, N. D. (1975). Evolution of the extant Mecoptera. *Journal of the Kansas Entomological Society* **48**: 331–50.

Ren, D. and Shih, C. K. (2005). The first discovery of fossil eomeropids from China (Insecta, Mecoptera). *Acta Zootaxonomica Sinica* **30**: 275–80.

Thornhill, R. (1980). Rape in *Panorpa* scorpionflies and a general rape hypothesis. *Animal Behaviour* **28**: 52–9.

Thornhill, R. (1981). *Panorpa* (Mecoptera: Panorpidae) scorpionflies: systems for understanding resource-defense polygyny and alternative male reproductive efforts. *Annual Review of Ecology and Systematics* **12**: 355–86.

Thornhill, R. and Sauer, K. P. (1991). The notal organ of the scorpionfly (*Panorpa vulgaris*): an adaptation to coerce mating duration. *Behavioural Ecology* **2**: 156–64.

Willman, R. (1987). The phylogenetic system of the Mecoptera. *Systematic Entomology* **12**: 519–24.

Zhong, W. and Hua, B. (2013). Mating behaviour and copulatory mechanism in the scorpionfly *Neopanorpa longiprocessa* (Mecoptera: Panorpidae). *PloS ONE* **8**: e74781.

Siphonaptera
(fleas)

Common name	Fleas	Metamorphosis	Complete (egg, larva, pupa, adult)
Derivation	Gk. *Siphon*–pipe, tube; *a+pteron*–wingless	Distribution	Worldwide
Size	Body length 1–8 mm. Mostly under 5 mm	Number of families	19
		Known world species	About 2,190 (0.19%)

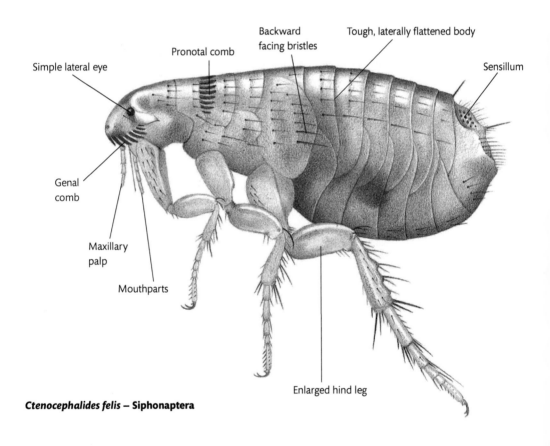

Ctenocephalides felis – **Siphonaptera**

Essential Entomology. Second Edition. George C. McGavin and Leonidas-Romanos Davranoglou, Oxford University Press.
© George C. McGavin and Leonidas-Romanos Davranoglou (2022). DOI: 10.1093/oso/9780192843111.001.0001

Key features

- small, wingless ectoparasites on mammals and some birds
- blood feeders
- characteristic jumping ability
- many species are vectors of disease

© Martin Gore

❯ Cat fleas do not live permanently on humans but will take blood meals. *Ctenocephalides felis*, UK.

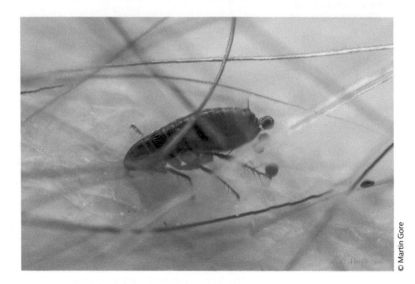

© Martin Gore

❯ While feeding, fleas excrete droplets of digested blood. These are among the principal food items for flea larvae. UK.

Found wherever there are suitable hosts, even in Arctic and Antarctic regions, fleas are a distinctive and readily recognizable group. The vast majority of flea species are parasitic on land mammals (well over 90%), the remainder being attached to bird species. Aquatic mammals are flea-free, as are roaming or thick-skinned mammals, such as elephants and rhinoceroses. As holometabolous insects, the larvae are very different from the adults, and have a different feeding ecology. Fleas have evolved as parasites of animals that regularly use lairs, burrows, dens, or nests, because it is here that the larvae feed and develop. There are, however, some strange and fascinating exceptions to this rule such as the Arctic Hare Flea, *Euhoplopsyllus glacialis* (Pulicidae), where the larvae live in the host's fur, and the Tasmanian Devil Flea (although it is found also in quolls and the extinct thylacine), *Uropsylla tasmanica* (Pygiopsyllidae), where the eggs are stuck to the hair of the host, and the larvae actually develop inside the host's skin. As adults, these blood-sucking insects are wingless and highly modified for an ectoparasitic life on the bodies of their hosts. The **tough, shiny, brownish or reddish body is streamlined and flattened from side to side with a variety of backward-pointing spines and bristles.** Comb-like structures on the cheeks on the posterior edge of the pronotum of many species, **genal and pronotal ctenidia**, respectively, are very useful identification features. It is thought that ctenidia prevent the flea from being pushed backwards and help it stay on the host's fur.

Fleas living on spiny hosts, such as porcupines and hedgehogs, have fewer, stouter, and more widely spaced comb teeth than those living on furry mammals, as their hosts cannot effectively scratch themselves. The head typically has very **short, three-segmented antennae**, which fit into special grooves, and **short, robust mouthparts** for piercing skin and sucking blood. A spring-like system allows the stylets to be rapidly and repeatedly driven through the skin to reach the blood vessels just below the surface. Fleas may have a **pair of simple, lateral eyes** similar to ocelli, and are negatively phototactic, but many nest-living species are blind. The legs are stout, and the **enlarged hind legs** are part of the flea's unique and powerful jumping mechanism. Before the flea jumps, the hind legs are flexed and rotated by the trochanter-levator muscles past a pivot point into a cocked position. In this position, a block of highly elastic rubber-like protein called resilin, located between the pleuron and the notum, is compressed and acts as an energy store. The jump is started by contraction of the trochanter-depressor muscles, which bring the legs back again, past the pivot point. At this stage, the energy from the resilin block is released and, together with the force generated from the contracting jumping muscles (notum-trochanter muscle), the legs are rotated rapidly and down against the substrate. Many fleas can jump considerable distances—the Cat Flea can achieve a high jump of over 30 cm (12 in) at up to 130 times the acceleration of gravity. Hungry fleas may jump hundreds of times an hour for several days to find a host.

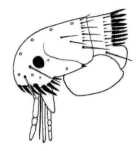

^ **Head of the Dog Flea** (*Ctenocephalides canis*).

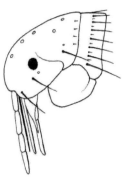

^ **Head of the Human Flea** (*Pulex irritans*) lacking genal and pronotal ctenidia.

The dorsal surface of the last visible abdominal segment bears a **conspicuous sensillum**, which can be flat or rounded in outline. This plate-like structure is covered with a variable number of sensory hairs arising from dome-like bases. Although it certainly has a sensory role, it is hard to believe that the exact function of such a large structure has remained unknown.

Unlike lice, which live permanently on their hosts, most fleas do not live on their host all the time. As adults, many species feed intermittently, and respond to the presence of an animal by sensing movement, temperature, humidity, carbon dioxide, and odours. Some species, collectively known as nest fleas (as compared with fur fleas which spend most of their time on the host), only move onto a host to feed and spend the rest of their time in the lair or burrow. Fleas regularly move from host to host, and in some species, can survive for well over a year without a blood meal. The life cycle is relatively simple in most species. The eggs are dropped from the fur or laid in the host's burrow, lair, or nest. The blind, pale, and worm-like larvae feed on a range of organic materials, which include dried blood and the excrement of the adult fleas, which they sometimes take directly from the anus. Incredibly, the larva of one species, The European Rat Flea (*Nosopsyllus fasciatus*), is known to beg for food by biting an adult on one of its posterior bristles. This action stimulates the flea to release a droplet of liquid excrement. Flea larvae do not have any thoracic legs or abdominal prolegs and move around by wriggling. There are three larval instars, and pupation takes place within a silken cocoon, which typically has bits of debris and small particles woven into its structure. Adult fleas will not emerge immediately once pupation is over; they will usually remain dormant inside until vibrations or other cues alert them to the presence of a host nearby. Workers clearing empty houses or new owners moving into property where pets have lived previously can be attacked by thousands of starving fleas that have matured in the absence of hosts. The life cycles of some fleas can be dependent on the development of their hosts, such as fleas whose hosts return annually to nesting sites. In some cases, the levels of hormones associated with the pregnancy of a host animal will stimulate the flea to reproduce in anticipation of new hosts.

Ceratophyllidae is the biggest of the 19 flea families, comprising about 20% of all flea species. More than 80% of the 455 species in this family are associated with small rodents, and the remainder with birds. Insectivorous bats are hosts to about 127 species that belong to the Ischnopysllidae. The species belonging to each genus of bat flea are confined to host species within a single bat family. Most bat fleas are found on cave-dwelling bats, where the environment provides a living for the larvae. Some specialized species in the family Malacopsyllidae use armadillos as hosts.

About 165 species make up the Pulicidae (common fleas), which, although a relatively small family, contains many species of medical or veterinary importance (see Box). This family includes species parasitic on dogs, cats, humans, hedgehogs, and porcupines, as well as many other carnivores and rodents.

Some pulicids are cosmopolitan in distribution, and among the best-known species are the Cat Flea (*Ctenocephalides felis*) (also a worldwide pest of dogs), the Dog Flea (*Ctenocephalides canis*), and the Human Flea (*Pulex irritans*). Apart from being very itchy and irritating, the bites of fleas can cause allergic reactions and transmit a variety of diseases and parasitic worms. Pet owners may only find a few dozen adult fleas on the bodies of their animals, but many tens of thousands of eggs, larvae, and pupae will be present in the home.

Fleas can act as vectors of a number of bacterial, rickettsial, and viral diseases of animals and humans (see Box). The best-known disease is Black Death (or bubonic plague) caused by the bacterium *Yersinia pestis*, and is actually primarily a disease of rodents, which can be passed to humans when the fleas' hosts die, and the fleas are forced to seek blood from other sources. A flea feeding on an infected brown or black rat (or other rodent) might pick up the bacteria in a blood meal. These bacteria multiply in the midgut of the flea, and, in some cases, may completely block the tract. When such a flea attempts to feed on the next host—rodent or human—the powerful pharyngeal pump tries to work but cannot pass the blood meal into the to gut. When it stops trying to feed, pressure and elastic forces cause the blood in the oesophagus, which is now mixed with some of the plague bacterial culture, to be pumped back into the host. In this way, a blocked flea may infect many hosts before it dies of starvation. There have been three great world outbreaks or pandemics of plague. The first is believed to have originated from Central Asia and spread through North Africa to Europe during AD 550–600. The second pandemic in the Middle Ages is thought to have caused 75 million human deaths throughout Europe and Asia. Outbreaks of the third plague pandemic in parts of India, Asia, Africa, and South America took place in the twentieth century, and plague can still occur today under certain conditions. The disease and its effects have been considerably reduced by the use of pesticides and modern drugs. Flea species in 50 genera across several families are vectors of plague in many parts of the world. Seventeen species in the genus *Xenopsylla* (Pulicidae) are disease vectors, the best-known of which is the cosmopolitan Rat or Plague Flea (*Xenopsylla cheopis*).

Stick-tight fleas belonging to the genus *Echidnophaga* (Pulicidae) are unusual in that they remain firmly attached to the skin of their host by their needle-like mouthparts. About 23 species are known, of which, one (*Echidnophaga gallinaceus*) is widespread and a serious pest of poultry. Even more unusual than stick-tight fleas are 13 species of sand-fleas, jiggers, or chigoes belonging to the Tungidae. Originally, an American species, *Tunga penetrans* is believed to have been spread to tropical Africa by the Atlantic slave trade. Another tungid species, common on rats, occurs in the eastern parts of China. Walking about in bare feet in dry, sandy areas of tropical Africa and Central and South America is not generally a good idea as, quite apart from the dangers of sharps and thorns, *Tunga penetrans* is a significant human parasite. The tiny females burrow into the skin of the foot, often under toenails, and remain encysted there, whereas males are free living. Gaseous exchange, excretion, mating, and egg-laying all take place through a small aperture at the rear of the embedded female. After fertilization, the abdomen swells enormously as the eggs develop, and the legs degenerate. Jiggers can grow to the size of pea causing pain, intense itching, and inflammation. In some cases, infection of the skin lesions can be serious and severe ulceration can lead to the loss of toes and even death through blood poisoning or tetanus infection.

Some fleas of medical or veterinary importance

Family	Species	Medical or veterinary importance
Pulicidae	*Xenopsylla cheopis*	Widespread vector of plague and murine or endemic typhus in rats and humans
	Ctenocephalides felis	Cat Flea
	Ctenocephalides canis	Dog Flea. Intermediate host for *Dipylidium caninum*, a tapeworm of cats, dogs and, sometimes, humans
	Pulex irritans	Human Flea*
Tungidae	*Tunga penetrans*	Jiggers, chigoes, sand fleas. Cause tungiasis in humans
Ceratophyllidae	*Ceratophyllus gallinae*	European Chicken Flea

* Although called the Human Flea, this species is also commonly found on goats, pigs, and some other animals such as foxes and badgers. Early humans would have quickly acquired this flea from another species, perhaps a previous occupant of a cave or shelter. *Pulex irritans* can act as a vector of bubonic plague and has also been implicated in the transmission of a number of other human diseases. Fleas have been implicated in the transmission of tularaemia and many other diseases than those already mentioned.

Host specificity in fleas is less strong than that of lice, and many species will readily feed on a range of related, or even unrelated, hosts. Some flea species may be found on more than 30 host species, and some unfortunate animals are host to more than 20 different flea species. The main reason for this is that, being holometabolous insects, where the larval stages have a totally different ecology, it is the nest or lair where the larvae develop, rather than the host, that is important to the flea's survival. Even though the host range may be broad, in some cases, where a flea is forced to take blood meals from a species that is not normally used as a host, they may lay fewer eggs. After killing and eating a rabbit, the ears of a domestic cat may become encrusted with dozens of feeding rabbit fleas, but even if they laid eggs, it is unlikely that the cat's bedding would provide as suitable a larval habitat as a rabbit burrow.

Insecticidal control of adults combined with treatment of the resting place and bedding of domestic animals is common, but the use of growth regulators, such as lufenuron, in feedstuffs is very effective. Adult fleas take up the substance from the blood and it is transferred to the eggs where it interferes with the formation of chitinous structures in the larvae.

Myxoma virus (*Fibromavirus myxomatosis*) causes the disease myxomatosis in rabbits and is spread by blood-sucking vector insects (e.g., mosquitoes). In Britain, the main vector in a control programme in the 1950s was the European Rabbit Flea (*Spilopysllus cuniculi*). This species was also introduced to Australia to try to control the huge rabbit populations, the rabbit having been introduced previously by European settlers for meat and skins.

The fossil record of fleas is sparse and challenging to interpret, as it is difficult to distinguish what is a true flea or an extinct unrelated ectoparasite that simply looks like siphonapteran without actually being one. The earliest unambiguous flea fossil is *Tarwinia australis* from the Early Cretaceous of Australia. Molecular studies also support a Cretaceous origin for fleas and suggest that they diversified on mammal hosts and infected birds and monotremes secondarily.

The relationship of fleas to other Antliophora has remained the last unsolved puzzle in reconstructing the insect tree of life. A close relationship of fleas to Mecoptera is supported by the vast majority of morphological and molecular studies. However, the great debate lies in whether fleas represent the sister group to Mecoptera, or a lineage within the latter order. If fleas are highly modified Mecoptera, then the rank of Siphonaptera as a distinct order is no longer necessary. However, phylogenomic data are currently conflicting. One study suggests with confidence that fleas originate within Mecoptera, and are most closely related to Nannochoristidae, while another study suggests that both phylogenomic and morphological data are inconclusive for a flea–nannochoristid relationship. At the moment, it might be prudent to maintain Siphonaptera as a distinct order until more data are available, and it is this approach we follow in this book.

Key reading

Bennet-Clark, H. C. and Lucey E. C. A. (1967). The jump of the flea: a study of the energetics and a model of the system. *Journal of Experimental Biology* **47**: 59–76.

Bibikova, V. A. (1977). Contemporary views on the interrelationships between fleas and the pathogens of animal and human diseases. *Annual Review of Entomology* **22**: 23–32.

Damgaard, P. D. B., Marchi, N., Rasmussen, S., Peyrot, M., Renaud, G., Korneliussen, T., Moreno-Mayar, J. V., Pedersen, M. W., Goldberg, A., Usmanova, E., Baimukhanov, N., Loman, V., Hedeager, L., Pedersen, A. G., Nielsen, K., Afanasiev, G., Akmatov, K., Aldashev, A., Alpaslan, A., Baimbetov, G., Bazaliiskii, V. I., Beisenov, A., Boldbaatar, B., Boldgiv, B., Dorzhu, C., Ellingvag, S., Erdenebaatar, D., Dajani, R., Dmitriev, E., Evdokimov, V., Frei, K. M., Gromov, A., Goryachev, A., Hakonarson, H., Hegay, T., Khachatryan, Z., Khaskhanov, R., Kitov, E., Kolbina, A., Kubatbek, T., Kukushkin, A., Kukushkin, I., Lau, N., Margaryan, A., Merkyte, I., Mertz, I. V., Mertz, V. K., Mijiddorj, E., Moiyesev, V., Mukhtarova, G., Nurmukhanbetov, B., Orozbekova, Z., Panyushkina, I., Pieta, K., Smrčka, V., Shevnina, I., Logvin, A., Sjögren, K. G., Štolcová, T., Taravella, A. M., Tashbaeva, K., Tkachev, A., Tulegenov, T., Voyakin, D., Yepiskoposyan, L., Undrakhbold, S., Varfolomeev, V., Weber, A., Wilson Sayres, M. A., Kradin, N., Allentoft, M. E., Orlando, L., Nielsen, R., Sikora, M., Heyer, E., Kristiansen, K., and Willerslev, E. (2018). 137 ancient human genomes from across the Eurasian steppes. *Nature* **557**: 369–74.

Hirst, L. F. (1953). *The conquest of plague. A study of the evolution and epidemiology.* Clarendon Press, Oxford.

Holland, G. P. (1964). Evolution, classification and host relationships of Siphonaptera. *Annual Review of Entomology* **9**: 123–46.

Huang, D. (2015). *Tarwinia australis* (Siphonaptera: Tarwiniidae) from the Lower Cretaceous Koonwarra fossil bed: Morphological revision and analysis of its evolutionary relationship. *Cretaceous Research* **52**: 507–15.

Meusemann, K., Trautwein, M., Friedrich, F., Beutel, R. G., Wiegmann, B. M., Donath, A., Podsiadlowski, L., Petersen, M., Niehuis, O., Mayer, C., Bayless, K. M., Shin, S., Liu, S., Hlinka, O., Minh, B. Q., Kozlov, A., Morel, M., Peters, R. S., Bartel, D., Grove, S., Zhou, X., Misof, B., and Yeates, D. K. (2020). Are fleas highly modified Mecoptera? Phylogenomic resolution of Antliophora (Insecta: Holometabola). *bioRxiv*. doi: http://dx.doi.org/10.1101/2020.11.19.390666

Kwak, M. L., Madden, C., and Wicker, L. (2017). The first record of the native flea *Acanthopsylla rothschildi* Rainbow, 1905 (Siphonaptera: Pygiopsyllidae) from the endangered Tasmanian devil (*Sarcophilus harrisii* Boitard, 1841), with a review of the fleas associated with the Tasmanian devil. *The Australian Entomologist* **44**: 293–6.

Linardi, P. M., Beaucournu, J. C., de Avelar, D. M., and Belaz, S. (2014). Notes on the genus *Tunga* (Siphonaptera: Tungidae) II—neosomes, morphology, classification, and other taxonomic notes. *Parasite* **21**: 68.

Rothschild, M. and Ford, B. (1964). Breeding of the rabbit flea (*Spilopsyllus cuniculi* Dale) controlled by reproductive hormones of the host. *Nature* **201**: 103–4.

Rust, M. K. and Dryden, M. W. (1997). The biology, ecology and management of the Cat Flea. *Annual Review of Entomology* **42**: 451–73.

Tihelka, E., Giagomelli, M., Huang, D.-Y., Pisani, D., Donoghue, P. C. J., and Cai, C.-Y. (2020). Fleas are parasitic scorpionflies. *Palaeoentomology* **3**: 641–653.

Zhu, Q., Hastriter, M. W., Whiting, M. F., and Dittmar, K. (2015). Fleas (Siphonaptera) are Cretaceous and evolved with Theria. *Molecular Phylogenetics and Evolution* **90**: 129–39.

Diptera

(flies)

Common name	True or two-winged flies	Metamorphosis	Complete (egg, larva, pupa, adult)
Derivation	Gk. *di*–two, *pteron*–a wing		
Size	Body length 0.5–60 mm. Wingspan up to 75 mm	Distribution	Worldwide
		Number of families	About 180
		Known world species	> 158,000 (13.96%)

Sarcophaga carnaria – **Diptera**

Essential Entomology. Second Edition. George C. McGavin and Leonidas-Romanos Davranoglou, Oxford University Press.
© George C. McGavin and Leonidas-Romanos Davranoglou (2022). DOI: 10.1093/oso/9780192843111.001.0001

Key features

- abundant and ubiquitous
- one pair of wings used for flight (although some wingless)
- some species have immense economic impact through disease transmission
- the fourth-largest order

> Flies dominate most ecosystems, being predators, herbivores, parasitoids, or decomposers. *Dolichopus ungulatus* feeding on an enchytraeid worm. UK.

© Anne Riley

> Nuptial gifts are frequently used to entice the opposite sex, as seen here in two mating empidid flies. While the female is eating a dead fly, the male mates with her. UK.

© Anne Riley

© Zestin Soh

❭ Dipterans are important pollinators. This *Stylogaster* sp. from Malaysia, is about to feed on nutritious nectar. Females of this genus are endoparasites, laying eggs into insects driven out by army ant raids.

Flies are one of the dominant insect orders in most habitats. Most of the species that make up this huge and diverse order are beneficial to the function of ecosystems as pollinators, parasites, and predators, and vital to the processes of decomposition and nutrient recycling. However, the word 'fly' usually conjures up images of dirt, disease, and death. The activities of relatively few species have a greater impact on humans and other animals than any other insect group. Enormous numbers of wild and domestic animals, and perhaps as many as one person in six, are affected by fly-borne diseases. Tremendous losses of crops and other important plants are also caused by flies.

Insects in other orders have '-fly' as part of their common names but, in these cases, the name is written as one word, for example caddisfly, scorpionfly, and mayfly. If the common name does refer to a dipteran, then two words should be used (e.g., horse fly, house fly, and hover fly). In keeping with the other very large orders, such as the Hymenoptera and Coleoptera, it is estimated that the number of species that have still not been described equals the number already named, if not considerably more.

Fly species are recognized by the possession of a **single pair of membranous, front wings**, although secondary winglessness occurs in some ectoparasitic species, such as the Sheep Ked, *Melophagus ovinus* (Hippoboscidae). The **hind wings are reduced to form a pair of 'gyroscopic' organs called halteres.**

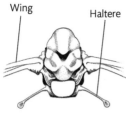

Wing   Haltere

⌃ Dorsal view of a tipulid thorax. The hind wings are reduced to form a pair of halteres.

These elegant gyroscopes, which beat at the same frequency as the wings, but out of phase, provide vital information to the flight system to keep the insect flying straight. Groups of surface stress receptors called campaniform sensillae are located in dorsal and ventral groups at the base of halteres and respond to the minute distortions in the cuticle created during flight. The sensillae are connected directly to the flight control systems by nerves and allow the fly to respond instantly to changes in pitch, roll, or yaw. Visual information from compound eyes and the brain is sent directly to the muscles that control the halteres, not to muscles that control the wings. In turn, signals from the halteres are relayed to the wing muscles. In this manner, the halteres, acting as flight stabilizers, can be fine-tuned by images from the eyes. The brain does not have a flight control centre, and the task is undertaken peripherally by a system of super-fast reflexes.

The thorax is modified with a very reduced front and hind segment. The middle segment, or mesothorax, is enlarged and packed with the flight muscles. The muscles operating the wings are not attached directly to the wing bases as in more primitive fliers, such as the dragonflies. Instead, the muscles alter the shape of the highly elastic, box-like thorax, and the wing hinges are arranged in such a way that rapid oscillations of the thoracic components are translated into wing motion (up to 1,000 wing beats per second). The flight systems that allow this phenomenal degree of aerial agility are complex and are not fully understood. Hovering, backward flight, 360 degree turns, and even upside-down flight and landing are nothing unusual to these insects. These aerial skills are essential to survival and are even brought into play during mating. Copulation in horse flies (Tabanidae), for example, takes place on the ground after bouts of hovering and high-speed interception manoeuvres by the males.

The order is divided into two suborders (see Box): the more plesiomorphic Nematocera (26 families) and the Brachycera (104 families). The Nematocera, or 'thread-horned' flies (the horn refers to the antenna), such as crane flies, mosquitoes, black flies, midges, and fungus gnats, are delicate and often have slender bodies, legs, and wings. Nematoceran antennae are elongate, threadlike, and multi-segmented. The larvae of many nematocerans are aquatic, and the adult females of several families are blood feeders. The Brachycera, or short-horned flies, are more robust with short, stout antennae of fewer than six segments.

A crane fly larva (or leatherjacket).

Within the Brachycera, two divisions are recognized (see Box). The Orthorrhapha (19 families) typified by the horse flies (Tabanidae), robber flies (Asilidae), and long-legged flies (Dolichopodidae), and a large group of 'higher' flies called the Cyclorrhapha (85 families), containing families such as the hover flies (Syrphidae), house flies (Muscidae), and blow flies (Calliphoridae).

Typically, flies have a **mobile head** with a pair of **large compound eyes**, and **three ocelli** are usually arranged in a triangular array on top of the head between the eyes. Dipteran mouthparts, which are designed for the ingestion of liquid foods, ranging from blood and nectar to the products of plant or animal decay, form a more or less elongate rostrum or proboscis. There are two basic types of mouthparts: biting/sucking or licking/sponging. The first type is typical of primitive flies like mosquitoes, while the second type is more characteristic of the higher Diptera, such as muscid or calliphorid flies. The rostrum usually incorporates the labrum, which may be small and flaplike, but is usually elongated to form the roof of the canal along which food is sucked. The floor of the food canal is made up of either the elongate hypopharynx or the overlapping mandibles (mainly in predatory species). The

A typical brachyceran fly.

hypopharynx is very deeply grooved along its length and carries the salivary secretions. The slender, elongate portions of the maxillae, which are not fused into the proboscis, are called the lacinia. The labium, which is usually the largest component of the mouthparts, terminates in two sensory labial palps (labella). In the higher flies, the labial palps, which are traversed by fine pseudotracheae, function as sponges. To suck up food, flies have two muscular pumps: the cibarial pump at the base of the food canal and the pharyngeal pump between the pharynx and the gut.

Principal dipteran taxa

Suborder (division)	Principal families	Larvae: pupae	Adults
Nematocera	Tipulidae Culicidae Chironomidae Ceratopogonidae Simuliidae Bibionidae Cecidomyiidae Mycetophilidae	Well-defined head capsule (eucephalic). Mandibles present moving in horizontal plane. Many species detritivorous or filter feeders, but some fungivorous, gall formers and predators. Pupa not enclosed in larval cuticle	Antennae thread-like with at least 6 segments (usually a lot more). Maxillary palps long (3–5 segments)
Brachycera (Orthorrhapha)	Tabanidae Stratiomyidae Asilidae Bombyliidae Empididae Dolichopodidae	Head capsule reduced (hemicephalic). Mandibles present, moving in vertical plane. Many species predacious or parasitic. Pupa not enclosed in larval cuticle	Antenna short, fewer than 6 segments. Last segment elongate or with bristle-like arista. Maxillary palps short (1–2 segments)
Brachycera (Cyclorrhapha)	Phoridae Syrphidae Conopidae Tephritidae Ephydridae Drosophilidae Chloropidae Anthomyiidae Muscidae Calliphoridae Sarcophagidae Tachinidae Oestridae Glossinidae Hippoboscidae	Head capsule vestigial, retractable (acephalic). Mandibles absent; function replaced by specialized mouth hooks. Larval feeding types very varied. Most species saprophagous (including dung and carrion feeders); others herbivorous, predacious, parasitic or parasitoids. Pupa enclosed in last larval cuticle (puparium)	Antenna short, fewer than 6 segments. Last segment elongate or with bristle-like arista. Maxillary palps short (1–2 segments)

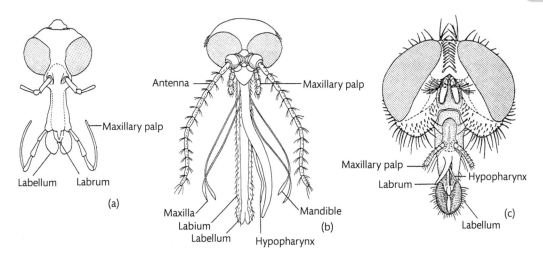

Mouthparts of some species of the Diptera.
(a) Tipulidae (crane fly) 'primitive' mouthparts;
(b) Culicidae (mosquito) piercing-sucking mouthparts;
(c) Muscidae (house fly) sponging-lapping mouthparts.

The mouthparts of a mosquito are made up of a bundle of slender, elongate stylets surrounded by the labium. Only females suck blood and have fully developed mouthparts. Before feeding, the female uses her labella to test the host's skin. To penetrate, the whole proboscis is pressed onto the surface, and the stylet bundle (made up of the labrum, mandibles, maxillae, and hypopharynx) is sawed through the tissue towards a blood vessel. The ends of the mandibles and maxillae have serrated edges. To allow deep stylet penetration the labium folds back on itself. Saliva is pumped down the hypopharynx and the blood meal is sucked back up the food canal.

The mouthparts of a house fly, which are made up from fusion of the clypeus, maxillae, and labium, can be extended by means of muscles and haemolymph pressure. Paired labial palps, located at the end of the proboscis, form a sponge-like structure (the labellum) traversed by grooves called pseudotrachea. Salivary secretions are pumped down the hypopharynx and flow on the undersides of the labellum. The pseudotracheae allow saliva to act on a large area of food, and the dissolved liquid flows into the pseudotracheae by means of capillary action to then be drawn back up the food canal. In these flies, mouthparts, such as the mandibles and maxillary lacinia, are lost.

Warning colouration and mimicry of other insects, particularly hymenopterans, is common in many dipteran families. Bees and wasps can be equipped with powerful stings, so the ability to look or behave like a bee or wasp affords flies protection from some of their enemies. Hover flies (Syrphidae) are especially well known for their very wasp or bee-like appearance. Sometimes flies imitate hymenopterans not only to avoid attack, but also in order to prey on their larvae. Species of African carpenter bees (*Xylocopa*, Anthophoridae) are perfectly mimicked by species of robber fly (*Hyperechia*, Asilidae). The female robber flies lay their eggs inside the nests of the carpenter bees. The fly larvae eat the bee larvae and their provisions.

Diptera are mostly all bisexual, and parthenogenesis is rare. The majority of adult flies do not live very long, and their main task is to find a mate

Larva and pupa of mosquito *Culex pipiens.*

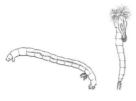

Larva and pupa of a chironomid midge.

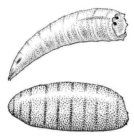

Larva and puparium of a cyclorrhaphan fly (e.g., blow fly).

> inquiline
a species that lives in the nest, gall, or home of another

and lay eggs. Courtship, which varies enormously between families and between species, can involve swarming displays, sound production, dances, nuptial gifts, pheromones and vibrational or acoustic songs. Drosophilid and tephritid flies (Drosophilidae and Tephritidae) are well known for the species-specific, short-range courtship songs that they make by vibrating their wings. The males in several families of flies fight for access to females and may have exaggerated physical traits that are sexually selected. The incredible stalked-eyed flies (Diopsidae) are an example. In several families, the males of some species have strange head modifications. In some, only the eyes, or the eyes and antennae, are placed on stalks, while, in others, the head bears bizarre antler-like outgrowths. In all these cases, the males display to females and engage in male-to-male ritual combat. Females seem to prefer to mate with larger males (indicated by the width between the eyes) and, like deer, smaller males lose out in any fights over territory.

Dance flies (Empididae) are predaceous on other insects, particularly other flies. The males offer nuptial gifts of food to their mates, and the size of the gift, and therefore how long the females feed, is related to how much sperm a male can transfer to the female during that time. Males have to be careful not to offer too large a prey item, as they fly in tandem with the female while she feeds, and thus have to support their own weight plus the combined weight of their mate and the nuptial gift. In some dance flies, the females form swarms and await the approach of a gift-bearing male.

Copulation often takes place on the wing, and eggs are laid in the vicinity of larval food, as they generally hatch quickly. In tsetse flies and some ectoparasitic fly families, the eggs hatch inside the mother and the larvae develop internally, feeding on special secretions through a pseudoplacenta. Fly larvae, or maggots as some are called, are cylindrical and often elongate, grub-like or worm-like, and, although they are highly variable in appearance, all lack thoracic legs. Nematoceran larvae go through four developmental stages or instars, and have a conspicuous, often dark, sclerotized head with biting jaws that move in a horizontal plane.

Brachyceran flies have between five and eight larval instars. Their larvae have a reduced head capsule with vertically biting mouthparts. Larvae of the cyclorrhaphan flies do not have any distinct head capsule. The tapered, anterior end is a dark, sclerotized cephalopharyngeal skeleton, of which the most conspicuous part is a pair of mouth hooks. The transformation from larva to adult takes place during the pupal stage and, in the higher flies (Cyclorrhapha), pupation occurs inside the skin of the puparium, the third and last larval stage.

The immense diversity of larval habitats and lifestyles is one of the major factors in the success of the order as a whole. Dipteran larvae can be herbivorous, predacious, parasitic, saprophagous, coprophagous, or fungivorous in all manner of freshwater and terrestrial habitats. The larvae of certain non-biting midges (Chironomidae) even live as **inquilines** inside the trap leaves of pitcher plants and are responsible for accelerating the breakdown of prey items, and the uptake of nitrogen by the host plant.

Species such as shore flies (Ephydridae) can even be found in some very harsh environments, such as pools of crude oil, hot springs, and the highly alkaline or saline water of marshes and mangroves.

Until the seventeenth century, it was thought that maggots in dead animals arise there by spontaneous generation. In 1668, Francisco Redi proved that maggots hatched from the eggs of flies. Flies and other insects feeding on carrion arrive and use the resources in a predictable succession associated with each stage of decay. Knowledge of the ecological succession under different conditions is of great use to forensic scientists, especially when deciding on the time of death in murder cases. The enormous scale of the recycling services carried out by flies should not be underestimated.

Although broad larval feeding patterns exist within each family, there are many exceptions. The larvae of most fungus gnats (Mycetophilidae and Keroplatidae) are fungivorous, but the larvae of *Arachnocampa luminosa* (Keroplatidae) catch prey in an unusual manner in New Zealand's Watiomo Cave. The larvae of these and other *Arachnocampa* species in New Zealand and Australia live in caves and catch flying prey on sticky threads that hang from the roof. The larvae lure their prey, which emerge from streams flowing through the caves, by being luminescent. The adults are also able to glow and use this for mate location in the darkness of the caves.

At least 20% of all fly species, representing more than 20 families, are parasitoids, that is, their larvae develop inside the bodies of hosts, which are killed in the process. Larvae of flies belonging to the family Conopidae are mostly endoparasitoids of aculeate Hymenoptera, although some species attack crickets, cockroaches, and other flies. Adult conopids are usually found on flowers, and often look very wasp-like, but this is for protection from predators and not to fool their hosts. In line with the findings in many parasite–host interactions, conopid larvae are able to manipulate the behaviour of their hosts. In one case, a conopid parasitoid of the bumble bee (*Bombus terrestris*) causes its host to burrow into the soil just before it dies. This is highly advantageous to the fly, because it now has a safe and protected site in which to **pupariate** and overwinter.

❯ **pupariate**
a term used of higher flies. The fly pupates inside a puparium which is the hardened skin of the last larval stage

The bumble bees can fight back a bit, and it has been shown that parasitized worker bees do not return to their nest as usual but stay outside all night. This slows up the growth of the parasitoid inside and may increase the bee's chances of survival by decreasing the chances of the parasitoid's successful development.

More than 8,200 species belong to the widespread family Tachinidae. These flies are typically stout bodied and many species resemble bristly house flies, but some can be much larger, very hairy, and almost bee-like. Adults feed on carbohydrate-rich food sources, such as honeydew, and males often congregate on hilltops, waiting for receptive females. Larval tachinids are parasitic on other insects and very rarely on other arthropods. The range of insect hosts used by these flies is extensive and includes the adult and immature stages of Lepidoptera, Coleoptera, Orthoptera, Hemiptera, Hymenoptera,

Diptera, and some other orders. The method by which the host is parasitized varies from the simple to the complex. Eggs can be laid directly on the host whereupon they hatch and burrow inside, or they can be laid through the skin of the host. The females of some species lay their eggs on an appropriate plant, which is then eaten by the host, and, in some cases, the tachinid female lays directly into the mouth of the host insect while it is feeding. While the larvae consume the host's tissues, eventually killing it, they obtain air from the outside through a hole in the host's body or by 'plugging in' to a trachea. Tachinids are of immense importance in the control of natural populations of pest insects. Many species have been used or are under investigation as biological control agents. For example, *Triarthria setipennis* was introduced to North America from Europe to control its host, the European earwig (*Forficula auricularia*). Another well-known species is *Cyzenis albicans*, which has been used against Winter Moth (*Operophtera brumata*), whose larvae can cause serious defoliation of a wide range of trees. The host-finding and egg-laying behaviour of some of these robust hairy flies is intriguing. Take, for example, *Ormia ochracea* whose larvae are parasitic inside the bodies of crickets. Mated female flies lay larvae on the outside of several species of *Gryllus* hosts, and the larvae burrow through the intersegmental membranes. Once inside, the young larvae feed on thoracic muscle tissue. The larvae then move to the abdominal cavity, moult, and feed on the fat body as well as other tissues. The host is killed when the fly pupariates and emerges. But how do the female tachinids find their host? The same way the female crickets find their mates—by the love songs of the male cricket. The flies locate the crickets by listening to the songs they emit, and as a result, they can parasitize male crickets and the females they attract. The story does not end there. It seems that both female crickets and female flies are preferentially attracted to the same qualities in the male song, and there is evidence that parasitoid pressure has shaped some of the characteristics of the cricket's songs. In fact, *Ormia ochracea* exerts such strong selection pressure on its hosts, that most individuals of a cricket species in Hawaii have lost their capacity to produce songs altogether, in order to escape the attention of the flies. Instead, they gather around the few males that are still capable of producing songs and steal the females that come by! The location of hosts by acoustic orientation also occurs in other species of tachinid that parasitize tettigoniids. Previously, it was thought that tachinid attack always resulted in the host's death, but recent work suggests that non-lethal parasitism might be quite common and might even explain the evolution of some host plant choices. In this study, the survival of a caterpillar (*Platyprepia virginalis*) to a tachinid attack (*Thelaira americana*) depended on what it ate. If parasitized, the caterpillars chose to eat poison hemlock, and if not parasitized, the caterpillars ate lupin. Parasitized caterpillars that ate poison hemlock survived better than those ate lupin, and, furthermore, the flies that emerged from the hemlock-fed caterpillars were heavier than those that emerged from lupin-fed hosts.

The range of feeding habits displayed by the larvae of hump-backed or scuttle flies (Phoridae) is very large; some feed on fungi, some on decaying matter or carrion, while others act as scavengers in the nest of ants and termites. Many are parasites on species in a number of insect orders, snails, and millipedes. There are even predatory adult phorids that eat on fig wasps, and whose larvae develop inside the fig fruit. The males and females of some ant-parasitizing species are very attracted to the alarm pheromones given off when their hosts fight. In the leaf-cutting ant *Atta cephalotes*, workers use tiny 'minim' workers to ride shotgun on top of their leaf loads to ward off other species of phorid flies that would otherwise find them easy targets. The minim workers are recruited by vibrational signals produced by the larger workers as they cut their leaf fragments. Other species of phorid fly who are specialist parasites of ant species can exert considerable influence of the behaviour of their host, causing them to forage at night, collect less food, and even defend themselves from enemies less effectively.

Many fly larvae are herbivorous pests. The larvae of the Hessian fly (*Mayetiola destructor*), a species of gall midge (Cecidomyiidae), burrow through the stems of grass crops like wheat, barley, and rye. *Oscinella frit* (Chloropidae) is another serious cereal pest in the northern hemisphere. The larvae of soldier flies (Stratiomyiidae) and rust flies (Psilidae) bore into root crops, and leaf mining species (Agromyzidae) attack a wide range of crops like alfalfa, cucumbers, and tomatoes. However, the major impact on humans relates to disease-carrying and blood-sucking dipterans (see Box). Many species of flies, such as muscids and calliphorids, are attracted to carrion, dung, and decaying matter, and can act as mechanical vectors of large range of pathogenic organisms that cause intestinal complaints, such as dysentery, and other diseases. The larvae of some flies invade human tissues, and others can act as vectors for human pinworm, hookworm, and roundworm, but the biggest threat by far comes from blood-feeding nematocerans that act as vectors for major diseases like malaria and yellow fever.

Many plants make use of flies as pollinators. The huge flower of the rare, parasitic tropical plant *Rafflesia* has evolved the scent of rotting meat to attract carrion flies. Although the flower is large, the flies are guided down tight channels around the central column of the male flower where they pick up sticky pollen on the back of their thorax. The flies are similarly confined to tight spaces in the female flower where they come into contact with the stigma.

Horse flies (Tabanidae) comprise more than 4,500 known species worldwide. Common names for these insects, which have large and often patterned eyes, include horse flies, deer flies, clegs, or gad flies. Adult females feed on mammalian blood while males feed exclusively on plant sugars. Tabanid larvae typically inhabit wet muddy areas by the margins of ponds, streams, or rivers, and eat small insects, worms, and other invertebrates. Most species are nuisance pests, and in the warmer parts of the world, they are vectors of several animal diseases. Two species of *Chrysops* are vectors of Loiasis, a filarial disease that affects humans, in west and central Africa.

˄ *Phlebotomus papatasi* is a vector of the viral disease sand fly fever in Mediterranean and Oriental regions.

The fly vectors and causative organisms of some human diseases

Disease	Vector	Organism
Anthrax	Horse flies. *Tabanus* spp. (Tabanidae)	Bacterium. *Bacillus anthracis*
Dengue fever	Mosquitoes. *Aedes aegypti* and *Culex* spp. (Culicidae)	Virus. (Flaviviridae)
Elephantiasis	Mosquitoes. *Culex fatigans* (Culicidae)	Nematode. *Wuchereria bancrofti*
Encephalitis (several)	Mosquitoes. *Aedes* and *Culex* spp. (Culicidae)	Virus. (Flaviviridae)
Filariasis (several)	Biting midges. *Culicoides* spp. (Ceratopogonidae)	Nematode. *Mansonella* and *Dipetalonema* spp.
Leishmaniasis (several)	Sand flies. *Phlebotomus* and *Lutzomyia* spp. (Psychodidae)	Protozoan. 12 species of *Leishmania*
Loiasis	Horse flies. *Chrysops* spp. (Tabanidae)	Nematode. *Loa loa*
Malaria (several)	Mosquitoes. *Anopheles* spp. (Culicidae)	Protozoan. 4 species of *Plasmodium*
Onchocerciasis	Black flies. *Simulium* spp. (Simuliidae)	Nematode. *Onchocerca volvulus*
Sleeping sickness	Tsetse flies. *Glossina* spp. (Glossinidae)	Protozoan. *Trypanosoma brucei*
Tularaemia	Horse flies. *Chrysops* spp. (Tabanidae)	Bacterium. *Francisella tularensis*
Yellow fever	Mosquitoes. *Aedes aegypti* and other spp. (Culicidae)	Virus. (Flaviviridae)
Zika virus	Mosquitoes. *Aedes aegypti* and *A. albopictus*	Virus. (Flaviviridae)

One of the most remarkable examples of insect engineering is the construction of mud cylinders by African species such as *Tabanus biguttatus*. The larvae live in pools, which eventually dry out. When the larva is fully grown, it first descends through the mud following a tight spiral path. It then spirals upwards, cutting across the descending channels thereby making a near-isolated cylinder of mud. Near the surface, it burrows straight up and enters the centre of the cylinder from the top. Once inside, the larva forms a sealed pupation chamber. As the mud dries out and bakes hard, cracks forming across the surface bypass the cylinders, leaving the pupating fly safe inside.

Although not vectors of any human disease, biting midges (Ceratopogonidae), such as the Scottish Biting Midge (*Culicoides impunctatus*), are well-known pests, and many parts of the northern hemisphere are rendered virtually uninhabitable due to these and other species of biting fly, like black flies and horse flies. *Culicoides* species of the *imicola* complex can, however, be vectors of serious viral animal diseases like African horse sickness and Bluetongue (Reoviridae).

Fly larvae in the family Oestridae are parasitic and eat the flesh of various animals. Species of calliphorid and sarcophagid flies lay their eggs or larvae in carrion, but some will also lay eggs in live tissue—a condition called a myiasis. Flies from all these families have been known to cause myiasis in humans. Although the principal hosts are rats, larvae of the Tumbu Fly (*Cordylobia anthropophaga*), a calliphorid species from tropical Africa, can burrow into human skin. Eggs can be laid on damp bedding or clothes, and, when hatched, a young larva can burrow under skin in less than a minute. A large painful swelling appears as the larva feeds and grows. The larva leaves the host to pupate in just over one week. Another calliphorid, the New World Screw-worm (*Cochliomyia hominivorax*), is an important pest of cattle, sheep, and horses in the North and South America, and will lay eggs on wounds, scratches, and even mucous membranes. The larvae burrow deeply into tissue and make a large boil or lesion, which can lead to disfigurement or even death. Control of this scourge has been achieved through the mass release of males, which have been sterilized by irradiation.

The Human Bot Fly (*Dermatobia hominis*) lives in Central and South America and uses a clever trick to provide their larvae with a suitable place to live. The females catch blood-feeding flies, lay their eggs on them, and then release them unharmed. When the eggs mature and the flies next feed on a mammalian host (including humans), the eggs hatch and the larvae burrow under the skin. In the cases where the larvae have been deposited on a human scalp, the afflicted person can hear the bot fly larvae munching through their flesh at night. Blow flies, such as *Lucilia* spp. (Calliphoridae), are responsible for flystrike of sheep. Traps baited with sheep offal can reduce local populations of these flies, but they are very significant pests in many parts of the world.

Warble flies or bot flies (Oestridae) live by laying live larvae in the noses or in the hair of their mammalian hosts. The larvae either feed in the nasal passage or burrow extensively under the skin. The cattle grubs, *Hypoderma lineatum* and *Hypoderma bovis*, are serious pests of domestic cattle in the northern hemisphere. Adapted to the migrations of their hosts, some species have an incredible ability to fly for long periods. A mated female Reindeer Warble Fly can cover up to 900 km during her life and may lose up to 40% of her initial body mass.

Gruesome as it might seem, sterile calliphorid larvae of certain species are sometimes used to clean infections which are difficult to treat with antibiotics. The grubs eat infected flesh and leave healthy tissue alone.

Mosquitoes (Culicidae) probably have a greater harmful effect on humans than any other insect family. They are slender, delicate flies with small, subspherical heads and elongate, slender, sucking mouthparts. The body is covered with tiny scales and appears pale brown to reddish brown. The wings are long and narrow with scales along the veins and margins. Adults belonging to the two important subfamilies, Anophelinae and Culicinae, can be recognized by the way they rest. Anopheline mosquitoes (*Anopheles* spp.) rest with their heads down and the body inclined at 45 degrees, whereas culicine (*Aedes* and *Culex*) species rest with their bodies held horizontally.

It is only the females that suck mammalian blood and therefore transmit disease. Males feed on nectar or honeydew. *Anopheles* eggs have small lateral air floats, and are laid singly on the surface of water, whereas *Culex* eggs are laid in floating rafts of 40–300 eggs. The larvae can be found almost anywhere, from rain-filled containers and tree holes to ponds and lakes, and are mostly saprophagous, although some are predacious. In most species, air is obtained from the surface. Anopheline larvae lie horizontally under the surface while culicine larvae hang at an angle from the surface film by a posterior siphon. The life cycle takes between 15–20 days.

Many mosquitoes are very common vectors of human diseases such as yellow fever, dengue fever, filariasis, and encephalitis, but malaria is by far the most serious. Malaria has killed people throughout the course of human history, and its effects are described in the Old Testament and even earlier texts. Causing more than 400,000 deaths every year, the majority of them pre-school age African children, malaria's impact continues to be immense. It is estimated that one-third of all human beings live in regions of the world where infection by the malarial parasite is likely, and 94% of all cases occur in Africa. Malaria is only carried by mosquitoes belonging to the genus *Anopheles*.

The discovery of the malarial parasite (*Plasmodium*) by Charles Laveran in 1880, and the elucidation of the role played by the mosquito in the life cycle of the parasite by Ronald Ross in 1898, made prevention a possibility. The disease had been treated with the bark of the cinchona tree from which quinine was eventually purified, and, to complete the victory, the breeding grounds of the mosquitoes were drained or covered with oil to suffocate the larvae and huge amounts of insecticides were used to kill adults. The development of DDT and its extensive use after the Second World War did, indeed, rid many countries of this terrible scourge, but it was not long before the mosquitoes developed resistance to insecticides, and the use of some were banned because of persistence and the harmful effects they had on non-target organisms. Today, there is clear evidence of a massive resurgence of the disease, partly due to resistance of the four species of *Plasmodium* parasites to drugs, such as chloroquine, and the resistance of the vector mosquitoes to insecticide, but also to the drug industry's lack of interest in producing new anti-malarial compounds for a long period of time. The hopes of a vaccine promised by molecular biology are diminishing, but new ideas involving genetic engineering may still win the war. Certain approaches include the release of millions of transgenically sterilized male mosquito vectors, which would mate with most females and outcompete fertile males, resulting in a huge drop in their overall numbers. Other approaches include the introduction of transgenic males that possess a gene that destroys X-chromosomes and results in male-only progeny. One thing is certain: the fight against malaria is ongoing and will likely continue for a long time. Recent evidence of global climate change suggests that the range of malaria may increase. The disease occurred in many parts of southern Europe (including England) in the past, and as the vector mosquitoes are still present, there is a real risk that reintroduction could take place.

Tsetse flies belong to the family Glossinidae and are dull brown or grey coloured, medium to large-sized flies (6–14 mm long). There are 21 known

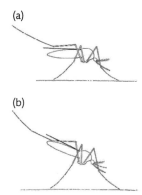

(a)

(b)

∧ Typical resting position of (a) the culicine mosquito and (b) the anopheline mosquito.

species, all found in subtropical and tropical parts of Africa, and their presence causes immense human suffering and renders large areas unfit for cattle rearing. In total, some 10 million km² of land are affected and 70 million humans are at risk from sleeping sickness—the disease carried by these flies. Both males and females feed on blood, and their needle-like mouthparts can penetrate the toughest skin or hide. Tsetse flies can be distinguished from other sorts of flies, such as horse flies, because some of the wing veins mark out a distinctive hatchet-shaped cell, and at rest, the wings are flat and crossed scissor-like over the abdomen. Different tsetse species are found in different types of habitat and prefer to bite different sorts of animals. Some are only found in east African countries, while others are more widespread. Some species breed in savannah woodlands, while others prefer coastal areas or damp, riverside forests.

Female tsetse flies are unusual in that they produce only one egg at a time, which hatches and stays inside the mother's body. Safe inside a structure called a uterus, the larva feeds on secretions produced by special nurse or 'milk' glands. When fully grown, the tsetse larva is large and may even weigh a little more than its mother. Larvae are deposited on the ground, or in the host's nest or resting place, where they pupate almost immediately. Adenotrophic viviparity, as it is called, also occurs in the Hippoboscidae (a family of more than 200 fly species, which are blood-sucking parasites of birds and mammals), the Nycteribiidae, and the Streblidae (two similarly sized families of blood-sucking parasites of bats). The advantages of being able to carry and feed young until they are just about to pupate are considerable and do away with the need for the larvae to fend for themselves. After a three- or four-week pupation, the adult flies emerge from the soil and find a mate within a day of two of emergence. Females store sperm inside a spermatheca, and only need to mate once in their lives. The first larva will be produced in 2–3 weeks and, depending on the species, female tsetse flies will produce more larvae every 8–10 days. Both sexes need blood meals every 2–3 days and different species exhibit preferences for cattle, horses, pigs, humans, and even crocodiles. In the wild, males live for about six weeks as adults, but females can live for more than three months.

One hundred years ago a Scottish microbiologist, David Bruce, serving in the Royal Army Medical Service, and his wife Mary Elizabeth Bruce (née Steele), went to Africa to investigate the cause of a disease of ruminant animals called nagana. In three years, they found that the disease was caused by protozoan organisms called trypanosomes (named *Trypanosoma brucei*, in their honour) that lived as parasites in the blood and tissues of affected animals. They also discovered that human sleeping sickness was caused by the same parasite, and that it was transmitted by tsetse flies when they fed on animal or human blood.

Once infected, a tsetse fly carries the parasite for the rest of its life and passes it on to the animals on which it feeds. The parasites are injected into the host animals along with the fly's saliva. Only about one in a hundred flies is infected with the parasite, so not every bite will result in the infection being passed on. People and animals get infected because they get bitten many times. In west and central Africa, domestic pigs are the main animals carrying the disease, whereas in east Africa it is mainly cattle and wild animals such as antelopes that are infected. Animals like antelope and buffalo are unaffected by the parasites, but they act as reservoirs from which the tsetse flies can pick up the infection. Although, in theory, all species of tsetse fly can transmit the trypanosome parasite to humans or domestic and wild animals, only a few—*Glossina palpalis*, *Glossina fuscipes*, *Glossina tachinoides*, *Glossina morsitans*, and *Glossina pallidipes*—are regular human-biting species associated with outbreaks of sleeping sickness.

In humans, the first sign of the disease is a local inflammation and swelling of the skin where the fly took a blood meal. Next, the victim will have influenza-like symptoms with a fever and chills, headache, and slight swelling of the hands and feet. Initially, only the bloodstream and the lymphatic system are infected, but damage to the brain and nervous system will also occur. This may take only a few weeks to happen, or it could take several years, but when it does, the victim feels sleepy and tired all the time, and may have convulsions and go into a coma. Untreated, the disease is almost always fatal, but some types of drug, if given early enough, can help. On a positive note, the number of cases is steadily plummeting, with about 1,000 new infections annually (compared to many tens of thousands in 1990s). Production of a vaccine is difficult as the parasite keeps changing slightly so that vaccine will not work properly.

In tsetse fly areas, prevention mainly involves trying to avoid being bitten, but this can be difficult because the flies are active during the hottest time of the day, not easily discouraged, and can bite through clothing. Many methods of control have been attempted, and these are aimed at trying to get rid of the flies themselves. This is not an easy job as the areas they live in are vast, and the flies can travel long distances. Tsetse fly control has included extensive habitat clearance, and the use of insecticides, but due to the low densities at which they exist, traps baited with animal dung and urine are more effective. The biconical trap is a clever device for catching tsetse flies. It works on the principle that tsetse flies are attracted to large dark shapes, which they mistake for animals. The bottom section, which has four oval holes around the outside, is made of dark blue cloth. The upper section is made of white cloth. Inside the bottom section, there are two pieces of black cloth at right angles, which makes the inside of the trap look very dark and inviting. Tsetse flies enter the trap through the holes, and then fly up where they enter a collecting bag. When the bag has a good catch, it is simply emptied of dead flies. The flies are killed by the heat that builds up inside the trap, and no insecticide is needed. It has been proposed that the patterning of some animals may be the result of selection to avoid tsetse fly bites. For instance, the characteristic vertical stripes of zebras may be important in protecting them, as tsetse flies may not see the zebras as well as they see non-striped hosts, and it has been shown that some blood-feeding flies have an aversion to landing on striped surfaces.

The origins of Diptera go very far back in time, with the oldest fossil flies dating to the Triassic (about 230 mya). As stated in previous chapters, the closest relatives of Diptera are the Mecoptera and the Siphonaptera.

Key reading

Armitage, P. S., Cranston, P. S., and Pinder, L. C. V. (eds) (1995). *The Chironomidae: the biology and ecology of non-biting midges*. Chapman and Hall, London.

Borkent, A. and Wirth, W. W. (1997). *World species of biting midges (Diptera: Ceratopogonidae)*. Bulletin of the American Museum of Natural History, 0003–0090. American Museum of Natural History, New York.

Centers for Disease Control and Prevention. Parasites—African trypanosomiasis (also known as Sleeping Sickness). Retrieved from: https://www.cdc.gov/parasites/sleepingsickness/

Chan, W. P., Prete, F., and Dickinson, M. H. (1998). Visual input to the efferent control system of a fly's gyroscope. *Science* **280**: 289–92.

Clements, A. N. (1999). *The biology of mosquitoes* (2 vols). Chapman and Hall, London (vol. **1**); CABI, Wallingford (vol. **2**).

Collins, F. H. and Paskewitz, S. M. (1995). Malaria: current and future prospects for control. *Annual Review of Entomology* **40**: 195–219.

Cranston, P. S. (ed) (1995). *Chironomids: from genes to ecosystems*. CSIRO Publishing, Victoria.

Crosskey, R. W. (ed) (1980). *Catalogue of the Diptera of the Afrotropical region*. British Museum (Natural History), London.

Denlinger, D. L. and Zdarek, J. (1994). Metamorphosis behavior of flies. *Annual Review of Entomology* **39**: 243–66.

Disney, R. H. L. (1994). *Scuttle flies: the Phoridae*. Chapman and Hall, London.

Eggleton, P. and Belshaw, R. (1992). Insect parasitoids: an evolutionary overview. *Philosophical Transactions of the Royal Society of London B* **337**: 1–20.

Erzinclioglu, Z. (1996). *Blowflies*. Naturalists' handbooks series **23**. Richmond Publishing, Slough.

Fabre, C.C., Hedwig, B., Conduit, G., Lawrence, P. A., Goodwin, S. F., and Casal, J. (2012). Substrate-borne vibratory communication during courtship in *Drosophila melanogaster*. *Current biology* **22**: 2180–5.

Feener, D. H., Jr. and Brown, B. V. (1997). Diptera as parasitoids. *Annual Review of Entomology* **42**: 73–97.

Foote, B. A. (1995). Biology of shore flies. *Annual Review of Entomology* **40**: 417–42.

Gibson, G. (1992). Do tsetse flies 'see' zebras? A field study of the visual response of tsetse striped targets. *Physiological Entomology* **17**: 141–7.

Gilbert, F. (1993) *Hoverflies*, (2nd edn). Naturalists' handbooks series **5**. Richmond Publishing, Slough.

Griffiths, G. C. D. (1972). *The phylogenetic classification of the Diptera Cyclorrapha*. Entomologica series **8**. Junk Publishers, The Hague.

Heinen-Kay, J. L., Urquhart, E. M., and Zuk, M. (2019). Obligately silent males sire more offspring than singers in a rapidly evolving cricket population. *Biology Letters* **15**: 20190198.

Karban, G. and English-Loeb, G. (1997). Tachinid parasitoids affect host plant choice by caterpillars to increase caterpillar survival. *Ecology* **78**: 603–11.

Kettle, D. S. (1990). *Medical and veterinary entomology*. CAB International, Wallingford.

Kirk-Spriggs, A. H. and Sinclair, B. J. (eds) (2017). *Manual of Afrotropical Diptera. Volume 1 Introductory chapters and keys to Diptera families*. Suricata series **4**. South African National Biodiversity Institute, Pretoria.

Kirk-Spriggs, A. H. and Sinclair, B. J. (eds) (2017). *Manual of Afrotropical Diptera. Volume 2. Nematocerous Diptera and lower Brachycera*. Suricata series **5**. South African National Biodiversity Institute, Pretoria.

Kirk-Spriggs, A. H. and Sinclair, B. J. (eds) (2021). *Manual of Afrotropical Diptera. Volume 3*. Suricata series **8**. South African National Biodiversity Institute, Pretoria.

Lane, R. P. and Crosskey, R. W. (1993). *Medical insects and arachnids*. Chapman and Hall, London.

Lehane, M. (1991). *Biology of blood-sucking insects*. Harper Collins Academic, London.

Lukashevich, E. D., Przhiboro, A. A., Marchal-Papier, F., and Grauvogel-Stamm, L. (2010). The oldest occurrence of immature Diptera (Insecta), Middle Triassic, France. *Annales de la Société entomologique de France* **46**: 4–22.

Mazzoni, V., Anfora, G., and Virant-Doberlet, M. (2013). Substrate vibrations during courtship in three *Drosophila* species. *PLoS ONE* **8**: e80708.

McAlpine, J. F. (ed) (1981/1987/1989). *Manual of Nearctic Diptera* (3 Vols). Monographs 27, 28, and 32. Research Branch, Agriculture Cananda. Canadian Government Publishing Centre, Hull.

Muller, C. B. (1994). Parasitoid induced digging behaviour in bumblebee workers. *Animal Behaviour* **48**: 961–6.

Muller, C. B. and Schmid-Hempel, P. (1993). Exploitation of cold temperature as a defence against parasitoids in bumble bees. *Nature* **363**: 65–7.

Nigam, Y. Dudley, E., Bexfield, A. E. Bond, Evans, J., and James, J. (2010). The physiology of wound healing by the medicinal maggot, Lucilia sericata. *Advances in Insect Physiology* **39**: 39–81.

Nilssen, A. C. and Anderson, J. R. (1995). Flight capacity of the reindeer warble fly, *Hypoderma tarandi* (L.) and the reindeer nose bot fly, *Cephenemyia trompe* (Modeer) (Diptera: Oestridae). *Canadian Journal of Zoology* **73**: 1228–38.

Papavero, N. (1977). *The world Oestridae (Diptera), mammals and continental drift*. Entomologica series **14**. Junk Publishers, The Hague.

Pascoal, S., Cezard, T., Eik-Nes, A., Gharbi, K., Majewska, J., Payne, E., Ritchie, M. G., Zuk., M., and Bailey, N. W. (2014). Rapid convergent evolution in wild crickets. *Current Biology* **24**: 1369–74.

Rogers, D. J. and Randolph, S. E. (1991). Mortality rates and population density of tsetse flies correlated with satellite imagery. *Nature* **351**: 739–41.

Rotenberry, J. T., Zuk, M., Simmons, L. W., and Hayes, C. (1996). Phonotactic parasitoids and cricket song structure: an evaluation of alternative hypotheses. *Evolutionary Ecology* **10**: 233–43.

Scholl, P. J. (1993). Biology and control of cattle grubs. *Annual Review of Entomology* **38**: 53–70.

Simoni, A., Hammond, A. M., Beaghton, A. K., Galizi, R., Taxiarchi, C., Kyrou, K., Meacci, D., Gribble, M., Morselli, G., Burt, A., Nolan, T., and Crisanti, A. (2020). A male-biased sex-distorter gene drive for the human malaria vector *Anopheles gambiae*. *Nature Biotechnology* **38**: 1054–60.

Skidmore, P. (1985). *The biology of Muscidae of the world*. Entomologica series **29**. Junk Publishers, The Hague.

Smith, K. V. G. (1986). *A manual of forensic entomology*. British Museum (Natural History) and Cornell University Press, Ithaca.

Snow, K. R. (1990). *Mosquitoes*. Naturalists' handbooks series **14**. Richmond Publishing, Slough.

Spencer, K. A. (1990). *Host specialisation in the world Agromyzidae (Diptera)*. Entomologica series **45**. Kluwer Academic Publishers, Dordrecht.

Stubbs, A. and Chandler, P. (eds) (1978). *A dipterist's handbook*. Amateur Entomologist's Society, London.

Tabachnick, W. J. (1996). *Culicoides variipennis* and bluetongue-virus epidemiology in the United States. *Annual Review of Entomology* **41**: 23–43.

Wagner, W. E., Jr. (1996). Convergent song preferences between female field crickets and acoustically orientating parasitoid flies. *Behavioural Ecology* **7**: 279–85.

Walker, S. M., Schwyn, D. A., Mokso, R., Wicklein, M., Müller, T., Doube, M., Stampanoni, M., Krapp, H. G., and Taylor, G. K. (2014). In vivo time-resolved microtomography reveals the mechanics of the blowfly flight motor. *PLoS Biology* **12**: e1001823.

World Health Organization. Malaria [version 1 April 2021]. Retrieved from: https://www.who.int/news-room/fact-sheets/detail/malaria

Wilkerson, R. C., Butler, J. F., and Pechuman, L. L. (1985). Swarming, hovering and mating behaviour of male horse flies and deer flies (Diptera: Tabanidae). *Myia* **3**: 515–46.

Wilkinson, G. S. and Reillo, P. R. (1994). Female choice response to artificial selection on an exaggerated male trait in a stalk-eyed fly. *Proceedings of the Royal Society of London B* **255**: 1–6.

Amphiesmenoptera

This superorder consists of caddisflies (Trichoptera) and butterflies and moths (Lepidoptera). Amphiesmenoptera means 'clothed-wing' in ancient Greek, referring to the hairs and scales that cover the wings of Trichoptera and Lepidoptera, respectively.

Trichoptera
(caddisflies)

Common name	Caddisflies	Distribution	Worldwide except Antarctica
Derivation	Gk. *trichos*–hair; *pteron*–a wing		
		Number of families	52
Size	Body length 2–38 mm	Known world species	16,000 (1.41%)
Metamorphosis	Complete (egg, larva, pupa, adult)		

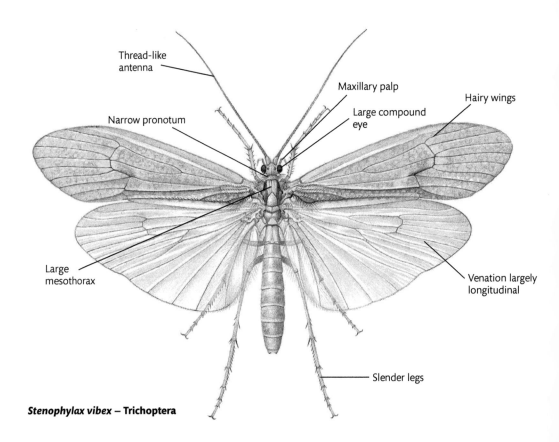

Thread-like antenna

Maxillary palp

Large compound eye

Hairy wings

Narrow pronotum

Large mesothorax

Venation largely longitudinal

Slender legs

Stenophylax vibex – **Trichoptera**

Essential Entomology. Second Edition. George C. McGavin and Leonidas-Romanos Davranoglou, Oxford University Press.
© George C. McGavin and Leonidas-Romanos Davranoglou (2022). DOI: 10.1093/oso/9780192843111.001.0001

Key features

- aquatic larvae, typically in self-constructed cases or shelters
- nocturnal and moth-like, with hairs and sometimes scales
- legs and antennae slender

A specimen of *Stenophylax permistus* from Leighton Moss, UK.

© Matt Doogue

Many caddisfly larvae construct a portable case that they carry with them wherever they go. The vast majority are aquatic, although the land caddis (*Enoicyla pusilla*) imaged here is exceptional in that it inhabits leaf litter. UK.

© Paul Brock

Caddisflies can be found almost everywhere that there is fresh water. Estimates point to there being as many as 50,000 species, the majority of which will be confined to the eastern hemisphere. The slender, elongate adults are sombre-coloured, and rather **moth-like** in appearance with **long, slender legs**. The body and wings, particularly the front wings, are covered with hairs. Caddisflies are most closely related to the Lepidoptera, but the latter are covered with scales and not hairs, and caddisflies lack the unique proboscis of most lepidopterans. The head has a pair of **large compound eyes**, and **long, thin, multi-segmented antennae**, which may be shorter or longer than the length of the body. **Two or three ocelli may be present.** The **weakly developed mouth-parts** allow adults to lick up water and nectar, but many do not feed at all. The head and basal part of the antennae may have scent-producing organs of various kinds, and the upper surface of the head and parts of the thorax may have small, hairy, wart-like bumps. **The front and hind pairs of wings are membranous, similar in size**, and held over the body in a characteristically tent-like manner. In flight, the hind wings are coupled to the front wings by means of special curved hairs or hooks. Two suborders are recognized: the Annulipalpia and the Integripalpia (see Box). In the Annulipalpia, the larvae construct stationary retreats, which gives them the common name 'fixed-retreat makers'. The Integripalpia, which is the most diverse caddisfly suborder, is characterized by larvae that make portable, often highly elaborate cases, while some carnivorous species are entirely free living. A third suborder is proposed from within the Integripalpia, the Spicipalpia, although whether this is a valid classification has remained controversial to this day. Both Annulipalpia and Integripalpia are split into several superfamilies (see Box).

Mostly nocturnal in habit, the adults hide in vegetation during the hours of daylight and are quite hard to find. Mating takes place in flight (sometimes in swarms), or on vegetation. In some species, when the females have coupled their genitalia with a male, the male alone flies the joined pair to the water margin where copulation occurs. Females are typically the bigger sex in caddisflies, but, in these particular species, males are larger to enable them to cope with the load imposed by the nuptial flight. Eggs, which are produced in masses or strings enveloped in a jelly-like substance, are either dropped onto or under water, or laid on overhanging plants from which they will eventually drop. The order is best known because of the aquatic larvae that construct impressive and elegant cases, inside which they live and finally pupate. A typical larva is caterpillar-like with a strong head and thorax and a soft abdomen. The head has chewing mouthparts and short antennae. There are three pairs of five-segmented thoracic legs, each of which the larvae use to move around, and, at the end of the abdomen, each has a pair of hooked prolegs to anchor it in its case. Larvae are apneustic (they have no spiracles) and oxygen is obtained from the water either by direct diffusion or through external gills located on the abdominal, and, sometimes, the thoracic segments. Caddisfly larvae may be herbivores, carnivores, or detritivores. The larvae of some species are free-living (Rhyacophiloidea), and some make silken nets to filter minute organic particles of food from the water currents (Hydropsychoidea). The majority of species make portable cases (e.g., Limnephiloidea) and can be found in flowing and still water. The cases, made from a great variety of materials, are held together with silk secreted from glands in the head and fed out through the labium. Case design and construction is very variable, tends to be specific to particular families, and in many instances, is a useful identification feature.

∧ **Limnephilid caddisfly larval cases vary enormously. Here, a species has used small snails shells as a construction material.**

Trichopteran suborders and some superfamilies

Suborder	Common families	Larval habits: food
Annulipalpia		
Hydropsychoidea (7 families)	Hydropsychidae	Net spinners: invertebrates, algae, particles
Philopotamoidea (2 families)	Philopotamidae	Net spinners: algae, particles
Integripalpia		
Limnephiloidea (9 families)	Goeridae	Case makers: algae, detritus
	Limnephilidae	Case makers: algae, detritus
Phryganeoidea (3 families)	Phryganeidae	Case makers: invertebrates, detritus
	Lepidostomatidae	Case makers: plant detritus
Leptoceroidea (7 families)	Leptoceridae	Case makers: invertebrates, algae, sponges, detritus
Sericostomatoidea (12 families)	Sericostomatidae	Case makers: detritus
'Spicipalpia'		
Hydroptiloidea (2 families)	Glossosomatidae	Saddle case makers: algae and detritus
	Hydroptilidae	Free-living then purse case: algae and diatoms
Rhyacophiloidea (2 families)	Rhyacophilidae	Free-living: invertebrates

The first four larval stages of purse casemakers (Hydroptilidae) have a short abdomen, and in addition to the thoracic legs, they have longish abdominal prolegs. They are active and free-living in water, and suck the juices of water plants, such as *Spirogyra*, through modified mouthparts. The last larval stage, which has a swollen abdomen, makes an open-ended purse or barrel-shaped case of silk, often with small stones attached. Larvae of the northern caddisflies (Limnephilidae) can be found in a wider variety of habitats than any other family of trichopterans, and exhibit variable preferences in terms of water current, temperature, and substrate texture. Their cases vary in size (6–65 mm in length) and have a circular, triangular, or irregular cross-section depending on species. The cases are constructed from a diverse range of materials including sand grains, pebbles, vegetation, sticks, and shells, or mixtures of more than one type of material.

In species that exclusively use twigs as construction materials, the cases are appropriately known as 'log cabins'. The larvae of some limnephilid species incorporate the occasional large twig in the case, which often projects sideways and deters fish from trying to swallow them. In some species, the cases made by the younger stages can be of a different design to those made by the older stages. Most limnephilid larvae are omnivorous, and feed on detritus or scrape microscopic algae and other organisms from the surface of stones and rocks. The larvae of giant casemakers (Phryganeidae) feed on plant debris or are predacious on small aquatic animals, and as a result, need a strong but light case that can be easily moved around. The beautifully regular, silk-lined, tapering cases are made of spirally arranged plant fragments with the occasional bit of gravel incorporated. Interestingly, left- and right-handed spirals occur within the same species, and the plant fragments used for the cases, which need to get longer as the case grows, are measured and expanded accordingly by the larva.

In some families, only bits of vegetation, such as cut leaves or sticks, are used. In others, there may be a high mineral content with everything from small sand grains to quite large pebbles being incorporated. Some species even use empty snail shells exclusively as their case-building material.

Cases of limnephiloid families may be long and thin, squat, or spiral in overall shape and several studies with various predators, such as fish and dragonfly nymphs, have shown that the design and construction of the case has a significant bearing on survivorship. A study of survivors of many species shows that they have stronger and sometimes wider cases than those that were eaten. The behaviour of the caddisfly larvae is also important. When confronted with a predatory fish, many caddisfly larvae will retreat inside their case and play dead. The fish may wait to see if the occupant moves again, but it is usually the insect that can wait the longest. If larvae are taken out of their cases and thus forced to rebuild them, the first priority is rapid protection, although some species will respond to the local conditions by making the new cases heavier in faster-flowing water. Net-spinning caddisflies, such as the Hydropsychidae, also alter their net construction, overall size, and mesh size in response to the amount of food present in the water and the speed at which it is flowing. They are therefore able to maximize the prey capture rate, while not risking damage to the net due to drag. There is evidence of larvae of algal grazing species defending patches of algae on the surface of rocks from each other, and even from other species of grazer, such as nymphal mayflies. The size of the patch is related to the size of the territory holder.

After between five and seven larval stages, pupation takes place within the silk-sealed larval case or in a specially made pupal case. The whole life cycle can take one year but is longer if the species occurs in mountainous or cold regions. When ready to emerge, the pharate adult (enclosed inside the pupal exoskeleton) cuts its way out of the case and swims to the surface or crawls out of water on plants where it emerges. The wasp *Argiotypus armatus* (Argiotypidae) is an unusual parasitoid that swims below the water surface to lay its eggs in pupal cases of the caddisfly *Silo pallipes* (Goeridae). The larvae are ectoparasites and, after five instars, the wasp pupates. It remains inside the pupal case of its host to overwinter and emerges in the following spring.

Many water courses become polluted by industrial effluent as well as by pesticides, fertilizers, and herbicides from farm run-off. As different caddisfly species vary enormously in their ability to tolerate various factors, such as acidity, oxygen content, and chemical composition of the water, they are invaluable indicators of changes in water quality. The levels of specific

Caddisfly larva use many types of material in the construction of their cases. Many designs are species- or genera-specific.

pollutants, such as pesticides and polychlorinated biphenyls (PCBs), can be analysed in adult body tissues.

Although the vast majority of species are freshwater dwellers, a few species from several families can be found in brackish water, and four species in the strange family Chathamiidae are marine (*Chathamia* and *Philansius*). Known only from parts of New Zealand and New South Wales, the larvae of these caddisflies make their cases from grains of sand and fragments of algae on which they feed. Bizarrely, the adult females of *Philansius plebius*, which have long and fairly tough ovipositors, are known to lay their eggs inside the coelomic cavity of two species of intertidal starfish. It is likely that females oviposit through pores on the aboral surface of the starfish. Inside the coelom, the eggs are protected from harmful waves and predation. After several weeks of incubation, the eggs hatch and the first instar larvae leave the host, either through the pores or, perhaps, through the stomach wall. They pass through seven instars feeding on calcareous and non-calcareous algae.

Some molecular studies estimate that caddisflies first appeared during the Permian (about 275 mya), which coincides with the earliest known fossils of early Trichoptera (or close relatives). Phylogenomic studies suggest a considerably younger date, between 215–154 mya. Additional data will further disentangle the origins of caddisflies.

Key reading

Andersen, D. T. and Lawson-Kerr, C. (1977). The embryonic development of the marine caddisfly, *Philanisus plebius* Walker (Trichoptera: Chathamiidae). *Biological Bulletin* **153**: 98–105.

Bournaud, M. and Tachet, H. (eds) (1987). *Proceedings of the Fifth International Symposium on Trichoptera*. Entomologica series **39**. Junk Publishers, The Hague.

Ivanov, V. D. (2002). Contribution to the Trichoptera phylogeny: new family tree with considerations of Trichoptera-Lepidoptera relations. *Nova Supplementa Entomologica (Proceedings of the 10th International Symposium on Trichoptera)* **15**: 277–92.

Macan, T. T. (1982). *The study of stoneflies, mayflies and caddisflies*. Amateur Entomologist's Society, London.

Mackay, R. J. and Wiggins, G. B. (1979). Ecological diversity in Trichoptera. *Annual Review of Entomology* **24**: 185–208.

Morse, J. C. (ed) (1983). *Proceedings of the Fourth International Symposium on Trichoptera*. Entomologica series **30**. Junk Publishers, The Hague.

Misof, B., Liu, S., Meusemann, K., Peters, R. S., Donath, A., Mayer, C., Frandsen, P. B., et al. (2014). Phylogenomics resolves the timing and pattern of insect evolution. *Science* **346**: 7639–67.

Morse, J. C. (1997) Phylogeny of Trichoptera. *Annual Review of Entomology* **42**: 427–50.

Nimmo, A. P. (1996). *Bibliographia Trichopterorum: a world bibliography of Trichoptera*. Pensoft, Sofia.

Otto, C. (1987). Behavioural adaptations by *Agrypnia pagetana* (Trichoptera) larvae to cases of different value. *Oikos* **50**: 191–6.

Petersson, E. (1995). Male load-lifting capacity and mating success in the swarming caddisfly, Athripsodes cinereus. *Physiological Entomology* **20**: 66–70.

Riek, E. F. (1976). The marine caddisfly family Chathamiidae (Trichoptera). *Journal of the Australian Entomological Society* **15**: 405–19.

Shields, O. (1988). Mesozoic history and neontology of Lepidoptera in relation to Trichoptera, Mecoptera and angiosperms. *Journal of Palaeontology* **62**: 251–8.

St. Clair, R. M., Dean, J. C., and Flint, O. S. (2018). Description of adults and immature stages of *Antipodoecia* Mosely from Australia and synonymy of the families Antipodoeciidae and Anomalopsychidae (Insecta: Trichoptera). *Zootaxa* **4532**: 125–36.

Thomas, J. A., Frandsen, P. B., Prendini, E., Zhou, X., and Holzenthal, R. W. (2020). A multigene phylogeny and timeline for Trichoptera (Insecta). *Systematic Entomology* **45**: 670–86.

Winterbourn, M. J. and Anderson N. H. (1980). The life history of *Philanisus plebius* Walker (Trichoptera: Chathamiidae), a caddisfly whose eggs were found in starfish. *Ecological Entomology* **5**: 293–303.

Lepidoptera
(butterflies and moths)

Common name	Butterflies and moths	**Metamorphosis**	Complete (egg, larva, pupa, adult)
Derivation	Gk. *lepidos*–scale; *pteron*–a wing	**Distribution**	Worldwide
Size	Wingspan 3–300 mm. Mostly under 75 mm	**Number of families**	About 141
		Known world species	> 180,000 (15.90%)

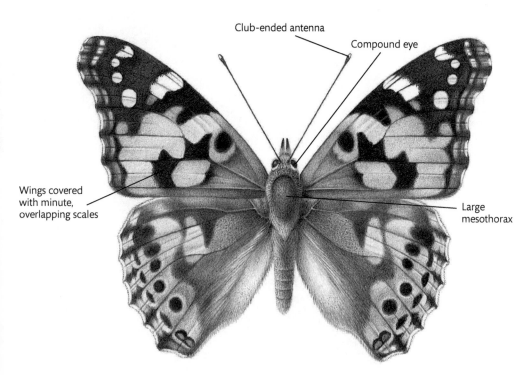

Club-ended antenna

Compound eye

Wings covered with minute, overlapping scales

Large mesothorax

Vanessa cardui – Lepidoptera

Essential Entomology. Second Edition. George C. McGavin and Leonidas-Romanos Davranoglou, Oxford University Press.
© George C. McGavin and Leonidas-Romanos Davranoglou (2022). DOI: 10.1093/oso/9780192843111.001.0001

Key features

- abundant and ubiquitous
- entire body and both sides of wings usually covered with minute, overlapping scales
- mouthparts usually in the form of a sucking proboscis (coiled at rest)
- some species are significant crop pests
- third-largest order

❯ The Dusky Thorn moth, *Ennomos fuscantaria*, feeds on European ash, as well as flowering plants and shrubs. UK.

❯ Many moths, such as *Speiredonia spectans* from Australia, possess distinctive eye spots on their wings that have evolved to startle predators.

> Plume moths are
unmistakable due to the
unique shape of their wings.
Buckleria paludum. UK.

© Paul Brock

Members of this readily recognizable order occur everywhere that there is vegetation. The first Lepidoptera appeared in the Jurassic, and then radiated greatly with the rise of the flowering plants in the Cretaceous.

The higher classification of the Lepidoptera has as much a confused history as that of many other large taxa, and there have been many attempts to divide the order into basic groups. A common theme among older classifications, and one that still pervades popular opinion, is that of a division of the order into moths and butterflies. It is often said that moths are nocturnal, drably coloured, hold their wings outstretched, and lack antennal clubs, whereas butterflies are diurnal, brightly coloured, hold their wings upright, and have marked terminal, antennal clubs; there are many exceptions to this rule of thumb. In fact, many moths hold their wings together in a roof-like manner over the body at rest, and quite a few are brightly coloured and diurnal. Many butterflies are very drab, but at least they do all have a distinctive antennal club. The skippers (Hesperiidae) with their odd, elongate antennal club are very closely related to the true butterflies, and both belong to the superfamily Papilionoidea, and together, they represent less than 15% of all lepidopteran species. The Hedylidae, with a single genus (*Macrosoma*) of 35 neotropical species, have a range of moth-like and butterfly-like characters, and were long-thought to represent a distinct superfamily. However, molecular data strongly support that they represent the sister-group to the skippers plus the true butterflies. These three groups can be considered as monophyletic.

❯ Butterflies in tropical climates frequently probe the ground for nutrients such as salt and other minerals, such as this aggregation of male Common Jays, *Graphium doson* in Sabah, Borneo.

© George McGavin

❯ The beautiful colours of butterflies, like those of this *Papilio machaon* from Greece, have long attracted the interest of entomologists, amateurs, and artists alike.

© Zestin Soh

'Moths' do not form a monophyletic group, but they are frequently referred to as such for convenience. Lepidoptera is divided into four suborders, three of them with only a handful of species; the fourth, the Glossata, contains the huge majority of the species (see Box).

⌃ The odd, elongate antennal club of hesperiid butterflies.

Lepidopteran suborders

Suborder	Superfamilies	Number of families: larval habit
Zeugloptera	1 Micropterigoidea	1 Micropterigidae: litter, fungal grazers
Aglossata	1 Agathiphagoidea	1 Agathiphagidae: seed borers
Heterobathmiina	1 Heterobathmioidea	1 Heterobathmiidae: leaf miners

Lepidopteran suborders *continued*

Suborder	Superfamilies	Number of families: larval habit
Glossata (non-ditrysians)*	11	23
	Acanthopteroctetoidea	1: leaf miners
	Andesianoidea	1: borers?
	Adeloidea	5: borers, gall formers, grazers, miners
	Eriocranioidea	3: leaf miners
	Lophocoronoidea	1: miners or borers?
	Neopseustoidea	2: miners
	Mnesarchaeoidea	1: grazers in litter, mosses, etc.
	Hepialoidea	5: stem and root borers etc.
	Nepticuloidea	2: leaf miners
	Palaephatoidea	1: leaf miners and then leaf tiers
	Tischerioidea	1: leaf miners
Glossata (Ditrysia)	30	114
	Tineoidea	5: detritivores, fungivores herbivores, etc.
	Yponomeutoidea	10: leaf miners and grazers, borers
	Galacticoidea	1: leaf tiers
	Gracillarioidea	3: leaf miners and grazers
	Gelechioidea	21: leaf miners, gall formers, rollers, detritivores, etc.
	Simaethistoidea	1: ?
	Cossoidea	7: wood borers, root feeders, stem borers
	Carposinoidea	2: borers, leaf tiers, gall formers
	Tortricoidea	1: borers, leaf tiers, and rollers
	Sesioidea	3: wood, stem, and root borers, leaf miners, and tiers
	Zygaenoidea	12: external plant feeders, ectoparasites
	Immoidea	1: external plant feeders
	Schreckensteinioidea	1: external plant feeders, grazers
	Urodoidea	2: external plant feeders
	Choreutoidea	1: external plant feeders, grazers
	Epermenioidea	1: external plant feeders
	Alucitoidea	2: borers in buds, fruit, and seeds
	Pterophoroidea	1: borers in buds, seeds, shoots, etc.
	Hyblaeoidea	1: external plant feeders, borers
	Thyridoidea	1: leaf tiers
	Pyraloidea	2: leaf rollers, tiers, or gallers
	Geometroidea	5: external-feeding herbivores, detritivores, coprophages, predators
	Drepanoidea	3: external plant feeders
	Calliduloidea	1: external plant feeders
	Papilionoidea	7: external plant feeders, leaf folders, tiers, and borers
	Lasiocampoidea	1: external plant feeders
	Mimallonoidea	1: external plant feeders, saprophages, predators
	Bombycoidea	10: external plant feeders, communal web makers, leaf tiers
	Noctuoidea	6: external plant feeders, some predators
	Whalleyanoidea	1: ?

* Here we have arranged all the non-ditrysian taxa together, although they are, strictly speaking, regarded as six separate infraorders.

The Micropterigidae comprise about 180 species of small, caddisfly-like moths whose adults have proper chewing mandibles, which they use to feed on pollen and spores. The larvae feed on leaf litter, fungal hyphae, or bryophytes. The two known species of the primitive Agathiphagidae occur in Queensland and Fiji, where the larvae feed on the seeds of Kauri pines (*Agathis* spp.). As in the Micropterigidae, the adults have articulated mandibles, but these are probably non-functional. About a ten species make up the Heterobathmiidae. The larvae mine inside the leaves of Southern Beech (*Nothofagus* spp.). The adults have functional mandibles and probably feed on *Nothofagus* pollen.

The Glossata, species with a proboscis of varying length (sometimes very reduced or vestigial), can be divided into two groups, the more advanced of which forms a taxon known as the Ditrysia. In these lepidopterans, the females have two genital openings, one of which is used by the male during copulation; the other is used for the passage of eggs during oviposition.

The most characteristic features of the order are the **minute overlapping scales or hairs that cover the entire body surface and wings**, and, typically, the possession of a **long, coiled proboscis**, which is used to suck liquids—usually nectar fruit juices or honey dew. The proboscis is formed from the very long maxillary galeae, which are deeply grooved and locked tightly together by means of opposing rows of hooks and spines. Despite the number of people who spend much of their lives studying these insects, the exact way in which the proboscis is extended and coiled is not yet fully understood. Another structure unique to most lepidopterans is the epiphysis—a spur-like structure located on the inner face of the front tibia, used in cleaning the proboscis.

The extraordinarily rich variety of pattern and colour seen in lepidopteran wings, must account, at least in part, for the enormous popular appeal of these insects. The wing scales that give these insects their colours and patterns can be pigmented or microsculptured with fine longitudinal ridges, such that they refract the light falling on them. The brilliant and iridescent sheen seen on the wings of many species are due to this physical colour production. It is well known that wing colours and patterns, whether warning or cryptic, protect from attacks by predators, such as birds, and, in many species, palatable species mimic the colours and patterns of non-palatable species. As long as the mimic is sufficiently rare, naive predators will always associate a particular colour pattern with non-palatability and the mimic will survive. There are two main types of mimicry: **batesian** and **mullerian**.

Wings are also of use in thermoregulation and courtship. Some butterfly wings have ultraviolet reflecting or absorbing patches, which are used in signalling. In some species, males have specialized wing scales called androconia that are responsible for the release of sexual odours. The head bears **well-developed compound eyes, the antennae**, and, **sometimes, two ocelli**. The antennae, which bear chemosensory receptors capable of responding to minute quantities of volatile compounds, such as sexual odours, are branched or feathery in male moths, and simple in female moths. The thorax carries three pairs of legs and two pairs of wings, which are mechanically coupled by means of a frenulum and retinaculum or overlapped during flight.

> batesian mimicry
where an edible or palatable species mimics the colours and patterns of an inedible or unpalatable species. The mimics are rare relative to the models

> mullerian mimicry
where several inedible or unpalatable species converge in colour pattern. As all the species are non-palatable, mimicry rings, as they are known, are enhanced by the frequency of all ring members and even distantly related species can be involved

In males, the frenulum is a single stiff bristle, whereas in females it is made up of more than one bristle. In butterflies, the base of the hind wings is enlarged and serves to couple the wings in flight by supporting the front wings.

Mate attraction and courtship in lepidopterans is highly variable, but typically involves displays, either singly or in groups, and the production of sexual pheromones. In many moths, resting females release a pheromone from abdominal glands. These scents are carried downwind and may be active over very long distances. Chemosensory sensilla on the antennae of males pick up the scent, often at very low concentrations. Males fly upwind to locate the female, court her, and copulate. The males of some species may gather in small groups or swarms near obvious features, such as a patch of sunlit foliage, and dance to attract females. Males also produce pheromones to attract females. In some arctiids and noctuids, males may have inflatable scent organs called coremata, which can be everted during courtship. The scent produced by these and other structures have several effects. They clearly act as a powerful aphrodisiac, but also keep the females motionless. There are instances where the production of sexual pheromones might be the downfall of a male moth. The bolas spider is able to mimic the scent of female moths, and, therefore, able to lure males within range of its sticky snare. Acoustic signals are used in the mating rituals of some Arctiidae, Pyralidae, and Sphingidae. In the Lesser Wax Moth, *Achroia grisella* (Pyralidae), males produce ultrasonic calls to attract mates over short distances. Louder ultrasound to attract females is produced by males of an Australian noctuid moth (*Hecatesia exultans*). The sounds carry up to 25 m, and males entering the territorial range of another male will be chased off. Some sounds produced by adult lepidopterans are defensive.

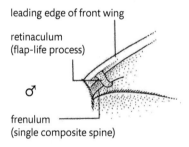

leading edge of front wing

retinaculum
(flap-life process)

♂

frenulum
(single composite spine)

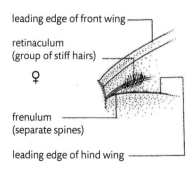

leading edge of front wing

retinaculum
(group of stiff hairs)

♀

frenulum
(separate spines)

leading edge of hind wing

The wing coupling mechanism of male and female moths—viewed from underneath where the wings join the thorax.

Butterflies that are diurnal rely more on visual courtship displays than most moths. The males of many species are territorial and fly along a regular patrol route looking for females to mate with or other males to chase away. A good way to lure the brilliant blue males of *Morpho* butterflies is to whirl a piece of metallic blue shiny paper on a weighted string around your head. The territories held by males may be determined by the presence of the larval food plant (and therefore oviposition site) or may be just easily recognizable features of the landscape, such as hilltops, clearings in woods, or patches of bare ground. The use of male pheromones, produced by hairs and scent scales on the wings or from inflatable abdominal hair pencils, is widespread and a variety of aerial manoeuvres performed by the males ensure that the females are liberally covered with the 'perfumes'.

After mating, the males of some species shower the female with a different scent, designed to put other males off. More commonly after mating, especially in the Nymphalidae and Papilionidae, the male will insert a sticky secretion produced by his accessory glands into the genitalia of the female. This substance hardens to form a plug, which prevents other males from mating. The eggs, of course, can still be laid because butterflies are ditrysians.

> **ditrysian**
having two distinct sexual openings: one for mating, and the other for laying eggs

Eggs are scattered in flight, laid singly, or laid in batches near or on the larval food. Leaf mining or boring species must burrow into the food plant. Lepidopteran larvae, or caterpillars, are elongate and cylindrical in shape, and may be smooth, spiny, or hairy. The toughened head has chewing mouthparts, very short, three-segmented antennae, and two laterally placed groups of about six ocelli. There are three pairs of thoracic legs, each with a single claw, and a variable number of abdominal prolegs. The abdomen typically has ten segments, and prolegs usually occur on segments 3–6 and 10, but there may be fewer than this number in certain families. The ends of the fleshy prolegs have many tiny hooks (crochets) arranged in circles or bands to help them hold on to the foodplant. The vast majority of caterpillars are herbivorous and often plant-specific, but there are species who eat fungi, dried organic material, and lichens. The caterpillars of the moth *Ceratophaga vastella* (Tineidae) burrow through the horns of dead animals, such as buffalo, and it is common to find festoons of pupating moths, or their empty cocoons, adorning skulls in the African savannahs. Nothing goes to waste in the natural world.

A few butterflies and moths have developed a carnivorous habit. The caterpillars of certain species in moth families such as the Psychidae, Tineidae, Pyralidae, and Noctuidae are predators of various types of scale insect. The Australian moth family Cyclotornidae (Zygaenoidea) contains a strange species, *Cyclotorna monocentra*, whose first instar caterpillars feed externally on the tissues of leafhoppers. The first instar caterpillars then build themselves a shelter and moult to the second instar. The leafhopper colonies are attended by ants, and interestingly, the second instar caterpillars are taken by the ants into their nest. Here the caterpillars eat ant larvae but are not attacked by their hosts on account of a sweet secretion they give in return. Also in the superfamily Zygaenoidea, the Epipyropidae harbours some more odd species. The caterpillars of small moths in the genera *Epipyrops* and *Agamopsyche* are parasitic. About 40 species are confined to the tropical and warm temperate regions of the southern hemisphere. Large fulgorid bugs are the preferred hosts, but some cicadellids and cicadas also serve as hosts for certain species. Typically, the caterpillars are external parasites, feeding either singly or gregariously in small numbers on the body fluids of their adult (and sometimes nymphal) host bug species. Once a young caterpillar finds a suitable host, it uses long, sharp mouthparts to penetrate the cuticle. When fully grown, the fifth instar caterpillar detaches itself and pupates inside a cocoon. The effects on the host are variable depending on size and the number of parasites. At one end of the spectrum, the host may show hardly any ill effects, while at the other end, the host may die soon after the caterpillars leave.

Even stranger are the ambush tactics employed by the caterpillars of some Hawaiian moth species belonging to the genus *Eupithecia* (Geometridae). These strange caterpillars strike in a snake-like manner at passing insect prey, mainly small flies such as *Drosophila*.

Carnivorous caterpillars may come as something of a surprise, but among the fruit-piercing moths that belong to the family Erebidae, there are species of vampire moth. *Calyptra eustrigata* and some other species have made the dietary switch from fruit juice to mammalian blood. It is interesting to note that trainee medics use oranges to practice giving injections, as the resistance offered by the peel is very similar to human skin.

Moths have other interesting relationships with mammals. In the family Pyralidae, five species of moth in two closely related genera, *Cryptoses* and *Bradypodicola*, have intimate associations with tree sloths. That the moths lived in the fur of the sloths was known for some time, and it was assumed that the caterpillars lived there as well, grazing on the algae growing on the hosts' fur. Sloths are celebrated for their lack of activity, but every week or so they climb down to the ground to defecate. Why they do not simply drop their dung from where they hang, high in the rain forest canopy, has more to do with sloth social life than having an easy life. At any rate, it is at this point that the sloth moths spring into action. Mated females leave the sloth while it is defecating and lay their eggs on the dung. The caterpillars are, of course, coprophagous. The female flutters back to her stately conveyance before it disappears up the tree again, and, in time, presumably, the newly emerged moths will find a sloth visiting the latrine.

Caterpillars may protect themselves by hiding in shelters or cases. The caterpillars of Psychidae, which are commonly called bagworms, make tough portable shelters from a variety of plant materials, including spines and leaf fragments. Shelter building by some caterpillars can greatly increase their survival by allowing them access to a food source not easily utilized by other insect herbivores. Such an approach is followed by a gelechiid moth in the genus *Polyhymno*, whose larvae live inside a shelter they create by joining separate leaflets of *Acacia cornigera* together with silk. This plant has no chemical defence, but instead relies entirely on patrolling ants (*Pseudomyrmex ferruginea*) for herbivore control. Unfortunately for the tree, the ants are unable to effectively rid it from this particular herbivore, as the latter constructs an effective and largely ant-proof shelter. In fact, *Polyhymno* infestations can be so severe that they often result in the death of the tree. The case building habit is also seen in other moth families, such as the Tineidae and Coleophoridae. A leaf mining or boring habit might be seen as a common defence from predators, but many parasitic wasps are specialized to seek out feeding caterpillars concealed in these ways. Parasitoids may be able to detect the movements of leaf miners, but the probing and insertion of the wasp's ovipositor can trigger avoidance response in some species.

Externally feeding herbivorous caterpillars employ a range of physical and chemical defences. They may possess a dense covering of poisonous or irritating hairs or spines, and some may be cryptically coloured, or resemble twigs or bird droppings. Others may be brightly coloured to warn birds of the presence of toxic substances within their body, or the ability to release noxious substances from eversible glands called osmeteria. Threat postures that display contrasting eyespots are fairly common, but some hawk moth caterpillars have evolved patterns that are believed to resemble the heads of snakes.

After anything between four and nine larval stages (typically five), pupation occurs. Lepidopteran pupae may be enveloped inside a silk cocoon spun by the mature larva, inside an underground silk-lined cell, or suspended in various ways from the foodplant. The term chrysalis is often used for lepidopteran pupae. In primitive moths, the pupa has moveable mandibles (decticous), and the appendages are not stuck to the body (exarate). In the majority of species, however, the pupa does not have moveable mandibles (adecticous), and the appendages are stuck to the body surface (obtect). The pupae of the higher Lepidoptera are not totally defenceless as some have special gin traps made of the opposable edges of adjacent abdominal segments. These can be snapped shut by wriggling body movements to deter egg laying by a pupal parasitoid.

In all insects air is distributed to the tissues through a system of spiracles, tracheae, and tracheoles. Tracheoles ramify profusely throughout the tissues, where oxygen and carbon dioxide are exchanged. However, not all tracheae end up in tissue. The purpose of unusual tracheal tufts in the rear end of many caterpillars' bodies was unknown for almost a century. Now, incredibly, we believe that they function pretty much in the same way as lungs. These thin-walled tracheal tufts arise from the posterior pair of spiracles. While some of these tuft branches are anchored in heart tissue, the majority of the tufts are free to wave to and fro in the haemolymph as the heart contracts and relaxes. In times of oxygen depletion, a certain type of haemocyte is attracted to these tufts and becomes closely apposed to the surfaces of the thin-walled tracheae in the tufts where they become oxygenated. Some of the tufts are concentrated in a special compartment called the tokus at the very rear of the caterpillar's abdomen. When oxygenated, haemocytes leave this compartment, and they are in exactly the right position to be picked up by the posterior ostioles of the dorsal heart and pumped forward to the anterior end of the body. Further studies may show that this system is common to the Lepidoptera and may even occur in other orders.

The following section is solely concerned with some of the most widespread ditrysian families and is not meant to imply that the early super-families of the Glossata do not contain very interesting species. The case of

Yucca plants and moths belonging to the non-ditrysian moth family Prodox-idae (Adeloidea) is worthy of mention as it is a classic case of an insect–plant mutualism. The female moths pollinate the *Yucca* flowers, but their larvae eat some of the plant's seeds.

The largest family in the Tineoidea is the Tineidae, and we have identified about 3,000 species of these drab, usually dull-brown moths. Tineid cater-pillars are best known for the destructive habits of some species that attack fur, woollen, and other textiles, and dry foods, but the majority are scav-engers and feed in fungi, decaying wood, or dry plant and animal matter. Very few species feed on plant foliage. The caterpillars make a tunnel or web of silk wherever they feed or construct a portable case from silk and debris. Many species are pests, and several have a worldwide distribution due mainly to commerce. The Webbing Clothes Moth (*Tineola bisselliella*), the Tapestry Moth (*Trichophaga tapetzella*), and the Case Bearing Clothes Moth (*Tinea pel-lionella*), are common household pest insects, whose caterpillars make holes in clothes and carpets. *Nemapogon granella* is a widespread pest of stored grain, and *Niditinea fuscipunctella*, the Poultry House Moth, can be found in chicken houses, where the caterpillars feed on fragments of feathers.

Within the Yponomeutoidea, the Plutellidae is economically significant. More than 100 species are found throughout the world, particularly in trop-ical and subtropical regions. The family includes the diamondback moths of the genus *Plutella*. Most species are cryptic, and many have bands or irreg-ular markings. When the wings are folded, many species appear to have a row of diamond-shaped marks along the back. The caterpillars mine, bore, or feed externally on a range of trees, shrubs, and plants, and often make a flimsy silk web. Many species are vegetable crop pests, especially on cabbage species. *Plutella xylostella* is a very serious pest worldwide, especially since it has developed resistance to *Bacillus thuringiensis* biopesticides.

More than 4,500 species of small or very small, grey or brown moths be-long to the family Gelechiidae (Gelechioidea). These moths can be found in a large range of habitats and the caterpillars feed externally, or inside leaf rolls, stems, leaf mines, or galls. Many species are serious agricultural pests and attack crops ranging from potato and peanuts to tomato and strawberries. Coniferous and fruit trees can also suffer great damage. The An-goumois Grain Moth (*Sitotroga cerealella*) is a serious, cosmopolitan pest of corn and stored grain, and the caterpillars of the pantropical Pink Bollworm (*Pectinophora gossypiella*) are major pests of cotton.

Carpenter moths (Cossidae) are medium or large, heavy-bodied moths, which usually have spotted, irregularly patterned, or mottled wings. The adults are nocturnal, and females lay their eggs in cracks in tree bark, or in the emergence tunnels of adult moths. Although the larvae of a few species bore into the pith of reeds, most attack trees. The caterpillars burrow into the wood of the branch or trunk on which they were laid, and, as they grow, the diam-eters of the tunnels inside the wood get bigger and bigger. Due to their large size and poor diet, cossids can take several years to reach adulthood. The fully grown larvae pupate within a cocoon made of silk and chewed wood fibres, usually in their tunnels close to the surface or in the soil. Characteristically, the empty pupal skin remains partly sticking out of the exit hole after the adult moths have emerged. Many species are pests of commercially important trees, such as hickory, maple, ash, oak, and pine, as well as fruit trees. The larvae of leopard moths, *Zeuzera pyrina* and *Zeuzera coffeae*, damage and may kill fruit and coffee trees, respectively. *Cossus cossus*, the European Goat Moth, whose larvae may be up to 100 mm long when fully grown, damage fruit and the softwood trees.

'Witchety' grubs, a favourite and highly nutritious delicacy of Australian aboriginal people, are the larvae of some cossid species that bore inside the roots and stems of certain *Acacia* (wijuti) species.

The Tortricoidea contains a single large family of smallish moths, the Tortricidae, with more than 6,340 species worldwide. Many species are cryptically coloured in brown, green, or grey, with patterns that resemble lichen, bark, or even bird droppings, but some species are brightly coloured. Most species have square-ended and broad front wings that are held roof-like over the body at rest. Tortricid caterpillars feed externally or internally on leaves, shoots, and buds, and many species tie or roll leaves together with silk. Some species bore into fruit, seeds, or stems. These moths eat a huge range of plants, and some species can be extremely polyphagous, having been recorded from almost 100 different host plants. Caterpillars of *Cydia saltitans* feed inside the seed of Croton plants in parts of southern USA and Mexico. If too hot, fully grown larvae wriggle violently inside the seed, causing it to jump, in an effort to move into shade. These 'Mexican jumping beans' are sold as curiosities. About one-third of all tortricid genera contain pest species that attack crops from avocado and lucerne to all manner of trees and soft fruit crops. A well-known and serious pest in apple orchards, the Codling Moth (*Cydia pomonella*), was introduced to North America from Europe over 200 years ago.

There are more than 1,360 species of clearwings or clear-winged moths belonging to the Sesiidae, the largest family in the Sesioidea. These day-flying moths are brightly coloured, black, bluish, or dark brown with yellow and orange markings, and many species mimic wasps or bees by having large areas of the wings clear of scales. The wings of many species are fully scaled at first, but scales are shed during the adult's first flight. Others may have no scales from the start or may even have transparent scales. The Hymenoptera-like appearance of these moths is accentuated by a banded abdomen and a buzzing flight. Caterpillars bore inside the branches and roots of trees and woody shrubs, and development of the caterpillars can take up to two years. *Sesia apiformis*, the Hornet Clearwing Moth, damages the trunks and roots of willows and poplars, and peach trees in North America suffer from damage by caterpillars of *Synanthedon excitosa*, the Peach Tree Borer.

Pyralid, snout, or grass moths (Pyraloidea) make up a very large superfamily of more than 15,600 species. They are very diverse varying enormously in size, colour and shape. The wings may be broad or narrow and have closely packed scales. In some species the front of the head has a small 'snout' formed by elongate palps held outstretched. These moths are very widespread, and the caterpillars can be herbivores or scavengers. Caterpillars of *Galleria mellonella*, the Wax Moth (Pyralidae), even have special enzymes for digesting the

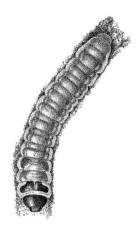
^ The caterpillar of the European Goat Moth (*Cossus cossus*) tunnels in the wood of tree trunks and branches.

honeycomb in bee colonies. The caterpillars of this and many other pyralid species make silk tubes or webs on their food. A few species are predators of scale insects or wood-boring larvae and a few feed in the roots of aquatic plants. A very large number of Pyraloidea are pests and the list of crops and stored products damaged by these small moths is very long indeed. The Cosmopolitan Meal Moth, *Pyralis farinalis* (Pyralidae), is a pest in stored grain and caterpillars of the European Corn Borer, *Ostrinia nubilalis* (Crambidae), damage young corn.

The Geometridae is an enormous moth family with more than 23,000 species. The wings of these moths often have patterns of fine lines and markings on a cryptic brown or green background. The bright green colour of some of these moths is due to a unique pigment called geoverdin, which is contained within the wing scales, and may be derived from chlorophyll. Geometrids can be found virtually everywhere there is vegetation, and the adults are usually nocturnal. The name geometrid, literally 'earth measurer', is derived from the characteristic looping motion of the caterpillars. Geometrid caterpillars, which are called inchworms or loopers, have prolegs on the sixth and tenth abdominal segments and move by drawing their hind end up to meet the front end. They then hold on with the abdominal prolegs and extend the head end forward and grip the substrate with the thoracic legs.

Caterpillars are protected by being cryptically coloured, and, at rest, many will assume a very twig-like appearance, sticking out at an angle from the host plant. The range of plants attacked by these moths is immense, and many species are serious pests of commercial deciduous and coniferous trees, being capable of causing severe damage and defoliation. The Winter Moth (*Operophtera brumata*) is a typical pest species, which was accidentally introduced to North America from Europe. The caterpillars feed on a wide range of deciduous trees, including apple and pear. Geometridae have a pair of well-developed tympanal organs located at the base of the abdomen. These organs give the moths bat-detecting ultrasonic hearing, which operates in the same range as the echolocating signals sent out by most bats (25–45 kHz). The flight behaviour of the moths depends on the intensity of the signal received. If the bat is perceived as very close, the moth will perform a spiral dive towards the ground; otherwise, the moth may just simply change direction.

The Papilionoidea or true butterflies comprise more than 17,500 species in seven families (including Hedylidae and Hesperiidae). The skippers (Hesperiidae) are a group of about 3,500 diurnal species of early butterflies, which until recently, were considered as related to, but outside, Papilionoidea. In general, they are moth-like and heavy-bodied with a wide head. Most species are drab brown with white or orange markings, but a few are pale with a blue, green, or purple sheen. The antennae terminate characteristically in an elongated club, which comes to a point and is often curved. The name 'skipper' refers to their rapid, darting flight pattern. The caterpillars, which are green, brown, or white, often with longitudinal stripes, mainly eat grasses and sedges, and the foliage of some herbaceous plants and trees. They usually feed at night and live within a shelter of silk-tied or rolled leaves during the day. A few grass-feeding species can be minor pests, as can *Urbanus proteus*, the Bean

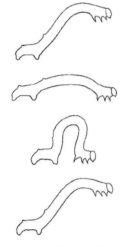

∧ A geometrid caterpillar moves in a characteristic looping manner using its thoracic legs and abdominal prolegs as alternate anchors.

Leaf Roller, which feeds on leguminous and cruciferous plants, and the Rice Leaf Roller (*Lerodea eufala*), which can attack sorghum and sugar cane.

Papilionids are often large and can be spectacularly attractive. Most of the 580 or so world species are swallowtails (Papilioninae), with short or long tails directed backwards from the ends of the hind wings. There are about 50 species of apollos (Parnassiinae), which do not have hind wing tails, and are white or grey in colour often with two red or yellow spots on the hind wings. Swallowtails are widespread, but apollos live in mountainous regions of the northern hemisphere. A characteristic feature of the caterpillars is a forked thoracic scent gland called an osmeterium. When threatened, the caterpillar everts the brightly coloured osmeterium, which gives off a repellent odour.

When fully grown, swallowtail caterpillars pupate on their host plant and the chrysalis is held upright by a silk belt around the thorax. Apollos pupate in plant litter or under stones surrounded by a loosely woven cocoon. These butterflies are of great aesthetic value, and the family contains the spectacular birdwing butterflies of south-east Asia. *Ornithoptera alexandrae*, from the south-east of Papua New Guinea, is the biggest butterfly in the world with females measuring 28 cm (11 in) across the wings.

∧ A characteristic feature of the Swallowtail Butterfly caterpillars is a forked thoracic scent gland called an osmeterium—here everted in defensive posture.

The Pieridae are among the commonest butterflies anywhere in the world. The common names, whites, sulphurs, and orange tips, refer to the wings of these butterflies, which are usually white or yellow, with orange, black, or dark grey markings. Pierids can often be seen feeding in groups at bird droppings, puddles, or wet patches on the ground. It is thought that this 'puddling' behaviour, which is common in males, provides dietary salts. Caterpillars of the best-known species feed on cabbage species and leguminous plants. The chrysalis, which is angular with a single head projection, is fixed upright to the host plant and held in place by a silk girdle around the abdomen. Many species are serious agricultural pests. The Cabbage White or Large White (*Pieris brassicae*) is a notorious pest of cabbage throughout Europe and the Mediterranean, and the Small White (*Pieris rapae*), which also attacks wild and cultivated cabbage, is perhaps the most destructive of all butterflies.

More than one-third of all butterflies belong to the family Nymphalidae, which is composed of several distinctive subfamilies with a total of about 5,700 species. Nymphalids are often called brush-foot butterflies because the front legs are very small and brush-like, and not used for walking. A good identification feature, therefore, is the presence of only four walking legs. The front legs have a sensory function, and, in females, at least, they are probably useful in host plant recognition. Nymphalid caterpillars are generally spiny and have many peculiar, branched projections, or have horns at either end of the body. They feed externally on the foliage of a huge range of host plants. The angular chrysalis is suspended, head down, from the host plant by a small group of terminal hooks called the cremaster. The spectacular owl butterflies (Brassolinae), the brilliantly metallic blue *Morpho* butterflies (Morphinae), and the passion vine-eating heliconiids (Heliconiinae) live in Central and South America. Browns and ringlets (Satyrinae) are pale yellowish-brown to dark brown or greyish coloured butterflies, which often have numerous eyespots around the wing margins. These butterflies are characteristic of heaths and meadows, particularly in upland areas, and adults fly in an erratic and bobbing manner. Satyrine caterpillars all feed on grasses or sedges and can be easily recognized by their forked tail segment. The subfamily Nymphalinae is the biggest in the family with more than 3,000 species. A well-known species, the Painted Lady (*Vanessa cardui*), is migratory and can be found all over the world. The caterpillars of this species usually feed on nettles and thistles, but it can occasionally be a minor pest of soya bean. The milkweed butterflies belong to the Danainae, and the best-known species are conspicuously coloured blackish-brown and reddish-orange with small white marks. The American Monarch Butterfly (*Danaus plexippus*) is typical. Adults and caterpillars are highly distasteful to predators on account of the toxic, cardiac glycoside compounds they sequester from their *Asclepias* host plants (Asclepiadaceae). The colouration of the adults and the yellow and black bands of the caterpillars warn predators of their non-palatability. The Monarch is a famous migratory species that has established itself all over the world. In North America it migrates from Canada to California and Mexico, where it overwinters in spectacular mass roosts.

More than 6,000 species of blue, copper, or hairstreak butterfly belong to the family Lycaenidae. Most species are slender-bodied and have a tropical or warm temperate distribution. Some species are very brilliantly coloured with iridescent scales, and others have slender tails on the hind wings. The caterpillars feed on the leaves, fruits, and flowers of a wide range of plants, and some species are predacious on aphids, coccids, and other small soft-bodied insects. Nearly one-third of the known larval life histories show complex, mutualistic associations with ant species. Lycaenid caterpillars are squat, and from above they look like a small slug or woodlouse. Caterpillars associated with ants secrete a special fluid containing sugars and amino acids from abdominal 'honey glands'. The ants allow the caterpillars to eat their larvae, or sometimes the aphids from which they harvest honeydew, in return for access to the honey gland fluid. The ants will even guard these caterpillars, and there is good evidence that the ants prevent parasitic wasps from attacking them as well.

The Lasiocampoidea comprise more than 1,500 species of eggar moths (Lasiocampidae), although lappet moths (Anthelidae) are sometimes included in this superfamily. Lasiocampids are common in wooded areas, hedgerows, and heathlands, where their host trees, grasses, or herbaceous plants grow. The caterpillars are often very hairy or have dorsal and lateral tufts and downward-pointing hair fringes. The caterpillars of some species live and feed communally inside tents or webs, which they spin across the foliage of their food plant. Pupation takes place inside a tough, papery, egg-shaped cocoon. Many lasiocampids, such as the tent caterpillars of the genus *Malacosoma*, are serious forest and orchard pests.

Moths belonging to the large superfamily Bombycoidea are stout-bodied and have broad wings. Although the silk of lasiocampid cocoons has been used by humans in the past, the main commercial silk-producing species belong to the Bombycidae. There are about 60 species of silk moths, the best known of which is the Silkworm, *Bombyx mori*. Silk production originated in

China more than 4,500 years ago, and the Chinese fiercely guarded their valuable secret (silk was worth more than gold) until AD 555, when two Roman monks smuggled some moth eggs and the seeds of the mulberry plant upon which the caterpillars fed to Constantinople and other parts of the Eastern Roman (Byzantine) Empire. The caterpillars are kept in large, airy, rearing drawers, and eat huge amounts of leaves in the course of their development. When fully grown, the caterpillars spin a cocoon in which to pupate. These are collected and boiled, and the silk is wound off onto reels (the hard bit is to find the end of the thread). A single cocoon may provide hundreds of yards of silk, but the silk of several cocoons has to be twisted together to make a single silk thread. Up to 2,000 cocoons may be needed to provide enough silk for a dress. Nothing is wasted as the boiled pupae are canned and sold as food.

The Saturniidae (atlas, emperor, and royal moths) are very large, heavy-bodied moths with very broad wings, which may span 200 mm or more. The wings of most of the 1,500 species have a characteristic eyespot near the centre and transparent patches, and hind wing 'tails' also occur in some species. The functional significance of the long hind wing tails of some species had long been debated. They are clearly important, as they have evolved multiple times independently, but what do they actually do? Saturniids fly at night, during the peak of bat activity, but they are at first glance defenceless against their attacks. Research has shown that these tails act as acoustic decoys to echolocating bats. The tails do not contain vital organs, nor are they crucial for flight. When in flight, these tails spin in a particular fashion, which lures bats to attack the tails and not the main body of the moth. In this way, moths survive more than half of bat attacks, and it is not infrequent to find saturniids with one or both tails missing.

^ A caterpillar of the Adonis Blue Butterfly (*Lysandra bellargus*) is attended by ants, which feed on the sweet secretions from the caterpillar's 'honey gland'.

Most adult saturniids inhabit wooded areas, and the caterpillars feed on a wide range of deciduous and coniferous trees, and shrubs. The body surface of caterpillars has fleshy protuberances called scoli, which bear spines and long hairs. Pupation takes place inside a very dense silk cocoon attached to the host plant. The biggest moths in the world belong to the genus *Attacus*, species of which can be found from South America to Mexico, and from Africa to the Orient. *Samia cynthia*, the Ailanthus Silk Moth or Eri Silk Moth, has been reared for the commercial production of silk. The green caterpillars of the spectacular pale green, long-tailed American Luna Moth (*Actias luna*) feed on deciduous trees, such as birch, hickory, and walnut.

Another charismatic family of the Bombycoidea are the hawk moths (Sphingidae), with just over 1,200 species. Hawk moths are a medium to quite large, heavy-bodied, powerful species with long, narrowish front wings. The proboscis, which is long, and in some species, several times longer than the whole of the body, is curled under the head when not in use. Some species (*Hemaris* and *Cephonodes* spp.) resemble bees with large transparent areas on the wings, and others (*Macroglossum* spp.) look like hummingbirds as they hover at flowers. Adults feed on nectar, which is sucked from the corolla tubes of flowers using a long proboscis. Hawk moth caterpillars, which can be brightly coloured or cryptic, have a characteristic dorsal spine-like process at the end of the abdomen. Most species pupate and overwinter in weak co-coons. The pupal stage of some species has a peculiar handle-shaped structure at the anterior end, which contains the proboscis. The Death's-head Hawk Moth, *Acherontia atropos*, is native to Africa, but migrates to Europe during the summer. The caterpillars eat potato and other related plants, and the adults raid bee colonies for honey. They make high-pitched piping noises similar to that made by queen bees, and it is thought that this calms the bees. *Xanthopan morgani* has an incredibly long proboscis reaching more than 25 cm (10 in) to feed from the very long nectaries of certain African orchids. Some hawk moths are serious crop pests. *Manduca quinquemaculata*, the Tomato Hornworm of North America, and *Manduca sexta*, the Tobacco Hornworm, damage tobacco, tomato, potato, and other plants in North America and elsewhere. The Striped Hawk Moth (*Hyles lineata*) is cosmopolitan, and its caterpillars attack a wide range of commercially important plants.

The Noctuoidea is the largest lepidopteran superfamily with well over 70,000 species. The largest family are the Erebidae, which after major reclassification based on molecular data now include 18 subfamilies which were previously considered distinct families, such as: the Lymantriinae (tussock moths), the Arctiinae (tiger and ermine moths), and the Herminiinae (litter moths), among others. Many of the erebid subfamilies were also part of the Noctuidae, the family which gives Noctuoidea its name. It is likely that the classification of this megadiverse lineage will change several times in the future, until multiple data types lead to a more stable phylogenetic system.

⌃ The larva of a typical hawk moth, *Sphinx ligustri*, showing the characteristic terminal horn.

Female tussock moths are weak fliers, and usually lay their eggs in quite large clumps on the bark of a great variety of host trees and shrubs, and often protect the mass by depositing irritant hairs from the end of their abdomens. The remarkably hairy and mostly brightly coloured caterpillars are phytophagous and feed externally, and often gregariously, on the foliage of their host plant. Most species have shaving brush-like tufts of hairs on the dorsal and lateral body surfaces. As well as thoracic legs, prolegs can be seen on five of the abdominal segments. The Gypsy Moth (*Lymantria dispar*) was introduced to North America in 1869, with the idea that its cocoons would yield silk. As has happened so many times with such introductions, it was a terrible idea as the moths escaped, and since then, have spread to cover large areas of the USA and Canada. In the early 1980s, this species defoliated an estimated 10 million acres of trees in eastern United States. The caterpillars of another pest, the Brown Tail Moth (*Euproctis chrysorrhoea*), feed communally inside silk webs on hawthorn and blackthorn trees. Humans can get incredible skin rashes from coming in contact with lymantriids, especially the caterpillars.

More than 11,000 species of tiger and ermine moths belong to the Arctiinae (formerly Arctiidae). The adults and their hairy caterpillars are poisonous, as many species feed on plants, such as potato, ragwort, and laburnum, which contain cardiac glycosides and other toxic substances. In addition to being distasteful, the adults have thoracic sound-producing organs, which emit trains of ultrasonic clicks that jam bat echolocation systems, enabling them to escape. The caterpillars, which are often as brightly coloured as the adults, feed at night and hide during the hours of daylight. Many species are polyphagous and normally feed on a wide range of low-growing herbaceous plants, although some species eat the foliage of deciduous and coniferous trees.

The Noctuidae are one of the largest lepidopteran families with more than 12,000 species, although this number may change with future reclassifications. Noctuid caterpillars, called cutworms, armyworms, and loopers, are among the world's most devastating agricultural pests, and affect a large range of crops, vegetables, and other important plants. Cutworms (Noctuinae, especially *Agrotis* spp.) generally chew through the base of the stem and kill the entire plant, and even a relatively small number of caterpillars can completely destroy a whole stand of corn or cotton.

> ∧ **Cutworms generally chew through the base of the stem and kill the entire plant.**

Many species are polyphagous, for example, the Old World Bollworm (*Helicoverpa armigera*) damages cotton, maize, and tomatoes. The huge list of names, which includes the Clover Cutworm, the Soybean Looper, the Corn Earworm, the Alfalfa Looper, the Celery Looper, the Tobacco Budworm, the Wheat Armyworm, the Lawn Armyworm, the Cotton Bollworm, and the Flax Bollworm, should convince anyone that nothing worth growing is going to be immune from these moths.

Phylogenomic studies estimate that the origins of Lepidoptera go back to the Late Carboniferous (about 300 mya), although the order started greatly diversifying much later, in the Middle Triassic (241 mya), when the evolution of the proboscis allowed lepidopterans to exploit nectar during the rise of the flowering plants (Angiosperms). The ancestor of true butterflies (Papilionoidea) was likely nocturnal (like most moths), but later descendants switched and became diurnal, most likely to exploit the abundant nectar of flowering plants that was available during daytime.

The oldest fossils of the order come in the form of lepidopteran wing scales, and date to the Jurassic (about 200 mya). They are 100 million years younger than the phylogenomic estimate of the origin of the order mentioned above, but this is likely a consequence of the scarcity of early Lepidoptera in the fossil record.

Key reading

Ackery, P. R. and Vane-Wright R. I. (1984). *Milkweed butterflies: their cladistics and biology.* British Museum (Natural History), London.

Barber, J. R., Leavell, B. C., Keener, A. L., Breinholt, J. W., Chadwell, B. A., McClure, C. J., Hill, G. M., and Kawahara, A. Y. (2015). Moth tails divert bat attack: evolution of acoustic deflection. *Proceedings of the National Academy of Sciences of the United States of America* **112**: 2812–16.

Bradley, J. D. (1982). Two new species of moths (Lepidoptera, Pyralidae, Chrysauginae) associated with the three-toed sloth (*Bradypus* spp.) in South America. *Acta Amazonica* **12**: 649–56.

Brehm, G., Fischer, M., Gorb, S., Kleinteich, T., Kühn, B., Neubert, D., Pohl, H., Wipfler, B., and Wurdinger, S. (2015). The unique sound production of the Death's-head hawkmoth (*Acherontia atropos* (Linnaeus, 1758)) revisited. *Science of Nature* **102**: 43.

Brunton, C. F. A. and Majerus, M. E. N. (1995). Ultraviolet colours in butterflies: intra- or inter-specific communication? *Proceedings of the Royal Society of London B* **260**: 199–204.

Carter, D. J. (1984). *Pest Lepidoptera of Europe: with special reference to the British Isles.* Entomologica series **31**. Junk Publishers, The Hague.

Carter, D. J. (1992). *Eyewitness handbooks: butterflies and moths.* Dorling Kindersley, London.

Cook, M. A. and Scoble, M. J. (1992). Tympanal organs of geometrid moths: a review of their morphology, function and systematic importance. *Systematic Entomology* **17**: 219–32.

Cook, M. A., Harwood, L. M., Scoble, M. J., and McGavin, G. C. (1994). The chemistry and systematic importance of the green wing pigment in emerald moths (Lepidoptera: Geometridae). *Biochemical Systematics and Ecology* **22**: 43–51.

Dickson, R. (1992). *A lepidopterist's handbook* (2nd edn). Amateur Entomologist's Society, London.

Davis, D. R. and Gentili, P. (2003). Andesianidae, a new family of monotrysian moths (Lepidoptera: Andesianoidea) from austral South America. *Invertebrate Systematics* **17**: 15–26.

Eubanks, M. D., Nesci, K. A., Petersen, M. K., Liu, Z., and Sanchez, H. B. (1997). The exploitation of an ant-defended host plant by a shelter-building herbivore. *Oecologia* **109**: 454–60.

Kawahara, A. Y., Plotkin, D., and Espeland, M. (2019). Phylogenomics reveals the evolutionary timing and pattern of butterflies and moths. *Proceedings of the National Academy of Sciences of the United States of America* **116**: 22657–63.

Larsen, T. B. (1996). *The butterflies of Kenya and their natural history.* Oxford University Press, Oxford.

Liu, Y., Yang, Z., Zhang, G., Yu, Q., and Wei, C. (2018). Cicada parasitic moths from China (Lepidoptera: Epipyropidae): morphology, identity, biology, and biogeography. *Systematics and Biodiversity* **16**: 417–27.

Locke, M. J. (1998). Caterpillars have evolved lungs for hemocyte gas exchange. *Journal of Insect Physiology* **44**: 1–20.

Mill, P. J. (1998). Caterpillars have lungs. *Nature* **391**: 129–30.

Montgomery, S. L. (1983). Carnivorous caterpillars: the behavior, biogeography and conservation of *Eupithecia* (Lepidoptera: Geometridae) in the Hawaiian Islands. *GeoJournal* **7**: 549–56.

New, T. R., Pyle, R. M., Thomas, J. A., Thomas, C. D., and Hammond P. (1995). Butterfly conservation management. *Annual Review of Entomology* **40**: 57–83.

Ohsaki, N. (1995). Preferential predation of female butterflies and the evolution of Batesian mimicry. *Nature* **378**: 173–5.

Pierce, N. E. (1989). Butterfly–ant mutualisms. In: *Toward a more exact ecology. British Ecological Society Symposium 30.* Grubb, P. J. and Whittaker, J. B. (eds), pp. 299–324. Blackwell Science, Oxford.

Pierce, N. E. (1995). Predatory and parasitic Lepidoptera: carnivores living on plants. *Journal of the Lepidopterist's Society* **49**: 412–53.

Pierce, N. E., Braby, M. F., Heath, A., Lohman, D. J., Mathew, J., Rand, D. B., and Travassos, M. A. (2002). The ecology and evolution of ant association in the Lycaenidae (Lepidoptera). *Annual Review of Entomology* **47**: 733–71.

Pollard, E. and Yates, T. (1993). *Monitoring butterflies for ecology and conservation.* Chapman and Hall, London.

Powell, J. A. (1992). Interrelations of yuccas and yucca moths. *Trends in Ecology and Evolution* **7**: 10–15.

Preston-Mafham, R. and Preston-Mafham, K. (1988). *Butterflies of the world.* Blandford Press, London.

Pullin, A. S. (ed) (1994). *Ecology and conservation of butterflies.* Chapman and Hall, London.

Renwick, J. A. A. and Chew, F. S. (1994). Oviposition behaviour in Lepidoptera. *Annual Review of Entomology* **39**: 47–79.

Rutowski, R. L. (1997). Sexual dimorphism, mating systems and ecology in butterflies. In: *The evolution of mating systems in insects and arachnids*. Choe, J. C. and Crespi, B. J. (eds), pp. 257–72. Cambridge University Press, Cambridge.

Scoble, M. J. (1992). *The Lepidoptera: form function and diversity*. Natural History Museum Publications. Oxford University Press, Oxford.

Scoble, M. J. (1996). In search of butterfly origins: morphology and molecules. *Trends in Ecology and Evolution* **11**: 274–5.

Spangler, H. G. (1988). Moth hearing, defense and communication. *Annual Review of Entomology* **33**: 59–81.

Stamp, N. E. and Casey, T. (eds) (1993). *Caterpillars: ecological and evolutionary constraints on foraging*. Chapman and Hall, London.

Talekar, N. S. and Shelton, A. M. (1993). Biology, ecology and management of the diamond-back moth. *Annual Review of Entomology* **38**: 275–301.

Thomas, J. A. and Wardlaw, J. C. (1992). The capacity of a *Myrmica* ant nest to support a predacious species of *Maculinea* butterfly. *Oecologia* **91**: 101–9.

Thompson, J. N. and Pellmyr, O. (1991). Evolution of oviposition behavior and host preference in Lepidoptera. *Annual Review of Entomology* **36**: 65–89.

Toussaint, E. F. A., Breinholt, J. W., Earl, C., Warren, A. D., Brower, A. V. Z., Yago, M., Dexter, K. M., Espeland, M., Pierce, N. E., Lohman, D. J., and Kawahara, A. Y. (2018). Anchored phylogenomics illuminates the skipper butterfly tree of life. *BMC Evolutionary Biology* **18**: 101.

van der Geest, L. P. S. and Evenhuis, H. H. (eds) (1991). *Tortricid pests. World crop pests 5*. Elsevier, Oxford.

van Eldijk T. J. B., Wappler, T., Strother, P. K., van der Weijst, C. M. H., Rajaei, H., Visscher, H., and van de Schootbrugge, B. (2018). A Triassic-Jurassic window into the evolution of Lepidoptera. *Science Advances* **4**: e1701568.

Vane–Wright, R. I. and Ackery, P. R. (eds) (1984). *The biology of butterflies*. Academic Press, London.

Young, M. (1997). *The natural history of moths*. T. & A. D. Poyser, London.

Zahiri, R., Holloway, J. D., Kitching, I. J., Lafontaine, J. D., Mutanen, M., and Wahlberg, N. (2012). Molecular phylogenetics of Erebidae (Lepidoptera, Noctuoidea). *Systematic Entomology* **37**: 102–24.

Zwick, A., Regier, J. C., Mitter, C., and Cummings, M. P. (2011) Increased gene sampling yields robust support for higher-level clades within Bombycoidea (Lepidoptera). *Systematic Entomology* **36**: 31–43.

Hymenoptera

(bees, wasps, and ants)

Common name	Sawflies, wasps, bees, and ants	**Metamorphosis**	Complete (egg, larva, pupa, adult)
Derivation	Gk. *hymen*–membrane; *pteron*–a wing	**Distribution**	Worldwide except Antarctica
Size	Body length 0.25–70 mm	**Number of families**	About 102
		Known world species	about 150,000 (13.25%)

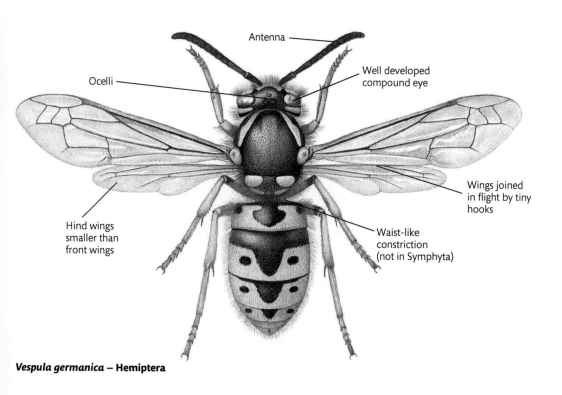

Vespula germanica – Hemiptera

Essential Entomology. Second Edition. George C. McGavin and Leonidas-Romanos Davranoglou, Oxford University Press.
© George C. McGavin and Leonidas-Romanos Davranoglou (2022). DOI: 10.1093/oso/9780192843111.001.0001

Key features

- abundant and ubiquitous
- body usually with a constricted waist
- some species live in social colonies
- ovipositor may be modified as a sting
- second largest order

❯ A female braconid wasp (likely *Diamblomera* sp.) probing a tree trunk with her ovipositor. Once she detects an insect larva, she will use her ovipositor to pierce it and insert an egg. Singapore.

❯ Many bees will rest on vegetation at night, as shown by this aggregation of *Nomia strigata* from Singapore.

Ants are the dominant predators in many ecosystems, both in terms of numbers and biomass. Here, a group of *Oecophylla* ants are dismembering an unfortunate *Tetragonula atripes* stingless bee. Malaysia.

Species within the Hymenoptera exhibit an incredible diversity of lifestyles: solitary or social; herbivorous, carnivorous, or parasitic. The number of un-described species may eventually be many times that already described (up to more than one million), and the order may be as big or bigger than the Coleoptera. If Diptera represents the order that has the most harmful effects on humans, then Hymenoptera must be regarded as the most beneficial. It would be difficult to picture a world without the strong control of natural in-sect populations exerted by parasitic and predatory wasp species, or without the flowering plants pollinated mainly by bees.

A nest of *Apoica pallens* wasps from Costa Rica. This species is highly unusual among wasps in that it forages almost exclusively at night.

❯ Closeup of the European social wasp, *Polistes nimpha*, drinking water. France.

The order is divided into two suborders: the Symphyta (sawflies) and the Apocrita (wasps, ants, and bees) (see Box). While Apocrita are monophyletic, the Symphyta are paraphyletic, and are therefore not a homogeneous 'natural' group. One group of sawflies, the Orussoidea, are parasitoids, and represent the sister group of Apocrita. Consequently, they are ranked together as the Vespina (Apocrita + Orussoidea) (see Box).

Although Apocrita are overall monophyletic, its megadiverse parasitoid lineages were once lumped in a now-obsolete infraorder known as the Parasitica. However, molecular and morphological studies have shown that this lineage is paraphyletic, and Parasitica are now split into many different superfamilies that largely represent monophyletic natural groups (see Box).

Another rank in the Apocrita is the Aculeata, which includes a monophyletic group of many related superfamilies that possess an ovipositor modified into a venom-delivering stinger. Three superfamilies of aculeate Apocrita—the Formicoidea (ants), the Vespoidea (social wasps), and the Apoidea (bees)—are considered, at least in behavioural terms, to be the most advanced of all insects.

Sawflies do not have a distinctly constricted waist, and females have a saw-like ovipositor for laying eggs into plant tissue. In the Apocrita, the first segment of the abdomen, called the propodeum, is fused to the thorax. The combined structure is called a mesosoma (in bees) or sometimes alitrunk (in ants). The second (and sometimes the third) abdominal segments are very

Symphyta

Apocrita

Parasitica

Aculeata

> The order Hymenoptera is divided into two suborders: the Symphyta (sawflies) and the Apocrita (Parasitica—parasitic wasps, and Aculeata—wasps, ants, and bees). Although Symphyta and Parasitica are paraphyletic, and therefore unnatural groupings, they are mentioned here due to their abundance in early classification schemes and studies of Hymenoptera.

narrow and form the petiole, which gives the distinctive wasp-waisted appearance. The swollen remainder of the abdomen behind the petiole is called the gaster or metasoma. Parasitic apocritans have a slender, and sometimes very elongate, ovipositor for penetrating and laying eggs inside other insects, and aculeate apocritans have a modified ovipositor in the form of a sting with an associated poison gland for defence. The presence of a waist in apocritans greatly increases the manoeuvrability of the abdomen and allows the sting or ovipositor to be brought into almost any position underneath the body, or even pointing forwards.

Hymenopteran superfamilies

Suborder	Superfamilies (no. of families)	Major larval habits
'Symphyta' (Sawflies)	Xyeloidea (1)	Phytophagous on coniferous and deciduous trees, some gall-formers
	Xiphydrioidea (1)	Borers on dead wood
	Pamphilioidea (2)	Phytophagous on coniferous and other vegetation, leaf rollers
	Cephoidea (1)	Borers in grass and pithy stemmed plants
	Siricoidea (2)	Mostly wood-boring, some may be parasitic
	Tenthredinoidea (7)	Phytophagous on trees, herbs, grasses
Vespina (Apocrita + Orussoidea)	Orussoidea (1)	Parasitoids of wood-boring beetles and sawflies
Apocrita (Wasps, ants and bees)	Stephanoidea (1)	Parasitoids of wood-boring beetles and sawflies
	Trigonaloidea (1)	Mostly hyperparasitoids of tachinids and ichneumonids
'Parasitica'	Megalyroidea (1)	Parasitoids of wood-boring beetles
	Ceraphronoidea (2)	Parasitoids
	Evanioidea (3)	Parasitoids
	Ichneumonoidea (3)	Parasitoids
	Proctotrupoidea (7)	Parasitoids
	Diaprioidea (5)	Parasitoids
	Platygastroidea (3)	Parasitoids
	Cynipoidea (5)	Gall formers or inquilines in galls, some parasitoids
	Chalcidoidea (22)	Parasitoids
	Chrysidoidea (7)	Mostly parasitoids
	Vespoidea (2)	Parasitoids, or feed on provisioned insect or spider prey
Aculeata*	Pompiloidea (4)	Parasitoids, kleptoparasites, or feed on provisioned insect or spider prey
	Thynnoidea (2)	Parasitoids
	Tiphioidea (2)	Parasitoids, or feed on provisioned insect prey

Continued

Hymenopteran superfamilies *continued*

Suborder	Superfamilies (no. of families)	Major larval habits
	Scolioidea (1)	Feed on provisioned insect prey
	Formicoidea (1)	Larvae feed on plant, fungal and animal material
	Apoidea (17)	Larvae feed on nectar and pollen, provisioned insects and spiders,

* The Aculeata are the sting-bearing apocritans.

The head carries a pair of **thread-like, multi-segmented antennae** of various designs, a pair of **well-developed compound eyes**, and, usually, **three ocelli**, which are positioned in a triangular array on top of the head between the eyes. The mouthparts are adapted for chewing and biting, but, in many species, liquids are ingested. In the bees (superfamily Apoidea), the maxillae and the labium are extended and modified to form a tongue of variable length through which nectar is sucked. The mandibles are versatile tools used for a variety of purposes, such as feeding, holding and moving prey, defence, and nest construction. In parasitic species, they are used to cut the adult free of the body of the host, and in other species, to leave the pupal cell.

Although many species lack wings, most members of the order are strong fliers with an enlarged meso- and meta-thorax, and **two pairs of membranous wings, which are strongly joined in flight by a row of small hook-like structures called hamuli**, located on the front edge of the hind wing. The front wings typically have a dark pterostigma on the front edge and are slightly bigger than the hind wings. Both pairs have a reduced pattern of venation marking out a number of large cells, but, in many species, particularly in the superfamilies Chalcidoidea and Proctotrupoidea, venation is almost lacking. Legs are generally not modified, except in Dryinidae (Chysidoidea), where the front legs of females are raptorial to grasp hosts, and in worker bees where they are modified for pollen gathering. Many male bees have secondary sexual modifications of the legs, some of them bizarre.

Hymenopteran larvae are either caterpillar-like (Symphyta) or grub-like (Apocrita). The larvae of sawflies have a well-defined head capsule, three pairs of thoracic legs, and abdominal prolegs, although prolegs are lost in species that bore into plant tissue. They can be separated from the caterpillars of moths and butterflies by a number of features. The prolegs do not have crochets, the eyes are simple with only one facet as opposed to several, and the antenna are reduced to small bumps, not three segmented as in lepidopteran caterpillars. Apocritan larvae tend to be simple and maggot- or grub-like with no legs and a reduced head capsule. The reasons for this dichotomy are related to the lifestyle of the larvae. Sawfly larvae are active herbivores that need to find food, whereas apocritan larvae develop from eggs laid inside the body of a host insect, or inside a cell that is already stocked or provisioned with food as the larva grows (only social hymenopterans practise progressive provisioning).

> parasitoid
a species that develops by consuming the body tissues of a single host, eventually causing its death

Parasitoids are insects with more than a certain amount of gruesome appeal. Most species of parasitoid are hymenopterans, the rest are tachinid, conopid, and some other flies. Parasitoids live and develop at the expense of another organisms. They are not parasites in the usual meaning of the word, in that the host never survives a parasitoid attack. It is an incredibly successful lifestyle, and one that, unseen, regulates natural populations of other species more effectively than any other single factor. Very many species are used in biological control programmes against pest insects.

Parasitoids can feed inside the host or outside the host's body, in which case they are termed endoparasitoids and ectoparasitoids, respectively. Another difference in parasitoid habits is important to bear in mind. If the female wasp kills or paralyses its host when laying an egg, and thus prevents the host from developing any further, it is called an idiobiont parasitoid. The larvae of idiobiont species are often ectoparasitoids. If, however, the female does not kill the host when egg laying, thus allowing the host to develop further, it is known as a koinobiont. Koinobionts tend to be endoparasitoids. In some cases, the development of the parasitoid can be delayed until the host is nearly fully grown. Species that are parasitoids of other parasitoids are called hyperparasitoids. Host-finding behaviour in parasitoids can involve a large number of factors. Many parasitoids seek hosts in specific locations, such as inside leaf mines or in rotten wood. The host itself may give out a chemical or vibrational signal of some sort that is intercepted by the parasitoid, but, in some cases, the signal may be produced by the parents of the host. For instance, *Leptopilina heterotoma* (Eucoilidae), a parasitoid of *Drosophila* larvae, uses the pheromones produced by mating adults at oviposition sites as a host-finding cue. Others, such as many members of the Dryinidae, locate their hemipteran hosts by listening to their vibrational songs.

Most female parasitoids laying a single egg in a host will try to ensure that the host is large enough to provide her offspring with sufficient resources to complete its development. In other species, many eggs may be laid, or many offspring will result from the repeated division of a single egg (polyembryony). In some parasitoids, the eggs are laid outside the host and the first instar larvae, which can move by wriggling, make contact themselves. The differences between sawfly larvae and those of Apocrita extend to the internal organs as well. In sawflies, the gut is normal, whereas in most apocritans the midgut is not joined to the hindgut until the last larval stage. Rather than foul its cell or the host's body, the larva stores its waste until just before it pupates. Parasitoid wasps locate hosts and lay eggs in them wherever they happen to be; inside wood, leaf mines, stems, or exposed. Many aculeate Hymenoptera (who have a sting) act as straightforward predators that catch and kill prey items to feed to their larvae, but, in many species, prey items are paralysed, and the developing larvae might be considered as parasitoids. The difference here is that, while some spider-hunting wasps do look for hosts in their own burrows, many hunt and transport the host either to a place where it is interred and an egg laid on it, or to an already-prepared chamber. Concealment is necessary to prevent the contents being raided or taken over by another wasp.

Sex determination in Hymenoptera is by haplodiploidy. Males, produced parthenogenetically from unfertilized eggs, are haploid, whereas females, produced from fertilized eggs, are diploid. Inseminated females can control the sex of their offspring by withholding sperm, which, in some species, can result in very biased sex ratios. There are many species in which males are rare or even unknown, and diploid females are produced parthenogenetically.

Herbivory is certainly the primitive and ancestral feeding habit of the order, and parasitism and carnivory may have arisen via inquiline species that may originally have been herbivores inside gall or other plant tissue but progressed to eating other small herbivores. With the exception of larvae of the Orussidae, which are parasitoids of wood-boring beetles, sawfly larvae are herbivorous on a wide range of plants. The Cephoidea contains a single family, the Cephidae (stem sawflies), with about 160 species. Stem sawflies are generally slim and cylindrical, and many species have yellow abdominal and thoracic markings. The larvae of all cephids are stem or twig borers of grasses, willows, raspberries, and other shrubs. Species in the genera *Cephus* and *Trachelus* can be pests in cereal crops, such as wheat, oats, rye, and barley. *Cephus pygmaeus*, the European Wheat Stem Sawfly was introduced to North America in the late 1800s and has become a pest in addition to the indigenous species *Cephus cinctus*.

In the superfamily Siricoidea, about 150 species of horntails or wood wasps make up the Siricidae. Siricids are large and stout and often strikingly coloured—black or metallic blue, or with yellow, hornet-like markings. The end of the abdomen has a distinctive terminal spine, which is short in males and spear-like in females. Females have a long, very strong ovipositor, which projects below the terminal spine. The ovipositor is used to drill through bark into the wood of diseased or dead trees where a single egg is laid at a time. There are two distinct subfamilies: the Siricinae, typified by species of the genus *Sirex* that attack pines or firs, and Tremicinae, for example, *Tremex* spp., that attack a range of deciduous trees. Siricid larvae bore into heartwood and eat the hyphae of wood-rotting fungus, which is introduced to the timber along with the egg at the time of oviposition. The larvae pupate in a cocoon made of silk and chewed wood just under the surface. Depending on the species, there can be between five and eleven moults, which can take two years or more. Sometimes, wood wasps have emerged from the timbers of new houses or furniture. In one instance, the almost simultaneous emergence of many large North American horntails from the wood behind the plaster of a new house in Oxford caused such alarm that the owners nearly moved out.

The Tenthredinoidea is the biggest sawfly superfamily containing more than 8,400 species, representing more than 85% of all sawfly diversity. Two families—the common sawflies (Tenthredinidae) and the conifer sawflies (Diprionidae)—contain significant numbers of pest species. Diprionids are stout, mainly dull brown or black species. They are distinguished by feathery or toothed antennae and a very broad abdomen. Larvae of the majority of species attack pine trees, but a few species are found on cedars, hemlock, firs, and spruce. The larvae feed externally on the needles and are often gregarious and warningly coloured when young. When attacked, gregarious larvae all simultaneously jerk their bodies upright, and have the ability to exude distasteful resins, which they have sequestered from their food. When fully grown, larvae pupate in a tough cocoon in the soil or glued to the host plant. Species of *Diprion* and *Neodiprion* can cause serious damage and defoliation in coniferous plantations. The Tenthredinidae, with more than 7,500 species, are variable in appearance and can be found in terrestrial habitats all over the world. Adult tenthredinids feed on pollen and nectar but many prey on other insects. Females lay their eggs in the tissues of the appropriate host plant, and the larvae usually chew or graze the foliage externally. Some species mine leaves, while others, such as species of *Pontania*, induce galls on the leaves of willow species. Many species are minor pests of garden plants, crops, and trees, but some can cause significant economic damage.

The Stephanoidea contains a single family (Stephanidae) of about 350 species whose larvae are parasitoids of wood-boring beetles and horntails living in dead wood. The adults locate their hosts by the sounds they make and reach them using a very long ovipositor. The Megalyroidea also contains a single family of about 50 species (Megalyridae), whose larvae, like stephanids, are parasitoids of the larvae of wood-boring beetles.

The superfamily Trigonaloidea contains a single family (Trigonalidae or Trigonalyidae) of around 90 species, in which life cycles can be very unusual. The females lay a great number of small eggs into the edges of leaves where they can be easily eaten by herbivorous insect larvae. A sawfly larva might become the host to one of these parasitoids, but, if consumed by a lepidopteran caterpillar, the egg hatches and the larva makes its way into the haemocoel of the host. Here it will wait until the host is parasitized again, this time by a tachinid fly or an ichneumonid wasp. The trigonalid larvae will then become a hyperparasitoid. Even if an infected caterpillar is seized by a vespid wasp, masticated, and then fed to the wasp's larvae, the trigonalid egg will still survive, hatch inside, and become a parasitoid of the wasp larva.

The Ichneumonoidea is dominated by two huge families, the Ichneumonidae and the Braconidae, with around 48,000 species collectively, and a tiny family, Trachypetidae, with just seven species. The Ichneumonoidea represents 33% of all Hymenoptera and 3% of all complex animal life, and there are likely as many, if not more, undescribed species. It is difficult to generalize about taxa this large, but the families differ in a number of respects. Most species in the family Ichneumonidae are ecto- and endoparasitoids of the larvae and pupae of holometabolous insects, although a few attack spiders and their eggs. The insects commonly used as hosts are sawflies, moths, beetles, and flies. Many ichneumonids can be hyperparasitoids, either of other ichneumonids or of tachinid flies. On the whole, the Braconidae tend to be parasitoids of hemimetabolous insects, such as bugs (including aphids), barklice, and termites, although a few do attack wasps and beetles. Nothing is known about the biology of Trachypetidae. A fascinating aspect of the life history of many Ichneumonoidea is their symbiosis with certain viruses known as polydnaviruses (PDVs). These viruses are integrated into the ichneumonoid genome, and they replicate in large numbers in the female reproductive tract. Each ichneumonoid species has its own unique type of

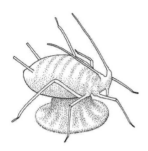

A mummified aphid stuck to foliage by a silk tent containing the cocoon of a pupating braconid wasp.

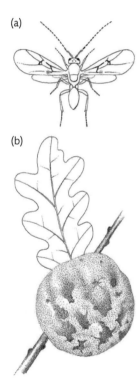

(a)

(b)

Gall-formers can have complicated life cycles involving sexual and asexual generations. (a) The winged sexual form of Biorhiza pallida having emerged from its oak apple gall. (b) The oak apple gall caused by oviposition of many eggs into a bud by the wingless asexual form that emerges from root galls.

PDV. When a female ichneumonoid oviposits inside an insect, she also injects huge numbers of PDVs, which are essential for successful parasitization, as they interfere with the host's immune response, granting the parasitoid larvae with a greater chance of survival. In this way, ichneumonoids have domesticated PDVs, in an insect–virus symbiosis, which is so far unique among arthropods (but may prove to be more common in the future).

Around 500 species belong to the seven families that make up the Proctotrupoidea. Most of the species in the Protoctrupidae are endoparasitoids of beetle larvae. The related Diapriidae (superfamily Diaprioidea) are endoparasitoids of larval or pupal Diptera. About 4,000 species belong to the three families that make up the Platygastroidea. These wasps are usually endoparasitoids of insect and spider eggs.

With the exception of some tenthredinid species, hymenopteran gall formers overwhelmingly belong to the family Cynipidae, the biggest of five in the Cynipoidea. Species in the other families are parasitoids of wood wasps and flies. It is likely that the gall-forming habit has evolved from an ancestral condition, perhaps a parasitoid of wood-boring larvae. Gall-forming species belong to the Cynipinae and species of the vast majority (*Andricus*, *Biorhiza*, *Callirhytis*, *Cynips*, *Neuroterus*, etc.) use oaks (*Quercus*) and related trees as host plants. A few gall-forming species (*Diplolepis* spp.) use members of the Rosaceae and Compositae as host plants. A few cynipids are not gall-formers themselves but live as inquilines inside galls of other species. Plant galls are a common sight and vary enormously in size, colour, texture, and location. Many are highly species-specific, and it is easier to identify the gall-causing species from the appearance of its gall than from its causer. Adult female gall wasps lay their eggs inside plant tissues, and the host plant responds by developing a gall around it. The mechanism of gall induction in the Cynipidae has remained elusive. The plant may isolate the gall wasp by growing a gall, but its structure can be complex, and as it is protective and has a nutritive layer to feed the developing larva or larvae, it is always beneficial to the gall former. Gall formers can have complicated life cycles involving sexual and asexual generations occurring either on different parts of the plant, or on two different host plants. When alternation of generations occurs, galls containing the developing asexual generations (all females) are larger and occur during the late summer and autumn. The females that emerge from these galls in the following spring lay eggs that give rise to galls containing males and females. These emerge, mate, and produce the next asexual generation. Gall wasps and their galls often support a rich and varied community of interdependent organisms, many of which are parasitoid wasps that attack the gall formers, their inquilines, and each other. Even when the galls are empty, they may provide shelter for other insects.

The Chalcidoidea comprises 22 families and more than 22,500 described species of small or very small parasitic wasp. They are a megadiverse lineage, and estimates suggest that up to 500,000 remain to be described. Some chalcidoids are herbivores and are seed feeders or gall formers, but the vast majority of species are parasitoids of the larval stages of insects, some spiders and

Chrysidid wasps often roll into a ball as protection against predators.

mites. The details of the life history of some of these species are truly remarkable. The females of many species in the family Chalcidae have very enlarged, strong hind legs, which they may use for fighting, but in a few they are used to hold the jaws of antlion larvae apart while the female stabs her ovipositor through the intersegmental membrane behind the head. The following is a brief description of the activities of some of the major families.

More than 850 species of tiny wasps belong to the Trichogrammatidae. These wasps parasitize the eggs of butterflies, beetles, bugs, thrips, flies, and other wasps. Some even cling to the body of host females to ensure that they are on hand when fresh eggs appear. The females of some species swim underwater to parasitize the eggs of water beetles and dragonflies. The entire larval development may be as short as three days. Trichogrammatids are so small (only 170 µm, or 0.17 mm—that is, smaller than many amoebae and protozoans) that, to save space in their miniature bodies, most of their brain neurons lose their nuclei. Even smaller than some trichogrammatids are the so-called fairyflies, the Mymaridae, among whose member species can be found the tiniest flying insects on Earth with wingspans of less than 0.25 mm. These tiny wasps are also egg parasitoids of other insect orders. Nearly half of all species use planthopper eggs and the eggs of related bug families, and the amazing females of *Caraphractus cinctus* swim underwater to locate the eggs of giant water beetles (Dytiscidae).

More than 650 species of fig wasp, Agaonidae, occur in subtropical and tropical regions where fig trees grow, and as there is probably at least one species of fig wasp specific to each of 900 or so fig species (and several species might inhabit a single fig species), there might be as many as 1,000 agaonids in total. Males and females are very different. The females are tiny, flattened, winged wasps whereas the males hardly resemble wasps at all, being wingless with odd-shaped heads, very weak middle legs, and a curled abdomen. Fig wasps and fig trees are totally dependent on each other for survival. The trees rely on the wasps for pollination and the wasps, which have complex lifecycles, can only reproduce inside figs. Both fig trees and Agaonidae play a vital role in tropical ecosystems worldwide and provide invaluable ecosystem services to traditional economies.

Nearly 3,500 species belong to the Pteromalidae. Most of these wasps are parasitoids or hyperparasitoids inside or outside the larvae or pupae of flies, beetles, wasps, butterflies, moths, and fleas. A few are gall-formers or herbivores. Females may lay one egg, while others may lay hundreds of eggs in a large host, which is paralysed and does not develop any further.

The Eulophidae, a family of more than 4,300 small species, are mainly parasitoids of leaf miner and gall-former larvae. The Torymidae comprise about 1,000 species that may be predacious or herbivorous in seeds as larvae. A large number of species are parasitoids of gall-forming Diptera and Hymenoptera, the females using their long ovipositors to drill through the tissue of the gall to reach the host larvae inside. Other torymid species attack caterpillars, the eggs of mantids, and the larvae of solitary bees and wasps that nest inside plant stems.

The Encyrtidae, comprising more than 3,700 species of small wasp, are mainly parasitoids of the immature stages and adults of scale insects, mealy bugs, aphids, and whiteflies, although they also have hosts in many other insect orders. Some species are hyperparasitoids of the larvae of pteromalids, braconids, and even other encyrtids. Many encyrtids are polyembryonic. The eggs divide repeatedly at an early stage of their development to produce anything from ten larvae in small hosts, to a couple of thousand larvae in large hosts.

Members of the seven families that make up the Chrysidoidea are parasitoids, with more than 6,000 species. Bethylids (Bethylidae) are a large group of about 2,000 species that attack caterpillars and beetle larvae. The female wasp locates a suitable host and stings it, causing paralysis or death. She then lays a number of eggs on it, and, in some species, she may give a certain amount of maternal care while her ectoparasitic larvae consume the host. An even larger family, with more than 3,000 species, is that of the jewel wasp (Chrysididae). The name of jewel or ruby-tailed wasp refers to the beautiful and strongly metallic coloured body of these wasps. The body surface is extremely hard to protect them from bee and wasp stings and the underside of the abdomen is flattened or concave, enabling the wasps to curl up into a ball.

Female chrysidids lay their eggs in the cells of nests of solitary bee or wasp species. The chrysidid larva eats the host larva or nest provisions left by the host bees or wasps. Some chrysidid species are parasitic on sawfly larvae or pupae, some on lepidopteran pupae, and a few eat the eggs of mantids. Sometimes the life histories can be complex, involving an intermediate host. When this host, carrying the jewel wasp's egg, is paralysed by a digger wasp and stored as food for the larvae, the egg hatches and the chrysidid larva feeds on the digger wasp egg and food supply.

There is only one other large chrysidoid family: the Dryinidae. The larvae of all 1,900 species are parasitoids of auchenorrhynchan nymphs or adults, particularly those belonging to the families Cicadellidae, Delphacidae, Cixiidae, Membracidae, Cercopidae, and Fulgoridae. Female dryinids locate their prey on vegetation and, when within close range, snatch their victim using their legs and, sometimes, jaws. Many female dryinids have a strange grasping or pincer-like front tarsi that are used to hold the bug pressed against the surface of the foliage or up in the air. The female then stings it and inserts an egg. In most species, the egg is placed between abdominal segments, in others the egg is inserted in the thorax or neck of the host bug. The effect of the sting lasts a short time, and the host revives and continues its life cycle normally. The hatched dryinid feeds, at first, on the host's internal fluids. The first instar is spent within the body of the host, but later instars develop inside a larval sac, which bulges through the intersegmental membranes. The mature larva will also eat the host's body tissues and the sac may become as big as the bug's abdomen. When mature, the larva leaves the host, almost always killing it, and moves towards the ground where it spins a cocoon.

The remaining superfamilies of the Aculeata contain a suite of fascinating life histories. Tiphiid wasps (Tiphiidae, superfamily Tiphioidea) are slender, shiny, yellowish-brown and black, or black wasps. Males of all species are fully winged, but females can be fully winged or wingless. The underneath of the middle thoracic segment has two lobes, which may cover the coxae of the middle legs. Tiphiids are parasitic on the larvae of beetles, bees, and wasps. Females of the genus *Tiphia* dig through soil to find the larvae of scarab beetles, and, when located, the female wasp will break open the cell, sting the beetle grub, and may chew it with her mandibles before laying her egg. The wasp then leaves and reseals the cell behind her. Females of the genus *Methocha* are very ant-like and specialize in parasitizing tiger beetle larvae within their burrows.

Velvet ants (Mutillidae, superfamily Pompiloidea), although velvety and often ant-like, are not ants. Nearly 7,000 species are found, mainly in tropical regions. These wasps are black or reddish-brown with spots or bands of red, yellow, or silver. They have short hairs and a head and thorax that has a sculpturing of coarse dimples. Males are fully winged, but the females are wingless and are most commonly seen running over the ground. Mutillid larvae are ectoparasitic on the larvae and pupae of bees and wasps, but also some dipterans and coleopterans. For example, species of the genus *Mutilla* specialize on bumble bees of the genus *Bombus*, while others attack a wider range of hosts, including halictid bees, and pompilid and sphecid wasps. When female mutillids find and bite open a host cell they check whether the larva inside is suitable. If too young, the cell is resealed, but if the larva is fully developed, an egg will be laid on it and the cell resealed. The mutillid larva hatches, eats the contents, and pupates within a tough cocoon inside the host's cell.

More than 5,000 species of spider-hunting wasps of the family Pompilidae (Pompiloidea) are known. Many pompilids are moderately sized, but the family includes some giants of the genus *Pepsis* measuring > 65 mm from head to tail. Spider-hunting wasps are generally dark-coloured, blue or black, usually with dark, yellowish, bluish, or black wings. Females are active hunters that fly and run rapidly over the ground in search of spiders. Females may make mud nests in crevices, dig branched burrows underground, or simply use the host spider's own burrow. After the spider is paralysed, the female makes a burrow or drags it to a prepared burrow, lays an egg on it, and seals the burrow. Some species are kleptoparasites, that is, they cheat by quickly laying their own eggs on a spider paralysed by another wasp before it is buried. Some even locate and break into an already-sealed nest.

More than 5,100 species in two families comprise the Vespoidea. Common or paper wasps belong the large family Vespidae. About 5,000 species are known and most live in tropical or subtropical regions. The family is divided into distinct subfamilies such as the Vespinae (yellow jackets and hornets), Eumeninae (potter wasps), and Polistinae (paper wasps). Vespine and polistine wasps are highly social insects living in colonies with a queen, males, and sterile worker females. Eumenine wasps and species belonging to the other subfamilies are solitary or very weakly social.

Yellow jackets and hornets make carton nests of paper made from chewed wood fibres. The nest, which may be underground, in natural cavities, or in bushes, may contain many thousands of workers. Developing larvae are fed on a nutritious paste of butchered or chewed-up insect prey. This sort of

Species of Eumenes make elegant vase-shaped nests and stock them with a number of paralysed caterpillars or beetle grubs, and an egg is suspended by a slender thread from the roof or walls of the nest before it is sealed.

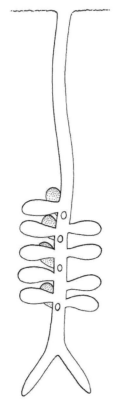

Section through a nest of a mining bee (*Halictus* sp.) showing numerous cells arranged around a central burrow.

feeding system is known as progressive provisioning, as opposed to the mass provisioning that occurs in spider-hunting wasps and others. Foraging may be opportunistic but there is evidence that vespids react to aggregations of suitable prey, such as mating swarms or leks, by picking up cues such as sexual pheromones. They are able to change their foraging strategy to capitalize on other prey items at different times of the day. The adults feed on nectar and other sugar rich materials. The nests of *Polistes* may be covered with a paper or mud carton, but many are uncovered in sheltered locations. Potter wasps make small mud-lined nests in the ground, in stems, and a variety of natural cavities. Species of the widespread genus *Eumenes* make elegant vase-shaped nests of mud or clay attached to twigs or rocks. The nests are stocked with a number of paralysed caterpillars or beetle grubs, and an egg is suspended by a slender thread from the roof or walls of the nest before it is sealed.

Ants all belong to one family, the Formicidae (superfamily Formicoidea), which are the sister group to Apoidea (bees, sand wasps, cockroach wasps, and aphid wasps). Just under 13,800 species are known and, although they are morphologically quite similar, their biology is very diverse. Ants appeared in the Early Cretaceous (145–65 mya) and radiated in the Late Cretaceous and Early Tertiary. With the exception of Antarctica and a few oceanic islands, ants are found everywhere on Earth, and have an immense impact on terrestrial ecosystems. In most habitats, ants are the major predators, and in parts of Central and South America, leaf-cutter ants (*Atta* spp.) are the major herbivores and the most serious insect pests. In many terrestrial ecosystems, ants move more earth than earthworms and are vital in nutrient recycling and plant dispersal. Like the termites and some of the bees and wasps, ants are eusocial and live in colonies ranging from a handful of individuals sharing a tiny shelter, to tens or hundreds of millions inhabiting underground structures reaching 6 m below ground level. Being eusocial means that two or more generations overlap at any one time, adults take care of the young, and there is a division of labour between castes (reproductive kings and queens and non-reproductive workers). Worker ants are wingless, sterile females, while the queens and males, who form mating swarms at certain times of the year, are winged. Mating may take place on the wing or on the ground; afterwards, the males die, and the females lose their wings. Chemical messages in the form of pheromones are important in colony regulation, trail making, raising alarms, and defence.

Caste is determined by a complex and poorly understood interplay of several factors such as diet, genetics, and epigenetics. The head is modified according to caste and species and may be very large with massive or otherwise specialized jaws. Some castes exhibit remarkable adaptations to the jaws and head for crushing seeds, blocking nest entrances, dismembering enemies, and catching fast-moving prey.

Ants can be herbivorous seed gathers or fungus growers, carnivorous or omnivorous, and some rely exclusively on the honeydew produced by sap-sucking bugs. Many species of insect and plant have evolved symbiotic and often highly complex relationships with ants. Many ants are associated with species of plants, which may provide them with homes in the form of galls or larger domatia (ant homes), often in return for protection of the plants from herbivores, encroaching vines, or fire.

The family is divided in about 16 distinct subfamilies, of which the biggest by far are the Myrmicinae and the Formicinae. Some myrmicine ants have stings, while formicines defend themselves by spraying formic acid. It is estimated that every year, the world's formicine ants release one million tonnes of formic acid into the atmosphere. The infamous driver and army ants of tropical regions belong to the subfamily Dorylinae. Colonies move in massive columns, several million strong, raiding termite or ant colonies, and kill and butcher anything that is unfortunate enough to get in the way (including tethered vertebrates). Some species can be pests in buildings, and the Red Imported Fire Ant (*Solenopsis invicta*) is an aggressive and abundant pest in the southern states of USA that causes great damage to soya bean crops.

The Apoidea contains 17 families, with about 26,700 species. The common names of these insects include solitary hunting wasps, digger wasps, and sand wasps, while the clade known as the Anthophila includes the honey bees, carpenter bees, sweat bees, bumble bees, etc. Most of the non-anthophilan Apoidea are solitary wasps that nest in soil, decaying wood, plant stems, in the burrows of insects, or in self-made mud nests. The females catch prey, paralyse it with their sting, and transport it back to the nest, where they will seal it in with an egg. Some species may use one single large prey item while others may use several small ones. Prey varies widely across many insect orders, as well as spiders, and many non-anthophilan Apoidea keep to a particular sort of prey item or a particular size of prey item. In most cases, the nest cells are mass-provisioned with enough food for the larva to complete its development, but the females of some species practice progressive provisioning and continue to supply their larvae with more prey as they grow. A few species are kleptoparasite and lay their own eggs in already provisioned nests of other sphecids. The European Beewolf, *Philanthus triangulum* (Philanthidae), provisions its brood cells with paralysed honey bees. When there are few honey bees around, the female chooses to lay more male eggs as sons are cheaper to produce. The female will also provide each offspring with less food than normal.

The Anthophila contains more than 16,000 species of bee, most of which are solitary. There are seven distinct families of which the larger ones deserve special mention. There are about 2,000 species of plasterer and yellow-faced bees (Colletidae). The species in this family are solitary and make very simple nests. Plasterer bees excavate nest burrows in the ground and use a special secretion from an abdominal gland to line the interior of their cells. This secretion dries to form a waterproof, cellophane-like film, which protects the interior. Yellow-faced bees nest in the pith of plant stems, the empty burrows of wood-boring insects, or even empty plant galls. Colletids transport pollen on their hind legs or carry it in their crops. The larval cells are provisioned with a sloppy mixture of regurgitated pollen and nectar.

The Halictidae, a family of more than 4,400 species, are called sweat bees because of the habit seen in some species of drinking perspiration. Many are solitary, while others are more social. Sweat bees mostly nest on the ground,

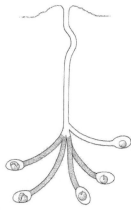

▲ Section through a nest of a solitary bee (*Andrena* sp.).

and the nests usually take the form of a single vertical burrow with or without lateral tunnels. Brood cells are arranged in dense clusters on the tunnels and are lined with a waterproof secretion. Some species are parasitic in the nests of other halictids. Many species are important crop plant pollinators.

More than 4,000 species of mining bees comprise the Andrenidae. Females of most species are solitary and make nests in soil burrows, although these often occur in large aggregations. Pollen from a few related plants is collected and transported in a brush of hair on the hind legs called a scopa. The larval cells are lined with a waxy substance secreted by an abdominal gland and are provisioned with a mixture of pollen and honey moulded into a small pellet.

Leaf-cutter and mason bees (Megachilidae) are solitary, nesting in natural cavities in the ground, dead wood, hollow stems, and snail shells. Megachilids do not line their brood cells with glandular secretions, but instead collect materials such as mud, resin, and leaf material, or animal or plant hairs to do the same job. Pollen, when collected, is carried in a brush of stiff hairs on the underside of the abdomen. Leaf-cutter bees are named for their habit of cutting neat circular pieces of leaves or petals to line their brood cells, which are arranged in rows inside the nest cavity. Other species are called carder bees because they use their jaws to strip the hairs from woolly leafed plants, which they tease out to line brood cells. Some species are parasitic in the nests of other bees, including other megachilids.

Just over 4,000 species of digger (Anthophorinae), cuckoo (Nomadinae), and carpenter bees (Xylocopinae) belong to the Apidae, the family that includes honey bees and bumble bees. Digger bees burrow in the ground and provision their brood cells with a mixture of pollen and honey. Cuckoo bees, which can be very wasp-like, are all parasitic in the nests of bees of all other families except the Megachilidae and Apidae. Carpenter bees make nests by excavating tunnels parallel to the grain of solid wood, in which a series of brood cells is made. Each cell is provisioned with a large mass of sticky pollen on which the female lays a single large egg before sealing the cell with chewed wood fibres. Female xylocopines will guard their nests against predators, parasites, and even female conspecifics that might attempt to take over the nest and lay their own eggs.

There are about 250 Bombini (bumble bees) species and 8 Apini (honey bees) species. Bumble bees are very hairy and stout-bodied, whereas honey bees are smaller and slenderer. The females of most species have a specialized pollen-carrying apparatus, the corbiculum or pollen basket, on the outer surface of each of the hind tibiae.

Members of these tribes are highly social and live in colonies with a queen, males, and sterile worker females. Bumble bee nests may be under or on the ground, and are made of grass with internal, wax-constructed brood cells. Many species make their nests in disused burrows and nests of small mammals. The larvae are fed on pollen and honey, which is stored in other cells. The queen rears workers first to build up the colony, with males appearing later in the summer. Some bumble bees behave like cuckoos, laying their eggs in nests of other bumble bees.

Corbiculum

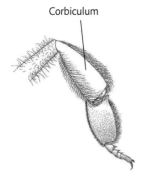

Location of the corbiculum (pollen basket) on the hind leg of a honey bee.

Early humans collected honey from wild bees and bee-keeping is documented from at least 9,000 years ago. Honey bees belong to the genus *Apis*, and the best known of the eight species is the Western Honey bee (*Apis mellifera*), which has been spread worldwide. Honey bees may possess the most complex social life of all eusocial insects. Each colony comprises a single egg-laying queen, many tens of thousands of workers, and up to 2,000 males or drones. The nest consists of a vertical array of double-sided wax combs divided into thousands of hexagonal cells. The hexagonal shape of the cells is the most economical shape and configuration, which will hold the most honey for the least amount of wax. Despite the fact that the cell walls are less than a tenth of a millimetre thick, a hexagonal array gives the comb great mechanical strength. Individual cells are built at a slope of 13 degrees upwards to keep the honey from running out. When honey bees start building a nest they gather together in a ball, the inside of which soon heats up to 35°C, the temperature needed for wax production. Wax is produced by glands on the underside of the worker bees' abdomens, and it is gathered and worked into shape with the front legs and mandibles. Wild honey bees need to start building their nest from scratch, but beekeepers provide their bees with a foundation sheet pressed out from pure beeswax, reinforced with wire and mounted in a frame. The modern type of hive known as the Langstroth hive was invented in 1851 in the USA and was a great advance over early bee skeps made out of straw. The queen is kept in the lower brood chamber by means of a queen excluder, a metal grille prevents her from laying eggs in the upper chambers. This means that the upper chambers are only used for storage, which makes the collection of honey and pollen very easy.

A colony has one queen, forty- to eighty-thousand workers, and a few hundred male or drone bees, whose only job is to fertilize new queens. Worker bees are sterile females, which are kept sterile by the queen who produces a pheromone, called queen substance, from small glands in her head. This substance is transferred from bee to bee within the hive and stops the workers' ovaries from developing. A queen can lay up to 1,500 eggs in a day, and in her four-year life she may lay one million eggs.

During the summer, an adult worker honey bee only lives for six weeks or so. When newly emerged, a young bee spends her first two days eating pollen and honey. After this she spends three weeks working inside the hive as a nurse feeding larvae with royal jelly, which she secretes from glands in her head. Bee larvae, which are going to become workers or drones (males), only get royal jelly for the first few days and then are fed on pollen and honey. Larvae that are going to become queens are fed on royal jelly throughout their development. When young, nurse bees produce wax, which is used to build new cells and repair the comb. For the last weeks of her life the worker is a forager. The exact age at which a worker becomes a forager varies according to colony condition. It seems that a high number of active foragers will delay the transition from nest bee to forager. Foragers gather nectar and pollen from flowers. The nectar is sucked up and stored in an anterior region of the gut called the honey crop. Enzymes in the gut act on the nectar and, when back in the nest, the bees 'pant' to evaporate water from the nectar. They then

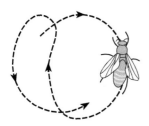

⌃ The Round Dance means the food supply is nearby (up to 100 yards).

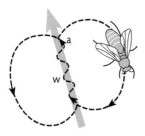

⌃ The Waggle Dance shows in symbolic form the direction and distance of the food source. Direction in relation to the sun's position is indicated by the angle (a) between the waggle run and the vertical. Distance is indicated by the duration of the waggle run (w).

regurgitate all of the nectar into storage cells. Here the nectar is further thickened by evaporation. Workers collect pollen and carry it in special baskets on their hind legs. Pollen, rich in protein, is used to feed worker and drone larvae. Foraging workers returning to the nest perform dances on the comb to tell other workers the direction and distance of the food source she has visited. A figure of eight waggle dance of varying speed is performed if the food source is more than 100 yards away. The angle of the waggle run in the middle of the dance relative to the vertical allows the bees to follow a course to the food source, relative to the position of the sun. If the food is less than 100 yards away, a round dance is performed.

Honey bee colonies attract a number of vertebrate and invertebrate intruders whose aim is to steal honey. Workers attack *en masse*, and many will die in the process. The sting of a honey bee is barbed, unlike that of a social wasp, which can sting repeatedly. An amazing thermal execution technique is used by honey bees in response to enemies such as hornets. The workers crowd tightly around the hornet and shiver their thoracic muscles. This generates a large amount of heat inside the ball of bees and the hornet is literally fried alive. Dying in defence of the colony has been pivotal for understanding the biological basis of altruism in animals. Due to the hymenopteran method of sex determination, haplodiploidy, females that are diploid share 75% of their genes with their sisters (50% comes from their haploid father and 25% from their diploid mother). Worker bees are thus more closely related to each other, than to their offspring, if they were able to produce any: if they had daughters themselves, they would only share 50% of their genes with their offspring, which is less than the 75% that they share with their sisters. Therefore, it makes evolutionary sense for them to help rear and defend their sisters and the queen, as the latter is essential in producing more worker bees.

When a queen ages, when the colony gets too crowded, or when there is not enough food, the workers construct large queen cells. Just before the first queen emerges, the old queen and about half the workers leave the colony as a swarm. The first queen to emerge stings any other queens to death and then leaves the hive for a mating flight. The drones compete with each other, and the queen may mate with several males. After mating the males die, but their sperm will be used by the queen to fertilize all the eggs she lays during her life.

Apids are major crop pollinators and the value of crops pollinated by honey bees alone greatly exceeds the value of the honey, wax, and other useful products they provide. Leaving aside the incalculable benefits that aculeate Hymenoptera bring, a significant number of people die every year from being stung. The Africanized 'killer bees' found in South America and in some southern United States of America, resulted from accidental mating between drones of *Apis mellifera mellifera* and queens of a very closely related African subspecies *Apis mellifera scutellata*. This hybrid is very industrious, but, unfortunately, very aggressive, and will readily attack anything that comes too near to the colony. The sting volume and the venom toxicity is the same as that of a normal honey bee, but, as victims may receive many hundreds of stings, the effect, particularly on the young and the elderly, can be fatal. Their behaviour is sufficient to identify them, but if you want to be really sure, you can measure the ratio of two wing veins, but it is best do it with a dead one.

The oldest hymenopteran fossils are known from the Triassic (around 224 mya). Phylogenomic work suggests that Hymenoptera arose considerably earlier, sometime between the Carboniferous–Triassic boundary (280 mya), rendering this order a very ancient lineage indeed. Hymenopterans are the sister group to all other Endopterygota.

Key reading

Andersson, M. (1984). The evolution of eusociality. *Annual Review of Ecology and Systematics* **15**: 165–89.

Beattie, A. J. (1985). *The evolutionary ecology of ant–plant Mutualisms*. Cambridge studies in ecology series. Cambridge University Press, Cambridge.

Blaimer, B. B., Gotzek, D., Brady, S. G., and Buffington, M. L. (2020). Comprehensive phylogenomic analyses re-write the evolution of parasitism within cynipoid wasps. *BMC Evolutionary Biology* **20**: 155.

Bohart, R. M. and Menke, A. S. (1976). *Sphecid wasps of the world*. University of California Press, Berkeley.

Bolton, B. (1995). *A new general catalogue of the ants of the world*. Harvard University Press, Boston.

Branstetter, M. G., Danforth, B. N., Pitts, J. P., Faircloth, B. C., Ward, P. S., Buffington, M. L., Gates, M. W., Kula, R. R., and Brady, S. G. (2017). Phylogenomic insights into the evolution of stinging wasps and the origins of ants and bees. *Current Biology* **27**: 1019–25.

Buffington, M. L., Forshage, M., Liljeblad, J., Tang, C.-T., and van Noort, S. (2020). World Cynipoidea (Hymenoptera): a key to higher-level groups. *Insect Systematics and Diversity* **4**: 1.

Burd, M. (1996). Foraging performance by *Atta colombica*, a leaf-cutting ant. *American Naturalist* **148**: 597–612.

Burke, G. R. and Strand, M. R. (2012). Polydnaviruses of parasitic wasps: domestication of viruses to act as gene delivery vectors. *Insects* **3**: 91–119.

Chittka, A., Wurm, Y., and Chittka, L. (2012). Epigenetics: the making of ant castes. *Current Biology* **22**: R835–8.

Chittka, L. (2022). *The Mind of a Bee*. Princeton University Press.

Cruaud, A., Jabbour-Zahab, R., Genson, G., Cruaud, C., Couloux, A., Kjellberg, F., Van Noort, S., and Rasplus, J. Y. (2010). Laying the foundations for a new classification of Agaonidae (Hymenoptera: Chalcidoidea), a multilocus phylogenetic approach. *Cladistics* **26**: 359–87.

Eggleton, P. and Belshaw, R. (1992). Insect parasitoids: an evolutionary overview. *Philosophical Transactions of the Royal Society of London B* **337**: 1–20.

Godfray, H. C. J. (1994). *Parasitoids: behavioural and evolutionary ecology*. Monographs in behavior and ecology series. Princeton University Press, Princeton.

Godfray, H. C. J. and Cook, J. M. (1997). Mating systems of parasitoid wasps. In: *The evolution of mating systems in insects and arachnids*. Choe, J. C. and Crespi, B. J. (eds), pp. 211–25. Cambridge University Press, Cambridge.

Goodman, L. J. and Fisher, R. C. (eds) (1991). *The behaviour and physiology of bees*. CAB International, Wallingford.

Heraty, J. M., Burks, R. A., Cruaud, A., Gibson, G. A. P., Liljeblad, J., Munro, J., Rasplus, J. Y., Delvare, G., Janšta, P., Gumovsky, A., Huber, J., Woolley, J. B., Krogmann, L., Heydon, S., Polaszek, A., Schmidt, S., Darling, D. C., Gates, M. W., Mottern, J., Murray, E., Dal Molin, A., Triapitsyn, S., Baur, H., Pinto, J. D., van Noort, S., George, J., and Yoder, M. (2013). A phylogenetic analysis of the megadiverse Chalcidoidea (Hymenoptera). *Cladistics* **29**: 466–542.

Herre, E. A., West, S. A., Cook, J. M., Compton, S. G., and Kjellberg, F. (1997). Fig-associated wasps: pollinators and parasites, sex-ratio adjustment and male polymorphism, population structure and its consequences. In: *The evolution of mating systems in insects and arachnids*. Choe, J. C. and Crespi, B. J. (eds), pp. 226–39. Cambridge University Press, Cambridge.

Hirota, T. and Mita, T. (2021). Role of host vibration and cuticular hydrocarbons in host location and recognition by *Haplogonatopus oratorius* (Hymenoptera: Dryinidae). *Applied Entomology and Zoology* **56**: 107–13.

Hölldobler, B. and Wilson, E. O. (1990). *The ants*. Springer-Verlag, Berlin.

Huber, J. T. (2009). Biodiversity of Hymenoptera. In: *Insect biodiversity: science and society*. Foottit, R. G. and Adler, P. H. (eds), pp. 303–23. Blackwell Publishing, London.

Itô, Y. (1993). *Behaviour and social evolution of wasps*. Oxford series in ecology and evolution. Oxford University Press, Oxford.

Jaffé, R., Kronauer, D. J. C, Bernhard, K. F., Boomsma, J. J., and Moritz, R. F. A. (2007). Worker caste determination in the army ant *Eciton burchellii*. *Biology Letters* **3**: 513–16.

Kimsey, L. S. and Bohart, R. M. (1990). *The chrysidid wasps of the world*. Oxford University Press, Oxford.

LaSalle, J. and Gauld, I. (eds) (1993). *Hymenoptera and biodiversity*. CAB International, Wallingford.

Malm, T. and Nyman, T. (2015). Phylogeny of the symphytan grade of Hymenoptera: new pieces into the old jigsaw(fly) puzzle. *Cladistics* **31**: 1–17.

Malyshev, S. I. (1968). *Genesis of the Hymenoptera and phases of their evolution*. Richards, O. W. and Uvarov, B. (eds). Methuen and Co. Ltd, London.

Matsura, M. and Yamane, S. (1990). *Biology of the vespine wasps*. Springer-Verlag, Berlin.

Michener, C. D. (1974). *The social behaviour of bees: a comparative study*. Harvard University Press, Cambridge, MA.

Michener, C. D. (2000). *The bees of the world*. Johns Hopkins University Press, Baltimore.

Moreau, C. S., Bell, C. D., Vila, R., Archibald, S. B., and Pierce, N. E. (2006). Phylogeny of the ants: diversification in the age of angiosperms. *Science* **312**: 101–4.

O'Toole, C. and Raw, A. (1991). *Bees of the world*. Blandford Press, London.

Peters, R. S., Krogmann, L., Mayer, C., Donath, A., Gunkel, S., Meusemann, K., Kozlov, A., Podsiadlowski, L., Petersen, M., Lanfear, R., Diez, P. A., Heraty, J., Kjer, K. M., Klopfstein, S., Meier, R., Polidori, C., Schmitt, T., Liu, S., Zhou, X., Wappler, T., Rust, J., Misof, B., and Niehuis, O. (2017). Evolutionary history of the Hymenoptera. *Current Biology* **27**: 1013–18.

Polilov, A. A. (2012). The smallest insects evolve anucleate neurons. *Arthropod Structure & Development* **41**: 29–34.

Prys-Jones, O. E. and Corbet, S. A. (1991). *Bumblebees* (2nd edn). Naturalists' handbooks series **6**. Richmond Publishing, Slough.

Quicke, D. L. J. (1997). *Parasitic wasps*. Chapman and Hall, London.

Quicke, D. L. J. (2015). *The braconid and ichneumonid parasitoid wasps: biology, systematics, evolution and ecology*. Wiley-Blackwell, Chichester.

Quicke, D. L. J., Austin, A. D., Fagan-Jeffries, E. P., Hebert, P. D. N., and Butcher, B. A. (2020). Recognition of the Trachypetidae stat.n. as a new extant family of Ichneumonoidea (Hymenoptera), based on molecular and morphological evidence. *Systematic Entomology* **45**: 771–82.

Roces, L and Holldobler, B. (1995). Vibrational communication between hitchhikers and foragers in leaf-cutting ants (*Atta cephalotes*). *Behavioural Ecology and Sociobiology* **37**: 297–302.

Roffet-Salque, M., Regert, M., Evershed, R., Outram, A. K., Cramp, L. J., Decavallas, O., Dunne, J., Gerbault, P., Mileto, S., Mirabaud, S., Pääkkönen, M., Smyth, J., Šoberl, L., Whelton, H. L., Alday-Ruiz, A., Asplund, H., Bartkowiak, M., Bayer-Niemeier, E., Belhouchet, L., Bernardini, F., Budja, M., Cooney, G., Cubas, M., Danaher, E. M., Diniz, M., Domboróczki, L., Fabbri, C., González-Urquijo, J. E., Guilaine, J., Hachi, S., Hartwell, B. N., Hofmann, D., Hohle, I., Ibáñez, J. J., Karul, N., Kherbouche, F., Kiely, J., Kotsakis, K., Lueth, F., Mallory, J. P., Manen, C., Marciniak, A., Maurice-Chabard, B., McGonigle, M. A., Mulazzani, S., Özdoğan, M., Perić, O. S., Perić, S. R., Petrasch, J., Pétrequin, A. M., Pétrequin, P., Poensgen, U., Pollard, C. J., Poplin, F., Radi, G., Stadler, P., Stäuble, H., Tasić, N., Urem-Kotsou, D., Vuković, J. B., Walsh, F., Whittle, A., Wolfram, S., Zapata-Peña, L., and Zoughlami, J. (2016). Widespread exploitation of the honeybee by early Neolithic farmers. *Nature* **534**: S17–18.

Ronquist, F., Klopfstein, S., Vilhelmsen, L., Schulmeister, S., Murray, D. L., and Rasnitsyn, A. P. (2012). A total-evidence approach to dating with fossils, applied to the early radiation of the Hymenoptera. *Systematic Biology* **61**: 973–99.

Roubik, D. W. (1989). *Ecology and natural history of tropical bees*. Cambridge University Press, Cambridge.

Schwarz, M. P., Richards, M. H., and Danforth, B. N. (2007). Changing paradigms in insect social evolution: insights from halictine and allodapine bees. *Annual Review of Entomology* **52**: 127–50

Shorthouse, J. D. and Rohfritsch, O. (eds) (1992). *Biology of insect-induced galls*. Oxford University Press, Oxford.

Skinner, G. (1996). *Ants*. Naturalists' handbooks series **24**. Richmond Publishing, Slough.

Strohm, E. and Linsenmair, K. E. (1997). Low resource availability causes extremely male-biased investment ratios in the European Beewolf *Philanthus triangulum* F. (Hymenoptera: Sphecidae). *Proceedings of the Royal Society of London B* **264**: 423–9.

Sullivan, D. J. (1987). Insect hyperparasitism. *Annual Review of Entomology* **32**: 49–70.

Sumner, S. (2022). *Endless Forms: The Secret World of Wasps*. William Collins.

Tang, P., Zhu, J. C., Zheng, B. Y., Wei, S. J., Sharkey, M., Chen, X. X., and Vogler, A. P. (2019). Mitochondrial phylogenomics of the Hymenoptera. *Molecular Phylogenetics and Evolution* **131**: 8–18.

Tautz, J. (2022). *Communication between honeybees: more than just a dance in the dark*. Springer.

Turillazzi, S. and West-Eberhard, M. J. (eds) (1996). *Natural history and evolution of paperwasps*. Oxford University Press, Oxford.

von Frisch, K. (1967). *The dance language and orientation of bees*. Harvard University Press, Boston.

Wharton, R. A. (1993). Bionomics of the Braconidae. *Annual Review of Entomology* **38**: 121–43.

Wiebes, J. T. (1979). Co-evolution of figs and their insect pollinators. *Annual Review of Ecology and Systematics* **10**: 1–12.

Winston, M. L. (1987). *The biology of the honey bee*. Harvard University Press, Boston.

Wiskerke, J. S. C., Dicke, M., and Vet, L. E. M. (1993). Larval parasitoid uses aggregation pheromone of adult hosts in foraging behaviour: a solution to the reliability-detectability problem. *Oecologia* **93**: 145–8.

Yeo, P. F. and Corbet, S. A. (1995). *Solitary wasps* (2nd edn). Naturalists' handbooks series **3**. Richmond Publishing, Slough.

SECTION 3

Fieldwork

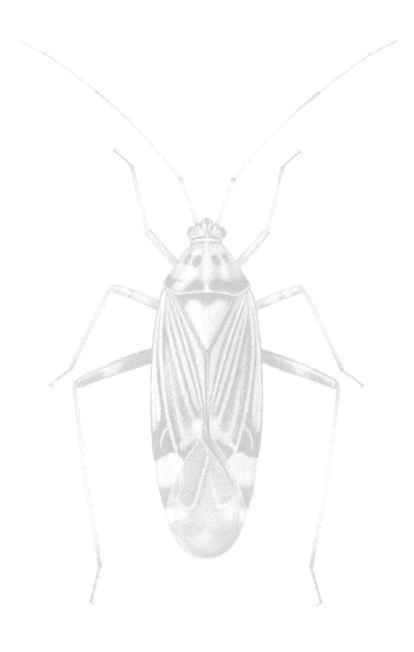

You can learn a certain amount from books, but in order to really learn about the secret lives of insects, you must study them at first hand. With organisms as small, numerous, and diverse as insects, you will not be able to get very far without collecting and killing them for identification. A few people profess to be able to recognize many species on sight, but for small and cryptic species their identifications will be questionable. In any case, an identification without voucher material carries little weight. Many species are distinguishable only by looking at their dissected genitalia or other minute details. Even with perfect eyesight and a hand lens it is impossible to count the number of bristles on the head or body of live flies, and most small moths rest with all their useful identification features obscured. To quantify biodiversity in any habitat, whether for a basic site description, environmental impact assessment, or pest management survey, it is necessary to find out what species are present, and in what numbers.

However, it is important to consider conservation issues. A growing appreciation of the importance of insects resulted in the protection of many species around the world, especially in the light of the growing destruction of natural habitats and other environmental changes. Among the influential international organizations that exist, the IUCN (the World Conservation Union, formerly the International Union of Conservation of Nature and Natural Resources) based in Gland (Switzerland) and the World Conservation Monitoring Centre based in Cambridge (UK) are particularly relevant for those interested in insects. Legal protection has been extended to particular species (Collins, 1987 & 1988; Heywood, 1995), and a list of threatened invertebrates is given by Wells et al. (1983), and of swallowtail butterflies by Collins and Morris (1985). An *IUCN Red List of Threatened Animals* (1996) lists 73 insect species as extinct or extinct in the wild, and a further 537 species are recorded as critically endangered, endangered, or vulnerable. In reality, this is the tip of an iceberg and little data are available. Whenever you collect, remember that there is a code of practice for collectors of biological material (Jermy et al.,1995, p. 57), and many countries have legislation that will affect your activities (e.g. Wildlife and Countryside Act, 1981).

Collecting

The indispensable reference work is *Ecological Methods* (Southwood and Henderson, 2000). Methods for collecting insects can be divided into two sorts—relative and absolute. Relative sampling methods provide only presence/absence data and indicate how abundant one species is relative to another, but do not allow you to measure the population sizes. Examples of relative sampling methods are pitfall trapping, flight intercept traps, and light trapping. In all these cases, it is unknown what area or unit of the habitat has been sampled. Using absolute sampling methods, the density of the species collected can be readily calculated (number per m^2 is a standard measure). Insecticidal knockdown techniques, suction sampling, hand searching (over a known area), and extraction from a known quantity of leaf litter are

Essential Entomology. Second Edition. George C. McGavin and Leonidas-Romanos Davranoglou, Oxford University Press.
© George C. McGavin and Leonidas-Romanos Davranoglou (2022). DOI: 10.1093/oso/9780192843111.001.0001

all examples of absolute sampling methods. No single method will collect all the species present in a sample area, but by using absolute techniques you will be able to make meaningful comparisons between different sites or studies.

Relative collecting techniques

There are many methods for collecting insects. The key to successful collecting is using the right equipment for the job, as well as the right timing. This section discusses relative collecting techniques.

Butterfly and aerial nets   Aerial or hand nets are lightweight nets designed for catching flying insects. They should be lightweight and strong with as large a mouth as is practicable. Quick reactions and balletic movements are needed to catch very fast-flying or highly manoeuvrable insects, such as dragonflies and skippers, but gentle passes over the top of vegetation will collect large amounts of small flies and wasps. It hardly needs saying that fine aerial nets should never be used for sampling insects from dense or thorny vegetation.

Sweep nets   Sweep nets are made from strong, close-woven material and are designed for thrashing through vegetation. The bag should be deep enough to allow you to trap the catch by gathering the material together in one hand below the frame. Apart from suction sampling, it is the only way to sample insects quickly from low-growing vegetation and grasses. A sweep net is considerably lighter and easier to use than any motorized suction sampler but there are, of course, drawbacks. Sweep sampling is a relative sampling method and is (at best) only semi-quantifiable. Sweeping will not catch leaf-mining and stem-boring species and insects that live close to the soil surface. You may also miss the larger, faster species, which, alerted by your vigorous progress through their habitat, will escape before you get to them. The sweep net can also be used to collect insects in a rough manner from trees and shrubs. It is impossible to use this technique if the foliage is wet as this will result in much of your catch being reduced to a crumpled mush.

To make your results comparable between different sites you will need to standardize your sampling effort in some way. You should use the same net (the mouth diameter of the net is an important factor) and do a fixed number of sweeps per sample (25, 50, or 100 sweeps are typical). How you sweep is important and people vary a great deal in their technique, so it is best to get the same people to do all the sampling. Another point worth remembering is that you should always sweep into the wind. Describing the proper technique is difficult, but bear in mind that the object is to sweep as evenly as possible from a particular area or volume of vegetation. Stoop down and hold the handle of the net firmly in both hands, and then sweep the mouth of the net to one side as far as you can comfortably reach. The essential point is that the net should be kept as low to the ground as possible at all times, and at the end of each pass the net is turned swiftly to begin the next pass or sweep to the other side. The movement of the net must be nearly horizontal and not parabolic. Repeat the sweeping action the required number of times as you walk through the vegetation, and at the end of the last sweep, turn the net mouth or grab the net above the catch so that you do not lose anything. The bigger the number of sweeps the more difficult it will be to sort out the catch from the plant debris that you will unavoidably collect as well. If you are sampling for a specific insect species or group, you might simply search through the catch *in situ*, collecting what you need with a pooter (see figure below), and returning the rest to the wild. Otherwise you need to bag up the material and sort it out later.

Transferring a sweep net catch to a strong polythene bag is easiest with two people. One person holds the plastic bag so that the opening assumes the shape of a slit. The other person grasps the net together above the catch. The contents of the sweep net are everted into the plastic bag by gradually drawing the material of the net gradually through and over the retaining hand. Expel most of the air from the plastic bag and tie the top (after putting a label inside).

There are two schools of thought as regards sorting sweep samples (and suction samples): the live sorters and the dead sorters. Killing the catch (for instance, with fumes of ethyl acetate) in the field is useful, as some of the catch might start to eat the others, especially if there is going to be a long delay between sampling and sorting. Also, heavier and stockier insects might trample over and damage more delicate species. Live sorting is also preferable, as dead insects, especially small ones, are ten times more difficult to sort out from debris than live ones and anything not required can be returned to the field. If you opt for live sorting, try to keep bagged catches out of the sun. Live sorting of sweep samples is greatly facilitated by using a sorting hood. This simple device, which can readily be made up in the field, employs the principle that most insects will move towards a light source. The sides, top, and bottom of the hood should be opaque with only the back or part of the back panel transparent. With a light placed behind or the hood simply angled towards the sun, your catch can be carefully tipped out and the live insects pootered or brushed into alcohol as they move towards the light. A simple hood can be made from a cardboard box lined with white paper with a cellophane sheet or fine netting taped to a hole in the back.

Beating trays   Victorian collectors were clever. You cannot easily collect insects in the rain so why carry a beating tray as well as an umbrella? Although better if made of a white material against which to see your catch, any umbrella held upside-down could serve as a reasonable beating tray.

Various designs of beating tray are available or you can make you own. The tray is held under the foliage to be sampled and the branch shaken or given sharp, jarring blows with a stick. Do not thrash the vegetation as you will damage the plants and end up with a large pile of leaves and debris on your tray, which will take longer to search through. Some insects are very good at holding on and you will not, of course, catch internal feeders, such as leaf miners and stem borers. If you are specifically trying to catch very active species, a modified beating tray in the form of a funnel-shaped net with a collecting bottle might be preferable. It is best to keep spiders and insects separate, as a spider silk quickly renders a pooter tube full of insects a tangled mess.

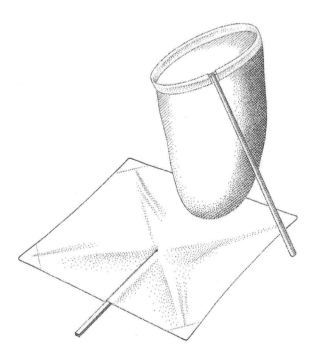

> A beating tray and butterfly net.

Like sweep netting, beating is a very simple and quick technique, but also suffers from being qualitative. It is also restricted to how high you can comfortably reach. You can, however, try to compare the beaten insect communities from different species of tree or shrub by replicating samples and standardizing the effort (the number and vigour of blows, perhaps) and area sampled as best you can. The beating of wet foliage is not recommended for the same reasons as sweeping wet vegetation.

Light traps   That some insects are attracted to light must have been noted shortly after humankind discovered fire. Eighteenth- and nineteenth-century entomologists used various forms of hurricane lamps and Tilley storm lanterns to go 'mothing', but, in recent decades, the use of light sources that produce large amounts of ultraviolet radiation has revolutionized light trapping. There have been many designs of light trap produced in the last 50 years, and all exploit the phenomenon that many insects (but by no means all) will fly towards bright lights at night. Moths, notably, are attracted in this way, but it should be remembered that it is mainly the males who respond, females preferring to sit in concealment among vegetation. The reason why moths are attracted to light is that in nature, they use the moon to navigate (transverse orientation). If they keep the angle that the light rays subtend to the ommatidia constant, they should fly in a more or less straight line. When confronted with a brighter light that is much nearer, the same behaviour will cause them to take a decreasing spiral path into the light source.

In tropical areas, lights of almost any kind can pull in fantastic numbers of beetles, bugs, and other insects, as well as moths. Around the world the installation of vapour discharge lamps for street lighting has reduced populations of some moths. Many predators have learned to cash in on this bonanza.

Bats will forage in the vicinity of bright lights, and birds will clean up the remainder in the morning. In Papua New Guinea, one of us (George McGavin) has often observed foraging columns of weaver ants (*Oecophylla smaragdina*) queuing up along lit fluorescent tubes to snatch insects as they fly towards the light.

The trouble with light traps is that you cannot be sure what will be attracted to them, and what will not: indeed, some insects are repelled by light. Many physiological, behavioural, and environmental factors are involved. The biggest problem with light trap data is that it is impossible to know over what area the trap is effective. In wooded habitats the range of the light may be quite small due to obstructions, but even in open habitats you cannot be sure. Experiments using a 125-watt mercury-vapour lamp have shown that the effective trap radius, far from being 90 m as originally thought, is rather less than 5 m.

Light traps are most useful to show the presence and perhaps relative abundance of particular species. Quantitative analysis is tricky, but with replications, using the same trap design, you might be able to get a rough comparison of different areas or an idea of how the populations of some species fluctuate over time. They are thus useful for general survey work but be very careful not to make more of the data collected than it warrants.

Large traps, such as the Robinson MV moth trap, which use mercury-vapour discharge bulbs, require connection to a mains supply or a generator. This type is not generally useful far from a field station or where you cannot drive a vehicle. Smaller variations of the Robinson trap, such as the Heath Trap that can be run from a 12-V car or motorcycle battery, can be used in more remote locations. There are a number of portable designs that use smaller batteries permitting their use in very remote or difficult terrain. Various sorts of light trap, their operation, and limitations can be found in Muirhead–Thomson (1991), Davies and Stork (1996), and Fry and Waring (1996). A very convenient light trap for use in the tropics is described by Robinson and Jones (1996). The heart of this cheap and simple trap is a 4-watt actinic tube powered by four standard D-size batteries. Larger-wattage tubes are available but need bulkier batteries.

Material that comes towards light can be collected by hand or by pooter and should be killed and preserved in a manner appropriate to the taxon in question. Most non-lepidopterans can be put into an ethyl acetate (avoid inhalation and contact with plastics, especially styrenes) or cyanide (avoid contact and inhalation) killing jar before being decanted into paper envelopes and stored dried or fresh in a sealed plastic tub with chlorocresol. Small or micro-moths should be taken from the killing jar and pinned using a minute or micro-pin vertically through the thorax. They are then pinned into small plastic boxes lined with Plastazote™ (or similar expanded polyethylene foam), and the wings should be set forward and held in place with additional micro-pins. In this manner a large number of fragile specimens can be kept and transported safely. Butterflies and robust moths, such as hawk moths, can be put in small envelopes or folded paper triangles inside a sealed plastic tub with chlorocresol as an antifungal agent.

Like most skills, the art of setting butterflies and moths is best learned by watching a professional, but instructions are given in a number of publications (Arnett, 1985; Dickson, 1992; Walker and Crosby, 1988).

Flight intercept traps   Any sort of static trap that catches insects in flight could reasonably be called a flight intercept trap (FIT). The term is used more specifically to describe a single, rather simple type of trap. A piece of material (usually black terylene) is stretched tightly between poles or trees, and plastic containers are arranged on the ground along its length. The containers are part-filled with water, preservative, and a little detergent to break the surface tension. Insects flying into a FIT will crawl up and escape from the top, drop downwards, or just fly off again. Some specialists use a roof over the FIT to prevent insects from flying away from the trap. If the traps are going to stay on site for more than a couple of days, placing crystals of chloral hydrate or pouring propylene glycol in the water containers will prevent decomposition of the catch. Although a remarkably effective technique to catch flying insects, one problem with FITs is that you cannot be sure whether they are acting purely passively or if some insects are attracted to them. Also, if you require directional data, such as whether or not insects are flying in a particular direction, for example, in or out of a woodland or other habitat, you will need to make modifications to available designs or make you own. Insects collected from the tubs can be strained out using a tea strainer or small sieve and transferred to alcohol in labelled tubes.

> A flight intercept trap, with collecting trays beneath.

Malaise traps, FITs of a particular design, are named after entomologist René Malaise, who noticed that insects that entered his tent always seemed to accumulate 'at the ceiling-corners in vain efforts to escape at that place without paying any attention to the open tent door'. His design, which was published in 1937, has spawned many variants over the years. Malaise commented on the main advantages of his trap; it is 'better than a man with a net' and 'it can catch all the time, by night as well as by day, and never be forced to quit catching when it was best because dinner-time was at hand'.

© George McGavin

❯ A Malaise trap.

Malaise traps are essentially open-sided tents made of material, which have a high point in the roof where the insects tend to collect and are funnelled into a container. To increase the efficiency of the Malaise trap, water-filled containers can be arranged on the ground under the central partition. In this way insects that do not crawl upwards in response to flight interruption will also be trapped. There is no doubt that they are effective in catching flying insects and, lashed to a light frame, they can even be pulled up into forest canopies. It is often said that Malaise-type traps are non-attractant, but anyone who has used them will know that this is not true. The colour of the material used to make them has a significant effect on the composition of the catch. In some habitats black malaise traps are very attractive to horse flies (Diptera; Tabanidae). This phenomenon has been put to good use in the design of traps for blood-feeding flies.

For the majority of survey-type work, the collecting bottle of the Malaise trap is filled one-third to one-half full with 70% alcohol, which serves to kill and preserve the catch. If, however, you are specifically collecting Diptera, most species are best collected dry with a killing agent, such as a piece of Vapona Fly Killer™, or a similar domestic, slow-release strip of insect killer containing dichlorvos wrapped loosely in tissue at the bottom of the collecting bottle.

Sticky traps and window traps are variants of FITs. Sticky traps are simply pieces of Perspex or glass coated with sticky material (e.g., Hyvis™) and fixed to a post. Window traps should be as transparent as possible and usually have a water-filled tray underneath to catch insects that hit the surface and drop down. The major problem with sticky traps is that the glue used to trap the catch is often difficult to remove and will almost certainly require the use of dangerous and/or inflammable solvents. Even if you get specimens off the traps in one piece, it is often very difficult to identify them. Sticky traps can be useful for the regular monitoring of populations of pest insects where identification does not pose a problem.

Pitfall traps   Pitfall traps are a very easy and cheap way to catch active ground-living arthropods of all kinds. A hole is dug with a trowel and a plastic container of some kind is sunk into the ground so that its rim is level or slightly below that of the surrounding ground. It is no good at all if there is even the slightest lip showing above ground, as small animals will simply walk around the circumference. Bits of debris and foliage that lie across the top of the trap will act like little walkways allowing insects to crawl across. Once properly set, pitfalls are left and animals will fall in, and the steep, slippery sides should ensure that most of them will not escape.

Pitfall traps are usually one-third filled with water or alcohol to trap and kill the catch. Very small insects can sometimes escape from the surface of water, so a few drops of detergent are added to lower the surface tension. When using water, the catch will start to decay in a day or two, faster in warm weather, so it is vital that the traps are emptied daily. If you are not able to service the traps regularly, or if you wish to leave them undisturbed for a week, you should use a mixture of 40% ethylene glycol (antifreeze) in the water as a preservative. Some older texts recommend a saturate solution of chloral hydrate, but it should be generally avoided as it is very toxic and needs to be handled with great care. Whichever technique you employ, the contents of the traps should be preserved as soon as possible in 70–80% alcohol. Of course, alcohol can be used in the traps in place of water, but it is expensive, will evaporate faster than water, and a few studies have suggested that the smell of alcohol can act as a repellent or attractant for some species. Do not use formaldehyde solutions—they are dangerous and a potential environmental hazard.

A well-tried and tested protocol for pitfall trapping any given area employs cheap, plastic drinks cups with a rim diameter of 7 cm (internal volume 200–250 ml). The standard unit of trapping effort could be, say 180 trap days; this could be 60 traps operated over a period of three days, or 30 traps over six days. Traps are placed in lines of ten with a spacing of 4–5 m between each trap and between adjacent rows. Each trap unit consists of two drinks cups, an outer one and inner one.

To set each trap, dig a hole larger than the drinks cups with a trowel and make sure it is just deeper than the rim of the inner cup. Put two cups inside each other into the hole and back-fill the space around the outer cup. Gently press down and smooth the soil around the rim of the inner cup. Do not worry if the inner cup is now partly filled with soil at this stage. The whole point of having two cups is that you can now take the inner one out and empty excess soil, preferably off-site. The inner cup is now replaced with a clean one. The outer cup protects the inner one and allows subsequent rapid emptying of the inner cup. The outer cup should have a small hole punched in the bottom to allow rain to drain away. If this is not done, rain can seep between the two cups and the inner one will float up, spilling its contents. If you are going to leave the traps for more than a day or two, or if there is the likelihood of very

① Dig hole slightly deeper than pitfall trap

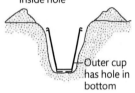

② Place two plastic cups inside hole

Outer cup has hole in bottom

③ Backfill cup and hole and tamp down

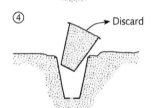

④

Discard

Using finger nail to hold outer cup in place, take out inner cup and discard soil

⑤ Carefully replace inner cup and check rim

Fill cup to $\frac{1}{3}$ with pitfall fluid (see text)

⌃ **Pitfall trap construction.**

heavy rain, it is a good idea to make a neat hole in the inner cup with a paper punch about 2 cm from the rim and glue a small piece of terylene or similar material over the hole. In this way, if the cup fills up, the excess water will drain away, and the catch will be retained.

When all the traps are dug in, the inner cups are one-third filled with water containing a few drops of detergent. To save time and depending on the nature of the habitat, the position of the grid can be marked with a cane. Catches from all the traps from one site and one day are collected by pouring them through a sieve and bulked together in 70% alcohol. If you really want to keep them, any vertebrates collected should be stored separately from the invertebrate material. As with keeping any biological material in tubes of alcohol, the total volume of the catch should never be more than half the volume of the storage tube. If you completely fill the tube there will be little room for the alcohol, which will become diluted anyway, resulting in the decomposition of the catch.

If you want to trap specific groups of insects, such as those attracted to carrion and dung, it is a simple matter to incorporate appropriate bait. You will, of course, still catch things that just blunder in but, with experience, these can be ignored.

The use of subterranean pitfall traps has produced some very interesting data on the occurrence, abundance, and distribution of species that spend most or part of their lives underground. Owen (1995, 1997) describes the construction and use of a simple pitfall trap for the repetitive sampling of underground faunas.

Pitfall traps can be used dry for live collection of specimens, in which case they need to be serviced much more regularly than those where the catch is killed in fluid. A large ground beetle in a dry trap will eat anything within it.

There are many drawbacks to using pitfalls. Sometimes, small mammals and amphibians get trapped and die unnecessarily. When pitfalling in dry areas, birds and other animals soon learn where to get a drink. Pitfall traps with a large surface area might act as a water or pan trap and attract flying insects as well. It would be impossible without watching it all the time to be sure whether an insect stumbled and fell into the trap as intended or flew into it. With large or long-term traps, it is often a good idea to arrange some sort of cover to keep rain out or mesh of various grades to exclude certain sizes of animal. Another problem is that you can never be sure whether even ground-living species might be attracted to them in some way, but as with light trapping, the biggest problem associated with pitfalls is that you cannot be sure over what area they are trapping, and they are greatly affected by vegetation structure. Despite these very serious disadvantages, pitfall trapping remains a standard qualitative sampling technique— it is cheap and produces large amounts of material with relatively little effort.

Water pan traps (or yellow pan traps)   A very simple trap can be made from almost any sort of shallow tray or container. The pan is partly filled with water and preservative to which a few drops of detergent are being added to reduce the surface tension. A bright yellow background colour has been found to be the most effective in attracting small flies, wasps, and bugs, such as aphids, although variants with blue colour may work better in certain areas. An approach that one of us (Leonidas-Romanos Davranoglou) and our colleagues frequently use is to create two-coloured traps, where one half of the trap is painted yellow, and the other half is blue—in this way, you can attract species that prefer either colour. To prevent overflow due to heavy rain, very small holes can be made on the sides of the trap. Pan traps have many advantages: you can put out scores of them very cheaply and you can use them virtually anywhere that you can reach, from ground level to the tops of trees. The main disadvantage, of course, is that the method is relative. Furthermore, they might dry up in very hot conditions, they need to be emptied frequently, and the catch preserved in alcohol. Lepidoptera caught will need careful drying.

© George McGavin

❯ A yellow pan trap.

Lures and baited traps   Insects will respond to all manner of environmental cues, and this can be used to locate food, mates, or egg-laying sites. Rather than employing general collecting techniques, specific insects can be attracted by various baits depending on their functional ecology. Indeed, many insects will only be caught using this sort of approach. For example, scarabs of the genera *Deltochilum* (subgenus *Aganhyboma*) and *Canthon* are highly unusual in that they frequently prey on millipedes. One of the main ways to collect them is by using pitfall traps baited with crushed or injured millipedes—the defensive chemicals they exude are irresistible to the scarabs and can attract them from far away. Baits can be divided into major groups: dung, carrion, rotting or fermenting plant material, sexual odours (pheromones), and live animals. There is plenty of scope for originality and ingenuity in the design and refinement of baited traps.

> A home-made bottle trap baited with a fish head.

Animal dung and carrion will attract specialist flies and beetles, and the baits can be incorporated into pitfall or FITs. The attraction of butterflies to putrid fluids is well known. A small dustbin or bucket partly filled with liquefying rotting fruit (add yeast to accelerate process) and/or carrion will attract certain nymphalid butterflies, particularly those of the genus *Charaxes*. The Purple Emperor butterfly, a woodland species found across Europe and Asia to Japan, was often caught by using 'the juices of a dead cat, stoat, or rabbit' or 'a seething mass of pig dung'. Males (mainly) of many species around the world also indulge in puddling. Large numbers of several species may gather to drink around the edge of puddles, urine patches, salt runs, and other damp areas. The physiological need for salt and other minerals has been put forward as an explanation for this phenomenon. Butterflies of the neotropical genus *Morpho* can be attracted to lures of bright flashing colour. Materials that reflect ultraviolet will be attractive to many sorts of insects, including water beetles.

The technique of 'sugaring', now not as fashionable as it once was, relied on the attraction of moths to sweet and fermenting fluids. Old books had many exotic recipes, but a good basic mixture would be a thick mush of rotting bananas, brown sugar, and dark beer boiled for ten minutes, to which is added a splash of rum or essence of pear drops (amyl acetate). The mixture is daubed on posts and tree trunks after dark. The mixture can also be soaked on to thick ropes, which are hung up.

Absolute collecting techniques

Hand searching Simply getting down on the ground and looking for insects is a much under-rated technique. There are some habitats where no other sampling method will yield results. The most valuable (and often neglected) aspect of hand searching is that you might find out what the animals are doing. Mass collection techniques such as sweeping, beating, or fogging let you know what is present and in what numbers, but that is all they tell you. Close observation will reveal much more.

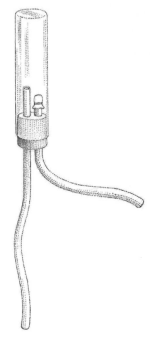

^ A pooter.

Imagine you need to look at grass tussocks where a particular plant bug lives. Sweeping and suction sampling may not provide much due to the dense growth and the foliage might be permanently wet in any case. The bugs in question live right on the surface of the soil and the only way to get them is go after them with a pooter. A good pooter or aspirator is one of the most useful bits of kit that you can have, and, although designs differ widely, they mostly rely on the operator sucking. There are special blow-pooters for medical, agricultural, and forensic sampling, which, sensibly, rely on the Bernoulli or venturi principle (you blow rather than suck). One of the simplest general purpose sucking designs uses a Sterilin™ universal specimen tube (approximately 39×25 mm and 30-cm^3 capacity). The tube is made from polystyrene so do not use solvents such as ethyl acetate near them. A bung with two holes drilled through it and short lengths of plastic tubing and pipe, and a small piece of terylene net secured over the end of the 'suck' tube to prevent insects from entering the operator's mouth, are all you need to complete the pooter. The diameter of the tubing is important. Too small and you will only be able to collect small species. Too big and you will need a lot of suck to get anything up the tube. Having collected what you want, the pooter tube can be tapped to shake all the insects to the bottom and then the bung pulled out and replaced with a screw-top. Large insects can be grasped using a pair of forceps, while very delicate insects, if not pootered, can be picked up on the end of an alcohol-moistened paint brush.

If pooters are used for long periods, there is a risk that minute particles, micro-organisms, and fungal spores could be inhaled, which might lead to allergies and chest infections.

Another situation where hand searching is useful is where arthropods have to be sampled from rock surfaces, particularly in caves or lava tubes. In these habitats, populations of animals are often sparse, and you can cover a lot of ground with a pooter and a pair of forceps. Different sites can be compared in a quantitative manner by introducing a time element.

Emergence traps and rearing Collecting insects by trapping them as they emerge, usually as adults from pupae or cocoons following larval development, diapause, aestivation, or overwintering, may be essential if you are going to do anything more than compile a simple species list. Rearing insects from leaf mines and stems, from under bark, from dead wood, from parasitized insects, from corpses, from macro-fungi, from the soil, and so on, may be the only practical way to collect many species. Different sorts of trap will be needed for each habitat/micro-habitat. At its simplest, an emergence trap or rearing container may be a cotton wool-plugged or muslin-covered tube or tub. Collecting insects in this way will provide much more useful biological information than general collecting techniques. For instance, beating the foliage of an oak tree may yield many species of gall wasp, but galls placed inside simple, muslin-covered containers will yield the gall causers, other gall occupants, and their parasites. It will also be possible to associate and preserve larval and pupal exuviae (shed skins) with the emerging adults. Deeming (1993) gives a good account of the simple techniques he uses to rear shoot flies belonging to several families. The same techniques could be equally well applied to the rearing of other insect groups, such as parasitic hymenopterans.

Many insect species emerge from soil or litter at the beginning of the growing season and are positively phototactic. Emergence traps are easily made from any suitable upturned opaque, usually black, container dug into the substrate and fitted with a transparent collecting bottle at the apex. Insects emerging from aquatic habitats can also be collected in emergence traps.

Emergence traps can be considered an absolute collecting method provided you sample from a known area or volume of substrate or habitat. Making simple and effective emergence traps to suit your particular purpose may be preferable (and certainly cheaper) to buying a ready-made design. A large, general-purpose model often used for processing large amounts of wood, litter, flood debris, compost, etc., is the Owen Emergence Trap. Measured quantities of material are placed inside the trap and left. Emerging insects should eventually find their way to the collecting bottle at the top. Some traps containing rotting wood have continued to yield material for up to two years.

Suction sampling   From the 1950s onwards, the need for more quantitative methods of sampling insects led to the development of a number of suction devices. The commonest was the petrol engine-driven D-Vac, or, more correctly, Dietrick suction insect sampler. Originally developed for use in agricultural crop systems, the D-Vac became widely used in ecological entomology, and virtually the regulation piece of equipment for sampling invertebrates in grassland habitats. The advantages of using a standard machine with a known sampling area and sampling effort were that different sites and different studies could be compared in a way that had been impossible before. The disadvantages, however, were many and varied. There was a tendency, among some users, to assume that everything was caught, but the machines were very variable in their efficiency and few calibration studies were ever carried out. Good practice involved D-Vac-ing for a number of timed sucks combined with careful hand searching afterwards to ensure that nothing had been missed. D-Vacs were very expensive, never very easy or comfortable to use, and needed constant maintenance. Despite modifications in the 1960s, these D-vacs are no longer available commercially. The advantages of suction sampling remain the same and some devices are still available. The high cost of commercial machines has resulted in a number of people making very simple modifications to leaf-blowers and similar items of garden machinery (Wilson et al., 1993; Stewart and Wright, 1995). Studies have shown that these home-made samplers do a very good job indeed, but, like all other designs, they will not sample everything with equal efficiency. You must calibrate your sampling in some way to estimate what percentage of organisms you are collecting. This percentage will vary greatly with insect body size, the sort of habitat you are studying, and the sampling time. Work out the cross-sectional area of the suction tube and you will be able to simply calculate the area you have sampled. For low vegetation a sampling unit might be 25 random timed sucks pooled from within a given area of the habitat. Alternatively, you could use a cylinder of known area to isolate the vegetation to be bug-vacuumed. Samples can be everted into plastic bags as described in the section on sweep nets.

Extraction techniques   Separating small invertebrates from the substrate can be a difficult and time-consuming process, but scores heavily on being an absolute method. Measured volumes, weights, or areas of sample will yield quantitative data. Different techniques are needed for different substrates, but the simplest form of extraction tool is a sieve. Here, gravity and agitation will only effect a partial separation. Different gauges of sieve (e.g., 10-, 5-, and 2-mm meshes) might help, but animals may hold on to or hide inside large debris and very small specimens will still need to be separated from fine debris. It is generally better to persuade the animals to leave the substrate under their own steam, and this principle is employed in a variety of extraction devices. The obvious advantage is that the samples obtained are relatively clean and require little further processing. A disadvantage is that the equipment can sometimes be bulky and, in cases where a heat or light source is required, difficult to arrange in the field. It is worth bearing in mind that the smaller and more delicate the animals of interest, the more difficult it will be to extract them efficiently.

Extraction funnels such as Berlese/Tullgren designs are suitable for the extraction of insects and other moderately robust specimens from leaf litter, crumbled soil, decaying wood, or even fresh foliage. There are very many variations of the same theme. The measured sample is placed on a mesh or sieve over a funnel, which leads to a collecting bottle containing alcohol (or dry for live collection). A (preferably boxed-in) source of light and heat source is placed directly above the sample causing the material to dry out. Animals migrate downwards through the sample. Rapid drying should be avoided, as this will kill smaller organisms before they escape from the sample.

An excellent piece of equipment for extracting invertebrates from moss, dead wood, litter, and even things like bird nests in the field, is the Winkler bag (also known as the Winkler extractor or Winkler apparatus). The bags are commercially available, and Owen (1987) gives a full account of the use and construction of this lightweight and highly effective apparatus. The Winkler bag is relatively cheap to make and, unlike other devices, does not require an artificial heat source to make it work. In essence, a wire-braced, box-shaped, calico bag is used to enclose measured samples of debris contained inside coarse netting. The bottom section of the bag tapers to a collecting bottle, which may be used dry or with alcohol. The top of the bag is tied up and the whole thing is hung up under cover to dry out gradually. Animals from samples that start fairly dry will be extracted in a few days, but wet samples may take more than a fortnight. To speed up the process it is advisable to occasionally take the samples from the inner bags and mix them to redistribute the wet bits. Winkler bags are particularly good at collecting small beetles, ants, hemipterans, parasitic wasps, flies, and cockroaches, and have resulted in the discovery of countless new species that would have been hard or impossible to collect with other methods.

Canopy techniques   All that is known to date about collecting arthropods from forest canopies can be found in two books: *Forest Canopies* (Lowman and Nadkarni, 1995) and *Canopy Arthropods* (Stork, Didham, and Adis, 1996). These two books, which have extensive bibliographies, provide up-to-date information on the biology of canopy arthropods, and how to sample and study them.

Essentially, you can either go up to collect material using a variety of approaches ranging from climbing, ladders, and cranes, to net rafts placed on the canopy by means of a dirigible (airship, balloon), or you can get the material to fall down using insecticides. Somewhere between these two extremes are things like composite traps, which are hoisted into the canopy (Basset, 1988).

The variety of insecticidal knockdown techniques that have been used in various parts of the world have a few things in common. Collecting trays of a known area are set up on the ground below or suspended from vegetation below the tree to be sampled. The trays, which can be conical or pyramidal, are best made from smooth, close-woven material such as rip-stop nylon. They should ideally have an integral collecting bottle at the bottom to speed up collection and minimize the loss of material. Under small trees or bushes, large plastic washing-up bowls can be used. A fast-acting insecticide, such as a synthetic pyrethroid, is blown into the foliage in the form of a fine mist or fog. Misting or fogging is commonly done early in the morning when there is no wind. Obviously, the longer the drop the greater the risk of a gust of wind blowing small insects clear of the trays. Wide treatment of the canopy around the area delimited by the trays below will reduce this error. Early samples are also thought to be more representative of the canopy fauna. Drop time is an important factor. Most studies have shown that 60–90 minutes is more than adequate. Samples should be preserved in alcohol as soon as is practicable.

Clearly, some techniques are better for some things than others. Hand searching in canopy foliage tends to miss small things, whereas insecticidal fogging or mist-blowing tends to under-record large things. Another problem with canopy sampling is that, although there have been a number of well-documented studies, they have used a wide range of sampling methods. Fogging, mist blowing, fumigation, and gassing are not comparable, and the use of different insecticides and even different formulations of the same insecticide have added to the difficulties.

Aquatic and taxa-specific sampling techniques

Chapter 5 in *Ecological Methods* (Southwood and Henderson, 2000) and chapter 10 in *Aquatic Insects* (Williams and Feltmate, 1992) are compulsory reading for anyone sampling freshwater habitats. In the main, insects can be simply collected from freshwater habitats using any number of hand net designs. Catches are simply tipped into white sorting trays or through sieves of various mesh sizes (0.5 mm or greater is fine for insects and larvae) and the species required removed for preservation in alcohol. Hand nets are long-handled, and for deep rivers or lakes should be able to take additional (screw-on) handle sections. The net bag should be made from strong polyester or nylon 1-mm mesh. The most common hand nets have circular or D-shaped mouths. Insects are caught by passing the net through the water or vegetation. Hand nets can also be used downstream to collect anything dislodged from vegetation or stones. Drift nets are rectangular-mouthed, long, tapering nets, which are staked or weighted down to collect organisms from shallow streams. A Surber sampler is a drift net with side screens fitted to a quadrat of known area (usually 0.1 m^2), which can be used to sample from shallow streams where the flow rate is less than 10 cm per second. Stones and plants within the quadrat are kicked, jarred, or brushed, the organisms dislodged, and are carried into the net by the current.

While it is very easy to collect insects in a relative manner from aquatic habitats, difficulties occur when trying to take reliable samples from a known volume or area of habitat. When collecting from open water you can easily calculate what volume of water has been sampled by your net. A number of sampling cages, cylinders, and grabs have been made which enclose a known volume of vegetation, bottom substrate, or sediment (Southwood and Henderson, 2000). In deeper water, dredges and trawls can be used to sample a known unit of substrate and open water habitat respectively.

Emergence traps, either fixed to the bottom or floating on the surface, can be used to collect insects that emerge as adults from freshwater habitats. Various sorts of artificial substrate (plastic mesh, discs, etc.) or natural substrate (leaf packs, stones) have been used to encourage colonization by invertebrates. These types of sampler are left in place for a specified time and then retrieved.

There are a number of useful techniques, mainly specific to particular taxa, which do not easily fit into the categories already mentioned.

Fleas   Fleas (Siphonaptera) are specialized ectoparasites of mammals and birds. Unlike parasitic lice (Phthiraptera), they do not spend their entire life attached to the host fur or feathers. Flea eggs are shed in the nest or lair of the host and the larvae are free-living and eat organic debris and the droppings of the adult fleas. Fleas and other ectoparasitic insects, such as lice and some Diptera, may be obtained by searching, combing, or fumigation (using chloroform or ether) of live or dead hosts. Remember to find out about any legislation in this area. You may need licenses to trap and handle live birds or mammals and it is essential to get training, especially if the use of anaesthesia is planned. Much larger numbers of fleas and other associated flies and beetles can be found in nests and lairs, which may be located visually, but mammals might only be active after dark or escape through vegetation after being released from live traps. Boonstra and Craine (1986) describe an inexpensive spool-and-line technique for following small mammals to their nests. Essentially, a thin thread is attached to a live trapped animals usually by gluing it to the fur. When released, the animal pulls thread from a spool, which allows the researcher to locate the nest. Early designs allowed tracking of up to 200 m, but improvements have increased this to more than 1500 m, depending on vegetation. Hawkins and Macdonald (1992) evaluate the technique for use with larger mammals.

Invertebrates living in nest material can be extracted using Berlese/Tullgren funnels or Winkler bags. These methods will also yield flea larvae, which would not be present on the body of the host. We are grateful to R. S. (Bob) George for the following description of his technique for collecting fleas from nest material. The equipment needed is a large, steep-sided bowl, specimen tubes containing 70% alcohol, a dissecting needle, and a large sheet of white paper. Before examination, nests must be kept separately in sealed boxes, tins, or polythene bags. The collector should wear a white shirt and work with rolled-up sleeves. A small amount of the nest material is placed in the bowl, which should stand in the middle of the white sheet. Adult fleas are picked up on the end of the alcohol-moistened dissecting needle and transferred to a tube. The material will also contain flea cocoons. These can be opened at the truncated end and, usually, adult fleas will emerge with a rush only to be collected. When no more adults can be found the material can be transferred to a sealed box either to breed out any remaining larvae or to allow a re-examination. Any fleas jumping free from the bowl are easily seen against the white paper, shirt, or bare arms and collected. This technique will produce many times more specimens than funnel extractions and is essential if you need quantitative data.

Butterflies   If you can recognize butterfly species you encounter, then you need not kill any. However, the butterfly transect walk techniques described by Pollard and Yates (1993) does need accurate identifications. Mark-capture-recapture techniques can be used to obtain absolute population estimates.

Bees   Bees can be most easily collected at flowers but try looking for nest sites in south or south-west facing exposures of bare earth or sand. In savannahs and deserts, some species will often nest in the vertical faces of dried-up riverbeds. These, and other species of bees that use the pith-filled cavities of plant stems, abandoned wood-boring beetle burrows, and empty snail shells, can be trapped in artificial nests. Trap nests can take the form of bundles of bamboo, pithy plant stems, or even wide drinks straws tied together; similarly, blocks of softwood drilled along the grain using a variety of hole sizes from 2 mm to 2 cm in diameter can be used. If appropriate trap nest designs are used, bees will start to colonize within hours of them being positioned in the field, and occupancy rates can be as high as 50–70%.

Termites   In tropical habitats, termites are vitally important as decomposers and soil conditioners, and are significant pests. There are also many interesting and largely unstudied species that live as inquilines in termite colonies. Numerically abundant (up to 10,000 individuals per m^2) the biomass of termites may outweigh all herbivorous vertebrate animal species put together. Despite this, termites do not appear a great deal in samples collected using standard entomological techniques. Even insecticidal techniques will not sample termites nesting in tree canopies. If sprayed or fogged, they will simply retreat inside their tunnels and nests. Sampling obvious structures, such as mounds, will provide only one-tenth of the termite abundance in some habitats. Eggleton and Bignell (1995) consider the various strengths and weaknesses of a variety of termite sampling procedures and develop a standard, stratified sampling protocol designed to take into account the patchy distribution and specialization of termite assemblages.

Collecting from live animals   Blood feeding and other types of insect may be collected by using live animals of various sorts, including humans, as bait. There are a number of obvious hazards. Large animals, even domesticated ones, can be dangerous at both ends. Tethered or not, approach from the front. Flies can be caught directly as they feed, but only a very tame animal will remain calm while an excited entomologist swipes at its hind quarters with a net. Small animals, such as rodents and bats, can give painful bites and may carry diseases such as rabies. Get training in handling live animals and wear protective clothing if required.

Killing methods and data recording

Insects should not be killed needlessly. However, the correct identification of many species may require the examination of minute body parts under a stereobinocular microscope or the dissection and preservation of the genitalia. This section discusses killing methods and data recording.

Killing methods

There are no clear guidelines about what constitutes the best or most humane way to kill insects, but common sense should tell you that the quicker the better is a good principle, as long as the specimens are intact and useable at the end (see Martin, 1977; Arnett, 1985, and Davies and Stork, 1996). When using chemicals, thoroughly familiarize yourself with the hazards and precautions. Appropriate safety procedures must be observed at all times, and bottles and jars should be labelled correctly.

A traditional method to kill insects is the cyanide killing jar. The technique is still favoured by some professional lepidopterists and orthopterists because it kills more quickly than other methods, keeps the specimens dry, and does not cause colour changes (ethyl acetate fumes, for instance, can cause some green pigments to fade). Glass or preferably high-density polyethylene jars for field use are made in advance and with careful preparation may give many months of service (Arnett, 1985). A typical jar uses a packed, 5-mm layer of powdered potassium cyanide crystals under a 1-cm-thick layer of sawdust and topped with a 1-cm layer of plaster of Paris. The plaster should be allowed to dry out thoroughly before the jar is ready for use. Some people cannot smell cyanide, and jars should be made under a fume hood or in a very well-ventilated space. Contaminated equipment should be immersed in dilute sodium hydroxide solution to which is added an excess of ferrous sulphate solution. After an hour, rinse and flush away with an excess of water. In its dry state, a cyanide jar will have a shelf life of one or two years. Once in the field, a few drops of water soaked into the plaster layer will react with the potassium cyanide to produce cyanide gas. Full production of the gas will build up over a day or two and should keep going for many weeks. An old-fashioned, but very effective, technique still used by some researchers today is the use of young, crushed laurel leaves wrapped in muslin inside a jar or bottle. Small quantities of cyanide gas are released, and the humid atmosphere keeps specimens soft and relaxed until they can be pinned.

Butterflies caught in a net can be quickly incapacitated by a gentle pinch across the thorax using the nails of the thumb and forefinger. The technique allows the transfer of specimens to a paper envelope with the risk of them flapping around and losing wing scales very much reduced. The envelopes can be stored flat in a plastic tub temporarily before being placed in an ethyl acetate jar. Pinching does not work very well on moths.

When using a Malaise trap the catch is generally going to be killed and stored in alcohol anyway. For many other mass collection techniques, such as suction sampling or sweeping (the catch will be live), or mist blowing (the catch will already be dead), the sheer volume of material means that simple immersion in alcohol is the only practicable combined killing and storage method.

Very large insects, such as dung beetles, can be killed very quickly by the injection of small quantity of saturated oxalic acid solution into the thorax or abdomen through the intersegmental membranes, but take care because, apart from being poisonous, oxalic acid is a white, crystalline powder which might arouse suspicion at customs. Dung beetles and other fairly robust species can also be killed very quickly by dropping them into boiling water (which also cleans them).

The vapours of various chemicals such as benzene, carbon tetrachloride, ether, and ammonia have been used in the past to kill insects. Many of them are dangerous and some highly inflammable, but probably the safest alternative to cyanide in the field is ethyl acetate. The fumes of ethyl acetate kill more slowly than cyanide and are still flammable. Killing bottles or tubes can be made with a layer of plaster of Paris at the bottom onto which a few drops of ethyl acetate can be dropped. More simply, a layer of cotton wool with a few pieces of filter paper on top can be arranged inside a polythene bottle. Bulk killing bottles should have a capacity of at least 500 ml and a wide top to allow you to get large specimens inside easily. Specimen tubes of strong glass or polythene are best for killing small insects. A wick of filter paper with a drop of ethyl acetate is held in place by the push fit top. Glass tubes are acceptable here and have the advantage of being transparent. It is best to have a small wad of tissue paper inside the killing chamber to soak up excess moisture. Ethyl acetate and acetone (sometimes used as a substitute and widely available as nail varnish remover) will dissolve styrene plastics. When dead, the insect should be removed and pinned or stored as required.

Data recording

Specimens collected must have adequate data associated with them. Information on where, when, who, and how any particular insect was caught should be as full as practicable. For instance, 'Tanzania, August 1996' is clearly inadequate as compared to 'collected at UV light, 12 August 1996, Ibaya Camp,

Mkomazi Game Reserve, Tanzania, G. C. McGavin'. In the field the data should be kept closely with the specimens (attached to it or inside the tube or envelope). Access to GPS (global positioning systems) enables collecting sites to be pinpointed as never before.

Do not write on the outside of tubes or on tube tops. Tube tops can get accidentally switched and labels can fall or get rubbed off. Labels for putting inside tubes of material stored in alcohol is best done on good-quality paper with pencil or waterproof drawing ink.

Never think that you will recall where or when a particular specimen was caught when you get back from the field. Specimens without proper data are completely useless.

Specimen preservation

In the field, the simplest techniques are drying or storing in alcohol. Most robust insects, such as beetles and bugs, can be simply dried in air and packed in layers of cellulose wadding or tissue. Delicate or soft-bodied species and spiders should be preserved in 70–80% alcohol (ethanol). Some coleopterists preserve small beetles in a 4% solution of glacial acetic acid until they are dried and set. If kept cool, the material should be fine for many months. Moths and butterflies should never be stored in alcohol and are best dried in small paper envelopes or folded paper triangles, or set and pinned directly into boxes. Set specimens take up a lot of room and are difficult to transport safely.

A problem can arise with large fat-bodied species, such as some grasshoppers and mantids. The large quantity of internal tissue can take a long time to dry properly, and decay may take place. A good technique is to stuff the abdominal and thoracic cavity with cotton wool. This is not as difficult as it may sound. Using a pair of fine dissection scissors make a slit in the ventral surface of the abdomen. You do not need to open the abdomen up from stem to stern. Remove the abdominal contents with a pair of fine watchmakers forceps. Be sure to reach up into the thoracic cavity to remove the front portion of the gut. Do not rip the guts from the rear end of the animal as you may damage or remove parts of the genitalia, which may be of value in identification at a later date. Use fine scissors to cut through the rectum close to the end of the abdomen. You do not have to remove every shred of soft material from the inside. Starting at the thorax, use fine forceps to push little bits of cotton wool into the cavity. When finished, the abdomen should be as close to the shape it is in real life as possible. You do not need to sew up the incision, as the cuticle will dry in position without being held in place.

The most useful field preservation technique for adult insects that we, and many others, have successfully used for years involves the use of chlorocresol (more correctly 4-chloro-m-cresol) (see Tindale, 1961). A small quantity (a level teaspoonful) is sprinkled in the bottom of an airtight polythene food storage box and covered with a layer of tissue or wadding. On top of this are laid specimens in labelled, paper envelopes or layered in tissue. Some crystals of chlorocresol should be sprinkled on top of every tenth layer or so. A crumpled tissue or two keeps the layers in place until the box is full, when a final generous sprinkle of chlorocresol should be added. The chlorocresol acts as a fungicide and bactericide, and keeps the insects moist and relaxed for months, so that when you come to mount them the job is very easy. Air-dried insects must be first relaxed in very warm humid conditions before they can be pinned and have their appendages arranged. The chlorocresol technique avoids all this, and even butterflies and moths treated in this way will remain 'settable' and in good condition for a long time.

Long-term preservation

For some studies involving large amounts of material, it might be simplest to leave specimens in alcohol until such times when they really need to be prepared in a more specific manner. However, spirit-preserved material does require maintenance. For long-term storage, critical specimens should, ideally, be kept in a freezer. This will reduce the need for topping up the alcohol and, as an extra bonus, keeps internal organs in absolutely first-class condition for decades.

Usually, identification and systematic study of specimens will require them to be mounted dry. Pinning insects enables them to be handled, examined, and stored safely. Sometimes inappropriate mounting can destroy the very features required for proper identification. General insect mounting, preservation, and curatorial techniques can be gleaned from a variety of sources, but among the best are Martin (1977) and Walker and Crosby (1988). Carter and Walker (1999) deal with all aspect of the physical care of botanical and zoological collections and give much practical guidance over the whole field of natural history curation.

Publications giving specialist techniques for particular orders of insects, such as the Dictyoptera (specifically, termites) (Krishna and Weesner, 1970), the Thysanoptera (Lewis, 1973), the Hemiptera (Dolling, 1991; McGavin, 1993), the Coleoptera (Cooter, 1991), the Diptera (Stubbs and Chandler, 1978), the Lepidoptera (Dickson, 1992), the Hymenoptera (Gauld and Bolton, 1988), and the stoneflies, mayflies, and caddisflies (Macan, 1982), should be consulted where appropriate. Many of these books also give details of collecting methods.

Most robust-bodied specimens can be simply taken out of alcohol, dried, and then mounted in the manner appropriate to the taxon concerned, but small or delicate specimens will need some treatment—otherwise, they will shrivel up and become useless. There are some simple techniques that will allow these specimens to be taken out of alcohol, dried, and mounted without bits or all of the body collapsing. We are grateful to Simon van Noort of the South African Museum in Cape Town for the following recipe, which can be used for a range of small or weakly sclerotized insects, especially parasitic wasps.

The quickest and easiest approach, particularly where there are large numbers of specimens, is to treat the whole sample in one go. Pipette the specimens in alcohol onto a square piece of fairly stiff card, the edges of which are folded up to make a miniature sort of tray. Arrange them in the position required for subsequent mounting at this stage. Place the card in a petri dish lid to keep the specimens saturated in alcohol while getting them into position. The specimens are best on their right-hand sides with wings folded flat above the body, which takes a bit of time, but is worth the effort, so that when it comes to mounting it is simply a matter of placing the specimen on a blob of glue on a card point. Once the specimens are arranged, drain off the excess alcohol by lifting up the card and letting it run off before placing the card tray into an acetone-saturated environment. Acetone is soaked liberally on to cotton wool at the bottom of a sealable acetone-proof container and the cards, bearing their arranged specimens that are still wet through with alcohol, should be placed horizontally or vertically, clear of the acetone, for a minimum of six hours. To dry the material, lift out the card tray, which is now saturated with acetone, and place it in a petri dish under a 60-watt desk lamp (close to the bulb) for 20–30 minutes. Once the specimens are dry, be careful not to breathe too hard over them, or they will end up all over the place. Once dry, they do not need to be mounted straight away and can be kept quite safe in a petri dish in a pest-free environment. The method generally gives good results but sometimes specimens may still collapse, even though others in the same sample are fine.

Another technique for the recovery of non-robust specimens from alcohol storage is particularly useful for small flies or bugs. Start by pinning the specimens direct from the alcohol. Stand the pinned specimens in a tube filled with 2-ethoxy-ethanol overnight (12 hours minimum). Transfer to a tube (glass) filled with ethyl acetate (1.5 hours for small species, 3 hours for larger species). When the specimen is removed, the wings, particularly, will stick together or crease up. Dab off excess ethyl acetate with a tissue and blow rapidly on the specimen to release the wings. If the wings do not immediately spring out, tease them out with a fine pin or forceps (J. Ismay, personal communication).

Bees that have been stored in alcohol cannot be simply taken out and dried, because the dense hair covering becomes matted and flat. A near life-like dried specimen can be prepared in the following manner. Remove the bee from alcohol, dab dry with tissue to get rid of most of the fluid, and immerse in 2-ethoxy-ethanol for 24 hours. Remove excess by dabbing with a tissue. Pin specimen in the normal manner, placing legs and wings in the desired positions. The male genitalia may be exposed at this stage using a small, hooked pin. Soak the pinned specimen in ethyl acetate for 24 hours. On removal, dry immediately with a small hair drier from behind. The hairs will fluff up and remain erect (C. O'Toole, personal communication).

Should important spirit-preserved material ever dry out, it can be rescued by soaking in a 2% solution of trisodium orthophosphate. The rehydration rate depends a great deal on the size and condition of the specimens, but it is best to check them every hour or so. A good indication that the process is complete is when the specimens sink to the bottom of the container. The specimens should then be thoroughly washed in distilled water before being put back into 70% alcohol.

Key reading

Aristophanous, M. (2010). Does your preservative preserve? A comparison of the efficacy of some pitfall trap solutions in preserving the internal reproductive organs of dung beetles. *ZooKeys* **34**: 1–16.

Arnett, R. H., Jr. (1985). *American insects: a handbook of the insects of America North of Mexico*. Van Nostrand Reinhold Company, New York.

Arnold, A. J. (1994). Insect suction sampling without nets, bags, or filters. *Crop Protection* **13**: 73–6.

Baňař, P. and Davranoglou, L.-R. (2018). Quantitative methods as a brilliant resource of seldom-collected heteropterans. In: *8th European Hemiptera Congress*, Zawiercie, Poland.

Basset, Y. (1988). A composite interception trap for sampling arthropods in tree canopies. *Journal of the Australian Entomological Society* **27**: 213–19.

Bedoussac, L., Favila, M. E., and López, R. M. (2007). Defensive volatile secretions of two diplopod species attract the carrion ball roller scarab *Canthon morsei* (Coleoptera: Scarabaeidae). *Chemoecology* **17**: 163–7.

Boonstra, R. and Craine, I. T. M. (1986). Natal nest location and small mammal tracking with a spool and line technique. *Canadian Journal of Zoology* **64**: 1034–6.

Carter, D. and Walker A. (eds) (1999). *Care and conservation of natural history collections*. Butterworth Heinemann, Oxford.

Collins, N. M. (1987). *Legislation to conserve insects in Europe*. The amateur entomologist's society pamphlet **13**. The Amateur Entomologist's Society, London.

Collins, N. M. (1988). Legislation on insects. *Amateur Entomologist's Bulletin* **47**: 53–5.

Collins, N. M. and Morris, M. G. (1985). *Threatened swallowtail butterflies of the world. The IUCN Red Data Book*. IUCN, Gland and Cambridge.

Cooter, J. (ed) (1991). *A coleopterist's handbook* (3rd edn). Amateur Entomologist's Society, London.

Davies, J. and Stork, N. E. (1996). Data and specimen collection: invertebrates. In: *Bio-diversity assessment: a guide to good practice*, vol **3**: *Field manual 2: data and specimen collection of animals*. Jermy, C. A., Long, D., Sands, M. J. S., Stork, N. E. and Winser, S. (eds), chapter 1. HMSO, London.

Deeming, J. C. (1993). A guide to the rearing of shoot flies. *Dipterist's Digest* **13**: 23–8.

Dickson, R. (1992). *A lepidopterist's handbook* (2nd edn). Amateur Entomologist's Society, London.

Dolling, W. R. (1991). *The Hemiptera*. Oxford University Press, Oxford.

Eggleton, P. and Bignell, D. E. (1995). Monitoring the response of tropical insects to changes in the environment: troubles with termites. In: *Insects in a changing environment. 17th Symposium of the Royal Entomological Society of London*. Harrington, R. and Stork, N. E. (eds), chapter 23. Academic Press, London.

Fry, R. and Waring, P. (1996). *A guide to moth traps and their use*. The Amateur Entomologist's Society, London.

Gauld, I. and Bolton B. (eds) (1988). *The Hymenoptera*. Oxford University Press, Oxford.

Hawkins, C. E. and Macdonald, D. W. (1992). A spool-and-line method for investigating the movement of badgers, *Meles meles*. Mammalia **56**: 322–5.

Heywood, V. (ed.) (1995). *Global biodiversity assessment*. United Nations Environment Programme. Cambridge University Press, Cambridge.

IUCN (1996). *The 1996 IUCN red list of threatened animals*. IUCN, Gland.

Jermy, C. A., Long, D., Sands, M. J. S., Stork, N. E., and Winser, S. (eds) (1995). *Biodiversity assessment: a guide to good practice* (3 vols). HMSO, London.

Krishna, K. and Weesner, F. M. (eds) (1970). *Biology of termites* (2 vols). Academic Press, London.

Krombein, K. V. (1967). *Trap-nesting wasp and bees: life histories, nests and associates*. Smithsonian Press, Washington, DC.

Lewis, T. (1973). *Thrips: their biology ecology and importance*. Academic Press, London.

Lincoln, R. J. and Sheals, J. G. (1979). *Invertebrate animals: collection and preservation.* Cambridge University Press, Cambridge.

Lowman, M. D. and Nadkarni, N. M. (eds) (1995). *Forest canopies.* Academic Press, London.

Macan, T. T. (1982). *The study of stoneflies, mayflies and caddisflies.* Amateur Entomologist's Society, London.

MacFadyen, A. (1962). Soil arthropod sampling. *Advances in Ecological Research* **1**: 1–34.

Martin, J. E. H. (1977). *The insects and arachnids of Canada. Part 1: collecting, preparing and preserving insects, mites and spiders.* Agriculture Canada, Hull.

McGavin, G. C. (1993). *Bugs of the world.* Blandford Press, London.

Muirhead-Thomson, R. C. (1991). *Trap responses of flying insects: the influence of trap design on capture efficiency.* Academic Press, London.

New, T. R. (1998). *Invertebrate surveys for conservation.* Oxford University Press, Oxford.

O'Toole, C. A. and Raw, A. (1991). *Bees of the world.* Blandford Press, London.

Owen, J. A. (1987). The 'Winkler extractor'. *Proceedings and Transactions of the British Entomology and Natural History Society* **20**: 129–32.

Owen, J. A. (1995). A pitfall trap for the repetitive sampling of hypogeal arthropod faunas. *Entomologist's Record* **107**: 225–8.

Owen, J. A. (1997). Observations on *Raymondionymus marqueti* (Aubé) (Col: Curculionidae) in north Surrey. *The Entomologist* **116**: 122–9.

Pollard, E. and Yates, T. (1993). *Monitoring butterflies for ecology and conservation.* Chapman and Hall, London.

Robinson, G. S. and Jones D. W. (1996). A cheap, low-powered insect light-trap for tropical use. *Tropical Biodiversity* **3**: 115–26.

Rodríguez-López, M. E., Y Gómez, B. G., Bueno-Villegas, J., and Malo, E. A. (2021). Attraction of *Canthon vazquezae* (Coleoptera: Scarabaeinae) to volatiles released by *Messicobolus magnificus* (Diplopoda: Spirobolida). *Journal of Insect Behaviour* **34**: 255–63. https://doi.org/10.1007/s10905-021-09785-x.

Shaw, M. R. (1997). *Rearing parasitic Hymenoptera.* Amateur Entomologist's Society, London.

Silva, F. A. B., Vidaurre, T., Vaz-de-Mello, F., and Louzada, J. (2012). Predatory behaviour in *Deltochilum*: convergent evolution or a primitive character within a clade? *Journal of Natural History* **46**: 1359–67.

Smith, K. V. G. (1986). *A manual of forensic entomology.* Cornell University Press, Ithaca.

Sokoloff, P. (1980). Practical hints for collecting and studying the Microlepidoptera. The Amateur Entomologist Series. Volume 16. The Amateur Entomologists' Society, London.

Southwood, T. R. E. and Henderson, P. A. (2000). *Ecological methods* (3rd edn). Blackwell Science, Oxford.

Stewart, A. J. A. and Wright, A. F. (1995). A new inexpensive suction apparatus for sampling arthropods in grasslands. *Ecological Entomology* **20**: 98–102.

Stork, N. E., Didham, R., and Adis, J. (eds) (1996). *Canopy arthropods.* Chapman and Hall, London.

Stubbs, A. and Chandler, P. (eds) (1978). *A dipterist's handbook.* Amateur Entomologist's Society, London.

Tindale, N. B. (1961). The chlorocresol method for field collecting. *Journal of the Lepidopterists Society* **15**: 195–7.

Walker, A. and Crosby, T. K. (1988). *The preparation and curation of insects.* DSIR information series **163**. SIPC, Wellington.

Wells, S. M., Pyle, R. M., and Collins, N. M. (1983). *The IUCN invertebrate red data book.* IUCN, Gland.

Williams, D. D. and Feltmate, B. W. (1992). *Aquatic insects.* CABI, Wallingford.

Wilson, E. O. and Peters F. M. (eds) (1988). *Biodiversity.* National Academy Press, Washington, DC.

Wilson, S. W., Smith, J. L., and Purcell, A. H. (1993). An inexpensive vacuum sampler for insect sampling. *Entomological News* **104**: 203–8.

Subject index

Notes Tables and figures are indicated by an italic *t* or *f* following the figure.